Genetic Diversity and Conservation of Woody Species

Genetic Diversity and Conservation of Woody Species

Editor

Gregor Kozlowski

MDPI • Basel • Beijing • Wuhan • Barcelona • Belgrade • Manchester • Tokyo • Cluj • Tianjin

Editor
Gregor Kozlowski
Department of Biology and
Botanical Garden
University of Fribourg
Fribourg
Switzerland

Editorial Office
MDPI
St. Alban-Anlage 66
4052 Basel, Switzerland

This is a reprint of articles from the Special Issue published online in the open access journal *Plants* (ISSN 2223-7747) (available at: www.mdpi.com/journal/plants/special_issues/genetic_diversity_conservation_woody_species).

For citation purposes, cite each article independently as indicated on the article page online and as indicated below:

LastName, A.A.; LastName, B.B.; LastName, C.C. Article Title. *Journal Name* **Year**, *Volume Number*, Page Range.

ISBN 978-3-0365-2056-8 (Hbk)
ISBN 978-3-0365-2055-1 (PDF)

Cover image courtesy of E. Kozlowski

Contents

About the Editor

Gregor Kozlowski

Prof. Dr. Gregor Kozlowski is a director of the Botanic Garden and group leader at the Department of Biology of the University of Fribourg. Furthermore, he is a conservator at the Natural History Museum in Fribourg (Switzerland). His main research interests are biogeography and conservation biology of relict, endemic, and threatened species.

Preface to "Genetic Diversity and Conservation of Woody Species"

Trees and other woody plants, such as shrubs and lianas, form the principal components in forests and many other ecosystems on our planet. Being among the largest and longest-living organisms, they support an immense share of the Earth's terrestrial biodiversity, providing food and habitats for innumerable microorganisms, epiphytes, invertebrate, and vertebrate species. Woody species are perfect study objects, giving us a link between the past, present, and future. Woody species have also accompanied our own species throughout its evolution. Even today, billions of people depend on trees and shrubs for fuel, medicine, food, tools, fodder for livestock, shade, and watershed maintenance. Woody species, therefore, have a high scientific, economic, social, cultural, and aesthetic value.

However, the future of many trees and shrubs is uncertain. Ten of thousands of species are threatened by overharvesting, non-native pests and diseases, changes in accelerated land use, and climate warming. Many aspects of their biology, ecology, and biogeography are still unexplored or insufficiently understood. These knowledge shortfalls, concerning their genetic diversity, for example, significantly hinder the development of protection strategies and the elaboration of efficient action plans. This book, dedicated to this very diverse group of plants, aims to encourage ongoing research and conservation efforts worldwide.

Gregor Kozlowski
Editor

plants

MDPI

Review

The Woody Planet: From Past Triumph to Manmade Decline

Laurence Fazan [1], Yi-Gang Song [2,3] and Gregor Kozlowski [1,3,4,*]

1 Department of Biology and Botanical Garden, University of Fribourg, Chemin du Musée 10, 1700 Fribourg, Switzerland; laurence.fazan@unifr.ch
2 Eastern China Conservation Center for Wild Endangered Plant Resources, Shanghai Chenshan Botanical Garden, Chenhua Road No.3888, Songjiang, Shanghai 201602, China; ygsong@cemps.ac.cn
3 Shanghai Chenshan Plant Science Research Center, Chinese Academy of Sciences, Chenhua Road No.3888, Songjiang, Shanghai 201602, China
4 Natural History Museum Fribourg, Chemin du Musée 6, 1700 Fribourg, Switzerland
* Correspondence: gregor.kozlowski@unifr.ch; Tel.: +41-26-300-88-42

Received: 6 November 2020; Accepted: 16 November 2020; Published: 17 November 2020

check for updates

Abstract: Woodiness evolved in land plants approximately 400 Mya, and very soon after this evolutionary invention, enormous terrestrial surfaces on Earth were covered by dense and luxurious forests. Forests store close to 80% of the biosphere's biomass, and more than 60% of the global biomass is made of wood (trunks, branches and roots). Among the total number of ca. 374,000 plant species worldwide, approximately 45% (138,500) are woody species—e.g., trees, shrubs or lianas. Furthermore, among all 453 described vascular plant families, 191 are entirely woody (42%). However, recent estimations demonstrate that the woody domination of our planet was even greater before the development of human civilization: 1.4 trillion trees, comprising more than 45% of forest biomass, and 35% of forest cover disappeared during the last few thousands of years of human dominance on our planet. The decline in the woody cover of Planet Earth did not decelerate during the last few centuries or decades. Ongoing overexploitation, land use and climate change have pushed ten thousand woody species to the brink of extinction. Our review highlights the importance, origin and past triumph of woody species and summarizes the unprecedented recent decline in woody species on our planet.

Keywords: biodiversity loss; global change; lianas; shrubs; trees; woody plant families

1. The Importance of Woody Species

Woody plants, such as trees, shrubs and lianas, form the principal components of forests and many other ecosystems on our planet [1] (Figure 1). Being among the largest and longest-living organisms, they support an immense share of the Earth's terrestrial biodiversity, providing food and habitats for innumerable microorganisms, epiphytes and invertebrate and vertebrate species [2,3]. They have also accompanied our own species throughout its evolution: billions of people still depend on trees and shrubs for fuel, medicine, food, tools, fodder for livestock, shade, watershed maintenance and climate regulation [4]. Woody species, therefore, have inestimable scientific, economic, social, cultural and aesthetic value [5].

Figure 1. Woody plants form the principal components in forests and many other ecosystems on our planet, providing food and habitats for innumerable organisms. (**a**) *Quercus bumelioides*, tropical rainforest, Tapanti National Park, Costa Rica; (**b**) *Salix polaris*, tundra, Svalbard, Norway; (**c**) *Rhizophora stylosa*, mangrove, Iriomote, Japan; (**d**) *Pinus cembra*, alpine forest, Prealps of Fribourg, Switzerland; (**e**) Mopane woodlands, Etosha National Park, Namibia; (**f**) Semidesert vegetation, Perekushkul, Azerbaijan. Pictures: E. Kozlowski (**a**,**c**,**e**,**f**), G. Kozlowski (**b**,**d**).

2. The Origin of Woody Plants

The earliest land plants had no vascular tissue and used ectohydric water transport by capillarity along their external surface [6]. Woodiness, which allowed endohydric, internal water transport, appeared approximately 100 million years after the conquering of land by plants [6–8]. The oldest wood tissue is derived from the Early Devonian, ca. 400 Mya [9]. These early woody plants were very small and used their woody structures as plumbing systems for taking up water rather than providing mechanical support [9]. Large woody plants evolved secondarily by taking advantage of the newly evolved support. Once woodiness evolved, however, the evolution of large woody trunks (so-called hyperstele) was extremely rapid [6]. In fact, it happened almost immediately, since tree habit and large trunks were already present between 385 and 390 Mya, mainly in progymnosperms [10] but also in cladoxylopsids [11]. Moreover, fossil records suggest that these organisms formed very early complex and dense forest ecosystems. These large woody organisms changed the evolutionary dynamics of terrestrial habitats, and their decaying branches, leaves and trunks significantly altered geochemical cycles all over the Earth [10,11]. Some other plant groups (e.g., arborescent lycophytes, horsetails and ferns) also evolved strategies to form tree-like organisms [6]. However, although they had some secondary xylem, their stems were determinate and short-lived, and thus, they should be denominated as "giant herbs" rather than true woody species or trees [6].

When gymnosperms, especially conifers, started to dominate our planet, they quickly left only epiphytic, aquatic or marginal niches for lycophytes and monilophytes [6]. Consequently, for more than 225 million years, the planet Earth was a true woody planet, since all seed plants were woody trees, shrubs or lianas [12]. Interestingly, the early history of angiosperms—the most diverse and dominant plant group today—also seemed to be woody [7,13,14]. The first angiosperms, however, were probably not large canopy trees but rather small shrubs or even lianas [6]. The lianous habit is potentially a driver of the evolution of vessels (one of the main novelties of angiosperms) due to the high hydraulic conductivity per unit area necessary in thin stems. Nevertheless, by the end of the Cretaceous (65 Mya), angiosperm trees, shrubs and lianas were predominant in the majority of forest ecosystems [6]. The gymnosperms, comprising a mere 1100 extant species, play nowadays an important role only in certain forest ecosystems, notably in the boreal zone of the Northern Hemisphere [15–17].

3. How Much of the Vegetal World Is Woody Today?

A recent report of the Royal Botanic Gardens in Kew (United Kingdom) estimated the total number of plant species to be ca. 380,000 worldwide [17]. Christenhusz and Byng [18] give the total number of ca. 374,000 officially described and accepted plant species, with approximately 308,000 vascular plants [18]. FritzJohn et al. [19] estimated that approximately 45% of vascular plants are woody species, and thus there might be ca. 138,500 species of trees, shrubs and lianas worldwide. According to the IUCN Global Tree Specialist Group, of this number, ca. 60,000 species are trees (43% of all woody species) [4]. The remaining 78,500 (57%) woody species are shrubs and lianas. There are no published syntheses or global data banks allowing the estimation of the proportion of lianas alone among all woody species. It is known, however, that especially in tropical forests, lianas represent ca. 35% of vascular plant diversity [20].

Among all 453 described vascular plant families, 191 are entirely woody (42%) (Appendix A, Figure 2). Gymnosperms are still (and as explained above, were in their long evolutionary history) clearly the woodiest plant group, with no herbaceous species at all among all known fossil and extant taxa [15,16]. The members of the other still extant ancient plant groups—e.g., lycopods and monilophytes (ferns and horsetails)—are entirely herbaceous today [19].

Figure 2. Selected members of the Fagaceae, a large and entirely woody plant family. (**a**,**b**) *Fagus sylvatica*, Ueberstorf, Switzerland; (**c**,**d**) *Quercus robur*, Enney, Switzerland. Both species belong to the most common trees in European forests; (**e**) *Castanopsis delavayi*, endemic to Southwest China. Yunnan, China; (**f**,**g**) *Quercus litseoides*, Shenzhen, Wutong Mountain, China. The species is one of the rarest and most endangered oaks worldwide. Pictures: E. Kozlowski (**a**–**e**), Y.-G. Song (**f**,**g**).

The overwhelming majority of trees, shrubs and lianas living today belong to angiosperms (Appendix A). Among the basal dicots, not less than 5 orders contain only woody plants (Amborellales, Austrobaileyales, Canellales, Magnoliales and Laurales) [18,21]. Many important for mankind, species-rich and entirely woody families, such as Myristicaceae (520 spp.), Magnoliaceae (294 spp.), Annonaceae (2500 spp.) and Lauraceae (2850 spp.), are basal dicots (Appendix A).

Among the eudicots, there are at least 6 entirely woody orders (Trochodendrales, Vitales, Fagales, Picramniales, Garryales and Aquifoliales). The ecologically most important, species-rich and entirely woody eudicot families belong to Proteaceae (1660 spp.), Vitaceae (910 spp.), Fagaceae (927 spp., Figure 2), Salicaceae (1220 spp.), Myrtaceae (5950 spp.), Anacardiaceae (860 spp.), Loranthaceae (1050 spp.), Sapotaceae (1273 spp.), Ebenaceae (800 spp.) and Oleaceae (790 spp.).

Additionally, several smaller woody plant families have played an important role in the historical biogeography and evolution of forests and woody flora during the past million years. These families, composed nearly exclusively of relict genera of trees and shrubs, possess enormous scientific value (for example, Altingiaceae with *Liquidambar*, Hamamelidaceae with *Parrotia* or Ulmaceae with *Zelkova*) [22]. Moreover, many of them have significant economic and cultural value, such as the Walnut family (Juglandaceae) [23]. Notably, practically all 60 species of Juglandaceae are important, useful trees for mankind. They are used for their high-quality wood, planted in parks and gardens as ornamental trees, and used for their edible fruits as well as for medicinal properties and are also used in tanning, staining, oil manufacturing and fish poisoning [24].

Interestingly, there are five taxonomic lineages across angiosperms that are not only exclusively herbaceous but that have in fact completely lost the vascular cambium during their evolution: (1) the whole order Nymphaeales, the plants of the genera (2) *Ceratophyllum* and (3) *Nelumbo*, (4) the family Podostemaceae and (5) all monocots [12]. For the first four plant groups, the convergent loss of the vascular cambium appears to be associated with the transition to an aquatic habitat [14]. Although not completely proven, this could also be the case for monocots due to the putative semiaquatic habitat of the last common ancestor of this group [25]. According to this argument, all monocot families, including those with common tree-like or shrub-like habits such as Arecaceae (2600 spp.) or Pandanaceae (982 spp.) should not be considered as woody taxa (Appendix B). According to Cronk and Forest [6], this group of plants should also be denominated as "giant herbs".

4. Domination of Woody Species

The domination of woody species on our planet was demonstrated recently by Crowther et al. [2]. According to their study, there are approximately 3.04 trillion trees worldwide, which grow mainly in tropical and subtropical forests (43%) and in boreal (24%) and temperate regions (22%) [2]. Even more impressive, however, are the estimations of the biomass of woody species [26]. The overall biomass composition of the biosphere was estimated at 550 gigatons of carbon (Gt C). Plants make up the majority of the biosphere (ca. 450 Gt C), and the stems and trunks of trees (wood) represent 70% of plant biomass (ca. 315 Gt C). Thus, only trees—not including other woody species—make up approximately 60% of the total biomass of our biosphere. In comparison, all animals (and humans) taken together make up merely 2 Gt C and thus less than 0.37% of the overall biomass of our planet [26].

Furthermore, according to estimations of Reichstein and Carvalhais [27], forests (e.g., the main plant community made of woody species) store close to 80% of all the biomass on Earth. However, even in sylvipastoral landscapes and in the tundra, the proportion of the biomass of woody species is between 50 and 75% [28,29].

5. The Manmade Decline

Ever since *Homo sapiens* started to increase in numbers and to colonize or discover new territories or islands, regional ecosystems have been altered by human presence and activities. Notable examples are the role of *H. sapiens* in the extinction of the Pleistocene megafauna which led to modification of ecosystems at a continental scale or the arrival of humans to large islands such as New Zealand or

Madagascar [30]. These human-influenced faunal extinctions are increasingly thought to have also affected plant communities at a large scale and increased biomass burning [30]. The discovery and spread of agriculture approximately 12,000–10,000 years ago led to unprecedented manmade ecosystem alterations (e.g., deforestation, domestication of plants and animals and irrigation) [30,31]. Although these human activities may have affected ecosystems on a local to continental scale, preindustrial societies lacked the necessary numbers, technology or organization to affect the environment at a global scale [31]. With the industrial development that started in ca. 1800, the global population increased more than six-fold. Deforestation and habitat losses for agricultural purposes, logging or urbanization prevailed and mankind transformed the environment at a global scale. Humans have changed the world's ecosystem during the past 50 years more rapidly and at a larger scale than ever before [32]. These changes have placed the woody domination of our planet into peril [2].

According to Crowther et al. [2], the global number of trees has fallen by approximately 46% since the start of human civilization (disappearance of ca. 1.40 trillion of trees). Thus, humans have reduced the number of trees by approximately half. In terms of biomass, if we extrapolate based on the estimation of Bar-On et al. [26], this would indicate that approximately 40% of the total planet biomass (ca. 220 Gt C) has been lost during only the last few thousand—or perhaps even the last few hundred—years. It is difficult to imagine what effect these dramatic changes have had on global biogeochemical cycles, carbon sequestration on our planet, as well as the decreases in other organisms depending directly or indirectly on woody species.

Humans dramatically reduced not only the biomass of woody species but also their diversity (number of species). The list of recently extinct woody taxa is long and covers all taxonomic groups and all continents and biomes, and it includes for example *Nesiota elliptica* (Rhamnaceae) from Saint Helena, *Ilex gardneriana* (Aquifoliaceae) from India and *Kokia lanceolata* (Malvaceae) from Hawaii (www.redlist.org). The IUCN Red List of Threatened Species enumerates 78 trees, 49 shrubs and one liana species, which are either globally extinct (EX) or extinct in the wild (EW). However, the global assessments of all woody species are by far not complete. If we take only the trees into consideration, the Global Tree Assessment (https://globaltreeassessment.org), coordinated by the Botanic Gardens Conservation International, estimates that approximately 40% of tree species are globally threatened. This means that more than 20,000 tree species are threatened with extinction. It is highly likely that similarly high numbers of shrub and liana species are threatened.

Since each of these endangered woody species is confronted with multiple threats, the elaboration of efficient conservation measures is extremely complex. For example, *Zelkova abelicea* (Ulmaceae) (Figure 3), which is a relict tree endemic to the Mediterranean island of Crete (Greece), has been classified as endangered (EN) according to the IUCN Red List (www.iucnredlist.org), and the species is mainly threatened by overbrowsing by goats [33]. There are only a few highly isolated populations composed of large fruiting trees, whereas heavily browsed dwarf individuals do not produce fruits at all [34,35]. The regeneration of populations via seedlings is nearly impossible due to the overgrazing, trampling and soil erosion caused by omnipresent large caprine and ovine flocks. Moreover, *Z. abelicea* stands are threatened by recurring droughts and wildfires as well as by ongoing climatic changes [36]. Like that of *Z. abelicea*, long-term conservation of the majority of endangered woody species will require an enormous investment of time and resources.

Recent conservation initiatives and research projects are attempting to change this dramatic loss of woody species diversity and deterioration or vanishing of the forest cover. On the one hand, there are several working groups associated with the International Union for Conservation of Nature (IUCN), which are conducting global inventories of the diversity and threats of woody species (e.g., Global Tree Assessment, https://globaltreeassessment.org), and developing solutions and specific conservation actions in order to save the World's threatened trees (e.g., Global Tree Campaign, https://globaltrees.org). On the other hand, recent literature from forest sciences offers silvicultural management solutions for maintenance and restoration of forest cover and woody species diversity [37], including old-growth forests [38] and endangered rare species that inhabit them [39].

Figure 3. *Zelkova abelicea* (Ulmaceae), an example of an endangered tree species endemic to the Mediterranean island of Crete (Greece). Only large trees can produce fruits (**a**,**b**). However, the majority of fruit is empty, which is probably due to unfavorable climatic conditions such as pronounced and recurrent droughts. The overwhelming majority of individuals, however, are dwarfed and nonfruiting due to overbrowsing by goats (**c**,**d**). The regeneration of populations via seedlings is nearly impossible due to the overgrazing, trampling and soil erosion caused by omnipresent large caprine and ovine flocks (**e**). Pictures: G. Kozlowski (**a**), H-R Siegel (**b**,**d**), L. Fazan (**c**,**e**).

6. Conclusions

The conquest of the Earth by woody species started approximately 400 million years ago. Woody species and their communities (e.g., forests, savannas and shrub ecosystems) dominated the biosphere of our planet in terms of both biomass and diversity for hundreds of millions of years. Despite the recent dramatic biomass depletion and diversity loss of this important plant group, the Earth is still a woody planet. Trees and other woody species are the living foundation on which most terrestrial biodiversity is built [1]. To efficiently protect threatened trees, shrubs and lianas and to stop the degradation of their declining ecosystems, existing international conservation programs must be supported, and new global protection initiatives must be developed [5].

Much more must also be done at the scientific level. Local, national and international research programs must be launched, and well-coordinated working groups dedicated to specific taxonomic or geographical groups must be formed to improve our knowledge of the most threatened and least known and studied among the ten thousand endangered woody species.

Author Contributions: Conceptualization, G.K.; writing—review and editing, G.K., Y.-G.S. and L.F. All authors have read and agreed to the published version of the manuscript.

Funding: This research was partially funded by Fondation Franklinia.

Acknowledgments: We would like to thank the research team of the Botanic Garden of the University of Fribourg (Switzerland) and of the Shanghai Chenshan Plant Science Research Center at the Shanghai Chenshan Botanical Garden (China) for their support and help in preparation of the manuscript.

Conflicts of Interest: The authors declare no conflict of interest.

Appendix A

Table A1. List of vascular plant families composed entirely of woody species with an approximate number of genera and species [15,17–19,21,23,40]. Asterisks (*) indicate orders entirely composed of woody taxa (all species in all families are woody). The life forms are listed in order of importance. T: trees; S: shrubs; L: lianas; O: other woody forms—e.g., palm-like cycads. Capitals: the indicated life form is very common within the family. Lower case letters: the indicated life form is rarely found within the family. The order of plant groups and families follows Christenhusz and Byng [18].

Plant Group	Family	No Genera/No Species	Life Forms
	GYMNOSPERMS		
Cycadales *	Cycadaceae	1/107	O
	Zamiaceae	9/230	O
Ginkgoales *	Ginkgoaceae	1/1	T
Welwitschiales *	Welwitschiaceae	1/1	O
Gnetales *	Gnetaceae	1/43	L, t
Ephedrales *	Ephedraceae	1/68	S, l
Pinales *	Pinaceae	11/228	T, s
Araucariales *	Araucariaceae	3/37	T
	Podocarpaceae	19/187	T, S
Cupressales *	Sciadopithyaceae	1/1	T
	Cupressaceae	29/149	T, S
	Cephalotaxaceae	1/8	T, S
	Taxaceae	5/20	T, S
	BASAL DICOTS		
Amborellales *	Amborellaceae	1/1	T
Austrobaileyales *	Austrobaileyaceae	1/1	L
	Trimeniaceae	1/8	T, S, L
	Schisandraceae	3/85	T, S, L
Canellales *	Canellaceae	5/23	T, s
	Winteraceae	5/65	S, T, l

Table A1. *Cont.*

Plant Group	Family	No Genera/No Species	Life Forms
	BASAL DICOTS		
Magnoliales *	Myristicaceae	21/520	T, s
	Magnoliaceae	2/294	T, S
	Degeneriaceae	1/2	T
	Himantandraceae	1/2	T
	Eupomatiaceae	1/3	S, t
	Annonaceae	105/2500	T, S, L
Laurales *	Calycanthaceae	3/10	S, T
	Siparunaceae	2/75	S, T, l
	Gomortegaceae	1/1	T
	Atherospermataceae	6/16	T, S
	Hernandiaceae	5/58	T, S, L
	Monimiaceae	24/217	T, S, L
	Lauraceae	45/2850	T, S
	EUDICOTS		
Ranunculales	Eupteleaceae	1/2	T, S
	Lardizabalaceae	7/40	L, s
Proteales	Sabiaceae	3/66	T, S, L
	Platanaceae	1/8	T
	Proteaceae	83/1660	S, T, o
Trochodendrales *	Trochodendraceae	2/2	T, S
Gunnerales	Myrothamnaceae	1/2	S
Saxifragales	Peridiscaceae	4/12	T, S
	Altingiaceae	1/15	T, S
	Hamamelidaceae	26/86	S, T
	Cercidiphyllaceae	1/2	T
	Daphniphyllaceae	1/30	T, S
	Iteaceae	2/18	T, S, l
	Grossulariaceae	1/150	S
	Aphanopetalaceae	1/2	L
	Tetracarpaeaceae	1/1	S
Vitales *	Vitaceae	14/910	L, S, T
Fabales	Quillajaceae	1/3	T
	Surianaceae	5/8	T, S
Rosales	Barbeyaceae	1/1	T
	Elaeagnaceae	3/60	S, T, l
	Ulmaceae	7/54	T, S
Fagales *	Nothofagaceae	1/43	T, S
	Fagaceae	8/927	T, s
	Myricaceae	3/57	S, t
	Juglandaceae	10/60	T, s
	Casuarinaceae	4/91	T, S
	Ticodendraceae	1/1	T
	Betulaceae	6/167	T, S
Cucurbitales	Anisophylleaceae	4/71	T, S
	Corynocarpaceae	1/5	T, S
	Tetramelaceae	2/2	T
Celastrales	Lepidobotryaceae	2/2	T
Oxalidales	Huaceae	2/4	T, S
	Connaraceae	12/180	T, S, L
	Cunoniaceae	27/330	T, S
	Elaeocarpaceae	12/615	T, S
	Brunelliaceae	1/60	T

Table A1. *Cont.*

Plant Group	Family	No Genera/No Species	Life Forms
	EUDICOTS		
Malpighiales	Pandaceae	3/17	T, S
	Irvingiaceae	3/13	T
	Ctenolophonaceae	1/2	T
	Rhizophoraceae	15/147	T, S
	Erythroxylaceae	4/242	S, T
	Bonnetiaceae	3/35	T, S
	Clusiaceae	13/750	S, T
	Calophyllaceae	14/475	T
	Caryocaraceae	2/26	T, S
	Lophopyxidaceae	1/1	S
	Putranjivaceae	2/216	T, S
	Centroplacaceae	2/6	T
	Balanopaceae	1/9	T, S
	Trigoniaceae	5/28	T, S, l
	Dichapetalaceae	3/170	T, S, L
	Chrysobalanaceae	18/533	T, S
	Humiriaceae	8/56	T, S
	Goupiaceae	1/2	T
	Lacistemataceae	2/14	S, T
	Salicaceae	56/1220	S, T
	Ixonanthaceae	3/17	T, S
	Picrodendraceae	25/96	T, S
Myrtales	Combretaceae	10/530	T, S, L
	Vochysiaceae	7/217	S, T
	Myrtaceae	132/5950	T, S
	Crypteroniaceae	3/13	T
	Alzateaceae	1/1	T, s
	Penaeaceae	9/32	S, T
Crossosomatales	Aphloiaceae	1/1	S, t
	Geissolomataceae	1/1	S
	Strasburgeriaceae	2/2	T
	Staphyleaceae	2/45	T, S
	Guamatelaceae	1/1	S
	Stachyuraceae	1/8	S, T, l
	Crossosomataceae	4/10	S, t
Picramniales *	Picramniaceae	3/49	T, S
Sapindales	Kirkiaceae	1/6	S, T
	Burseraceae	19/615	T, S
	Anacardiacaeae	83/860	T, S, L
	Simaroubaceae	22/108	T, S
	Meliaceae	53/600	T, S
Huerteales	Petenaeaceae	1/1	T, s
	Gerrardinaceae	1/2	S, t
	Tapisciaceae	2/6	T
	Dipentodontaceae	2/20	T, S
Malvales	Muntingiaceae	3/3	T, S
	Sphaerosepalaceae	2/18	T, S
	Sarcolaenaceae	10/71	T, S
	Dipterocarpaceae	16/695	T
Brassicales	Akaniaceae	2/2	T
	Caricaceae	6/35	T, S
	Setchellanthaceae	1/1	S
	Koeberliniaceae	1/2	S, T
	Bataceae	1/2	S
	Salvadoraceae	3/11	S, T

Table A1. *Cont.*

Plant Group	Family	No Genera/No Species	Life Forms
		EUDICOTS	
	Tiganophytaceae	1/1	S
	Pentadiplandraceae	1/1	S, l
	Capparaceae	30/324	S, T
Berberidopsidales	Aextoxicaceae	1/1	T
	Berberidopsidaceae	2/3	L
Santalales	Olacaceae	29/180	T, S, L
	Opiliaceae	11/33	T, S, l
	Misodendraceae	1/8	S
	Loranthaceae	76/1050	S, t, l
Caryophyllales	Tamaricaceae	4/78	S, T
	Dioncophyllaceae	3/3	S, L
	Ancistrocladaceae	1/21	L, S
	Rhabdodendraceae	1/3	S, T
	Simmondsiaceae	1/1	S
	Physenaceae	1/2	S, T
	Asteropeiaceae	1/8	T, S
	Achatocarpaceae	2/11	S, T
	Stegnospermataceae	1/4	T, S
	Barbeuiaceae	1/1	L
	Sarcobataceae	1/2	S
	Didiereaceae	7/22	T, S, l
Cornales	Nyssaceae	5/37	T
	Curtisiaceae	1/1	T
Ericales	Marcgraviaceae	7/120	L, S, t
	Tetrameristaceae	3/5	T, S
	Fouquieriaceae	1/11	S, T
	Lecythidaceae	25/355	T, s
	Sladeniaceae	2/3	T
	Pentaphylacaceae	12/330	T, S
	Sapotaceae	54/1273	T, S, L
	Ebenaceae	4/800	S, T
	Theaceae	9/240	T, S
	Symplocaceae	2/260	T, S
	Styracaceae	11/160	T, S
	Roridulaceae	1/2	S
	Actinidiaceae	3/360	L, T, S
	Clethraceae	2/75	S, T
	Cyrillaceae	2/2	S, T
Icacinales	Oncothecaceae	1/2	T, S
	Icacinaceae	25/165	T, S, L
Metteniusales	Metteniusaceae	11/50	T, S, l
Garryales *	Eucommiaceae	1/1	T
	Garryaceae	2/25	T, S
Gentianales	Gelsemiaceae	3/11	S, T, L
Solanales	Montiniaceae	3/5	S, T
Lamiales	Plocospermataceae	1/1	S, T
	Oleaceae	26/790	S, T, L
	Schlegeliaceae	4/37	T, S, L
	Thomandersiaceae	1/6	S, T
	Paulowniaceae	3/8	T, L
Aquifoliales *	Stemonuraceae	12/90	S, T
	Cardiopteridaceae	5/43	T, S, l
	Phyllonomaceae	1/4	T, S
	Helwingiaceae	1/4	S, t
	Aquifoliaceae	1/500	T, S

Table A1. *Cont.*

Plant Group	Family	No Genera/No Species	Life Forms
	EUDICOTS		
Asterales	Rousseaceae	4/6	S, T, l
	Alseuosmiaceae	5/13	S
	Phellinaceae	1/12	S, T
	Argophyllaceae	2/21	S, T
Bruniales	Bruniaceae	6/81	S, t
Paracryphiales	Paracryphiaceae	3/36	S, T
Apiales	Pennantiaceae	1/4	S, T, l
	Torricelliaceae	3/10	T, S
	Griselinaceae	1/7	S, L, T
	Pittosporaceae	7/245	T, S, L
	Myodocarpaceae	2/15	T, S

Appendix B

Table A2. List of monocot plant families composed entirely of tree-, shrub- and liana-like organisms [21]. Monocots have completely lost the vascular cambium during their evolution and thus should be denominated as "giant herbs" rather than true woody species [6]; therefore, were not included in the Appendix A. The life forms are listed in order of importance. T: trees; S: shrubs; L: lianas (capitals: often within the family; lower case letters: rarely). The order of plant families follows Christenhusz and Byng [18].

Plant Group	Family	No Genera/No Species	Life Forms
	MONOCOTS		
Pandanales	Pandanaceae	5/982	T, S, L
Liliales	Philesiaceae	2/2	S, L
	Ripogonaceae	1/6	S, L
Arecales	Dasypogogonaceae	4/16	S, t
	Arecaceae	181/2600	T, s

References

1. Brodribb, T.J.; Powers, J.; Cochard, H.; Choat, B. Hanging by a thread? Forests and drought. *Science* **2020**, *368*, 261–266. [CrossRef] [PubMed]
2. Crowther, T.W.; Glick, H.B.; Covey, K.R.; Bettigole, C.; Maynard, D.S.; Thomas, S.M.; Smith, J.R.; Hintler, G.; Duguid, M.C.; Amatulli, G.; et al. Mapping tree density at global scale. *Nature* **2015**, *525*, 201–205. [CrossRef] [PubMed]
3. Rogers, P.C.; McAvoy, D.J. Mule deer impede Pando's recovery: Implications for aspen resilience from a single-genotype forest. *PLoS ONE* **2018**, *13*, e0203619. [CrossRef] [PubMed]
4. Beech, E.; Rivers, M.; Oldfield, S.; Smith, P.P. GlobalTreeSearch: The first complete global database of tree species and country distributions. *J. Sustain. For.* **2017**, *36*, 454–489. [CrossRef]
5. Watson, J.E.M.; Evans, T.; Venter, O.; Williams, B.; Tulloch, A.; Stewart, C.; Thompson, I.; Ray, J.C.; Murray, K.; Salazar, A.; et al. The exceptional value of intact forest ecosystems. *Nat. Ecol. Evol.* **2018**, *2*, 599–610. [CrossRef] [PubMed]
6. Cronk, Q.C.B.; Forest, F. The evolution of angiosperm trees: From palaeobotany to genomics. In *Comparative and Evolutionary Genomics of Angiosperm Trees*; Grooves, A.T., Cronk, Q.C.B., Eds.; Springer: Berlin/Heidelberg, Germany, 2017; pp. 1–17.
7. Philippe, M.; Gomez, B.; Girard, V.; Coiffard, C.; Daviero-Gomez, V.; Thevenard, F.; Billon-Bruyat, J.P.; Guiomar, M.; Latil, J.L.; Le Loeuff, J.; et al. Woody or not woody? Evidence for early angiosperm habit from the Early Cretaceous fossil wood record of Europe. *Palaeoworld* **2008**, *17*, 142–152. [CrossRef]

8. Morris, J.L.; Puttick, M.N.; Clark, J.W.; Edwards, D.; Kenrick, P.; Pressel, S.; Wellman, C.H.; Yang, Z.; Schneider, H.; Donoghue, P.C.J. The timescale of early land plant evolution. *Proc. Natl. Acad. Sci. USA* **2018**, *115*, 2274–2283. [CrossRef]
9. Gerrienne, P.; Gensel, P.G.; Strullu-Derrien, C.; Lardeux, H.; Steemans, P.; Prestianni, C. A simple type of wood in two early Devonian plants. *Science* **2011**, *333*, 837. [CrossRef]
10. Meyer-Berthaud, B.; Scheckler, S.E.; Wendt, J. *Archaeopteris* is the earliest known modern tree. *Nature* **1999**, *398*, 700–701. [CrossRef]
11. Stein, W.E.; Mannolini, F.; VanAller Hernick, L.; Landing, E.; Berry, C.M. Giant cladoxylopsid trees resolve the enigma of the Earth's earliest forest stumps at Gilboa. *Nature* **2007**, *446*, 904–907. [CrossRef]
12. Povilus, R.A.; DaCosta, J.M.; Grassa, C.; Satyaki, P.R.V.; Moeglein, M.; Jaenisch, J.; Xi, Z.; Mathews, S.; Gehring, M.; Davis, C.C.; et al. Water lily (*Nymphaea thermarum*) genome reveals variable genomic signatures of ancient vascular cambium losses. *Proc. Natl. Acad. Sci. USA* **2020**, *117*, 8649–8656. [CrossRef] [PubMed]
13. Carlquist, S. More woodiness/less woodiness: Evolutionary avenues, ontogenetic mechanisms. *Int. J. Plant Sci.* **2013**, *174*, 964–991. [CrossRef]
14. Crane, P.R.; Friis, E.M. Water lilies, loss of woodiness, and model systems. *Proc. Natl. Acad. Sci. USA* **2020**, *117*, 9674–9676. [CrossRef] [PubMed]
15. Fragnière, Y.; Bétrisey, S.; Cardinaux, L.; Kozlowski, G. Fighting their last stand? A global analysis of the distribution and conservation status of gymnosperms. *J. Biogeogr.* **2015**, *42*, 809–820. [CrossRef]
16. Kozlowski, G.; Stoffel, M.; Bétrisey, S.; Cardinaux, L.; Mota, M. Hydrophobia of gymnosperms: Myth or reality? *Ecohydrology* **2015**, *8*, 105–112. [CrossRef]
17. *SOTWP—State of the World's Plants*; Royal Botanic Gardens, Kew: London, UK, 2017; Available online: https://stateoftheworldsplants.org (accessed on 14 October 2020).
18. Christenhusz, M.J.M.; Byng, J.W. The number of known plants species in the world and its annual increase. *Phytotaxa* **2016**, *261*, 201–217. [CrossRef]
19. FitzJohn, R.G.; Pennell, M.E.; Zanne, A.M.; Stevens, P.F.; Tank, D.C.; Cornwell, W.K. How much of the world is woody? *J. Ecol.* **2014**, *102*, 1266–1272. [CrossRef]
20. Meyer, L.; Kissling, W.D.; Lohmann, L.G.; Hortal, J.; Diniz-Filho, J.A.F. Deconstructing species richness-environment relationships in Neotropical lianas. *J. Biogeogr.* **2020**. [CrossRef]
21. Christenhusz, M.J.M.; Fay, M.F.; Chase, M.W. Plants of the world. In *An Illustrated Encyclopedia of Vascular Plants*; University of Chicago Press: Chicago, IL, USA, 2017.
22. Kozlowski, G.; Gratzfeld, J. *Zelkova—An ancient tree. Global Status and Conservation Action*; Natural History Museum Fribourg: Fribourg, Switzerland, 2013; p. 60.
23. Song, Y.G.; Fragnière, Y.; Meng, H.H.; Li, Y.; Bétrisey, S.; Corrales, A.; Manchester, S.; Deng, M.; Jasinska, A.K.; Van Sam, H.; et al. Global biogeographic synthesis and priority conservation regions of the relict tree family Juglandaceae. *J. Biogeogr.* **2020**, *47*, 643–657. [CrossRef]
24. Kozlowski, G.; Bétrisey, S.; Song, Y.G. *Wingnuts (Pterocarya) and Walnut Family. Relict Trees: Linking the Past, Present and Future*; Natural History Museum Fribourg: Fribourg, Switzerland, 2018; p. 128.
25. Givnish, T.J.; Zuluaga, A.; Spalink, D.; Soto Gomez, M.; Lam, V.K.Y.; Saarela, J.M.; Saas, C.; Iles, W.J.D.; Lima da Sousa, D.J.; Leebens-Mack, J.; et al. Monocot plastid phylogenomics, timeline, net rates of species diversification, the power of multi-gene analyses, and a functional model for the origin of monocots. *Am. J. Bot.* **2018**, *105*, 1888–1910. [CrossRef]
26. Bar-On, Y.M.; Phillips, R.; Milo, R. The biomass distribution on Earth. *Proc. Natl. Acad. Sci. USA* **2018**, *115*, 6506–6511. [CrossRef] [PubMed]
27. Reichstein, M.; Carvalhais, N. Aspects of forest biomass in the Earth system: Its role and major unknowns. *Surv. Geophys.* **2019**, *40*, 693–707. [CrossRef]
28. Wielgolaski, F.E. Vegetation types and plant biomass in tundra. *Arct. Alp. Res.* **1972**, *4*, 291–305. [CrossRef]
29. Zomer, R.J.; Neufeldt, H.; Xu, J.; Ahrends, A.; Bossio, D.; Trabucco, A.; Van Noordwijk, M.; Wang, M. Global tree cover and biomass carbon on agricultural land: The contribution of agroforestry to global and national carbon budgets. *Sci. Rep.* **2016**, *6*, 29987. [CrossRef] [PubMed]
30. Braje, T.J.; Erlandson, J.M. Human acceleration of animal and plant extinctions: A Late Pleistocene, Holocene and Anthropocene continuum. *Anthropocene* **2013**, *4*, 14–23. [CrossRef]
31. Steffen, W.; Crutzen, P.J.; McNeill, J.R. The Anthropocene: Are humans now overwhelming the great forces of Nature. *AMBIO J. Hum. Environ.* **2007**, *36*, 614–621. [CrossRef]

32. Reid, W.V.; Mooney, H.A.; Cropper, A.; Capistrano, D.; Carpenter, S.R.; Chopra, K.; Dasgupta, P.; Dietz, T.; Duraiappah, A.K.; Hassan, R.; et al. *Ecosystems and Human Well-Being—Synthesis: A Report of the Millennium Ecosystem Assessment*; Island Press: Washington, DC, USA, 2005; p. 137.
33. Kozlowski, G.; Frey, D.; Fazan, L.; Egli, B.; Pirintsos, S. *Zelkova abelicea*. IUCN Red List of Threatened Species. 2012. Available online: www.iucnredlist.org (accessed on 19 July 2020).
34. Fazan, L.; Stoffel, M.; Frey, D.J.; Pirintsos, S.; Kozlowski, G. Small does not mean young: Age estimation of severely browsed trees in anthropogenic Mediterranean landscapes. *Biol. Conserv.* **2012**, *153*, 97–100. [CrossRef]
35. Kozlowski, G.; Frey, D.; Fazan, L.; Egli, B.; Bétrisey, S.; Gratzfeld, J.; Garfì, G.; Pirintsos, S. Tertiary relict tree *Zelkova abelicea* Ulmaceae): Distribution, population structure and conservation status. *Oryx* **2014**, *48*, 80–87. [CrossRef]
36. Kozlowski, G.; Bétrisey, S.; Song, Y.G.; Fazan, L.; Garfì, G. *The Red List of Zelkova*; Natural History Museum Fribourg: Fribourg, Switzerland, 2018; p. 32.
37. Höhl, M.; Ahimbisibwe, V.; Stanturf, J.A.; Elsasser, P.; Kleine, M.; Bolte, A. Forest Landscape Restoration—What generates failure and success? *Forests* **2020**, *11*, 938. [CrossRef]
38. Keren, S.; Medarević, M.; Obradović, S.; Zlokapa, B. Five decades of structural and compositional changes in managed and unmanaged montane stands: A case study from South-East Europe. *Forests* **2018**, *9*, 479. [CrossRef]
39. Sabatini, M.F.; Keeton, W.S.; Lindner, M.; Svoboda, M.; Verkerk, P.J.; Bauhus, J.; Bruelheide, H.; Burrascano, S.; Debaive, N.; Duarte, I.; et al. Protection gaps and restoration opportunities for primary forests in Europe. *Divers. Distrib.* **2020**. [CrossRef]
40. Swanepoel, W.; Chase, M.W.; Christenhusz, M.J.M.; Maurin, O.; Forest, F.; Van Wyk, A.E. From the frying pan: An unusual dwarf shrub from Namibia turns out to be a new brassicalean family. *Phytotaxa* **2020**, *439*, 173–182. [CrossRef]

Publisher's Note: MDPI stays neutral with regard to jurisdictional claims in published maps and institutional affiliations.

plants

MDPI

Article

Biogeographic Overview of Ulmaceae: Diversity, Distribution, Ecological Preferences, and Conservation Status

Yann Fragnière [1], Yi-Gang Song [2], Laurence Fazan [1], Steven R. Manchester [3], Giuseppe Garfì [4] and Gregor Kozlowski [1,2,5,*]

[1] Department of Biology and Botanic Garden, University of Fribourg, Chemin du Musée 10, CH-1700 Fribourg, Switzerland; yann.fragniere@unifr.ch (Y.F.); laurence.fazan@unifr.ch (L.F.)
[2] Eastern China Conservation Centre for Wild Endangered Plant Resources, Shanghai Chenshan Botanical Garden, 3888 Chenhua Road, Songjiang, Shanghai 201602, China; ygsong@cemps.ac.cn
[3] Florida Museum of Natural History, University of Florida, 1659 Museum Rd, Gainesville, FL 32611, USA; steven@flmnh.ufl.edu
[4] Institute of Biosciences and BioResources—National Research Council, Corso Calatafimi 414, 90129 Palermo, Italy; giuseppe.garfi@ibbr.cnr.it
[5] Natural History Museum Fribourg, Chemin du Musée 6, CH-1700 Fribourg, Switzerland
* Correspondence: gregor.kozlowski@unifr.ch; Tel.: +41-26-300-88-42

Abstract: The elm family (Ulmaceae) is a woody plant group with important scientific, societal, and economic value. We aim to present the first biogeographic synthesis investigating the global diversity, distribution, ecological preferences, and the conservation status of Ulmaceae. A literature review was performed to explore the available data for all extant species. Our study made it possible to map the actual global distribution of Ulmaceae with high precision, and to elucidate the centers of diversity, located mainly in China and in the southeastern USA. A detailed comparative analysis of the macroclimatic niche for each species was produced, which shows the general biogeographic pattern of the family and pinpoints the outlier species. The results corroborate recent molecular analyses and support the division of Ulmaceae into two taxonomically, biogeographically, and ecologically well-differentiated groups: the so-called temperate clade with 4 genera and 43 species and the tropical clade with 3 genera and 13 species. The elm family is often described as a typical temperate plant group, however the diversity peak of all Ulmaceae is located in the subtropical zone, and a non-negligible part of the family is exclusively distributed in the tropics. We also noticed that a high proportion of Ulmaceae is linked to humid macro- or microhabitats. Finally, we highlighted that nearly 25% of all Ulmaceae are threatened. Fieldwork, conservation efforts, and research activities are still necessary for this family, particularly for the tropical members and the most endangered species.

Keywords: climatic niche; diversity centers; elm family; Köppen–Geiger climate classification; latitudinal diversity gradient; relict trees

Citation: Fragnière, Y.; Song, Y.-G.; Fazan, L.; Manchester, S.R.; Garfì, G.; Kozlowski, G. Biogeographic Overview of Ulmaceae: Diversity, Distribution, Ecological Preferences, and Conservation Status. *Plants* **2021**, *10*, 1111. https://doi.org/10.3390/plants10061111

Academic Editor: Stefan Zerbe

Received: 6 April 2021
Accepted: 27 May 2021
Published: 31 May 2021

1. Introduction

Although ranking among the smallest families in the plant kingdom in terms of the number of species, the elm family (Ulmaceae) has important scientific, economic, societal, and conservation value [1,2]. Ulmaceae is an ancient and exclusively woody plant group consisting of deciduous, rarely evergreen trees and shrubs [3–5]. It is an extremely interesting plant family with respect to different scientific issues, such as paleobotany, biogeography, systematics, plant evolution, or species diversification. However, though considered a relatively well known plant group, many of its representatives, especially from the tropical regions, still remain insufficiently investigated, e.g., [6,7].

Ulmaceae includes many relict trees. Fossil records date the origin of the family to the early Cenozoic Era. By the early Paleocene, members of the elm family were already widespread throughout the entire Northern Hemisphere [8,9], but the oldest confirmed records of extant genera, such as *Ulmus* and *Zelkova*, are from the Eocene. At least one widespread genus went extinct, i.e., *Cedrelospermum*, from the Paleogene and Neogene of Europe, Asia, and North America [10,11]. Other genera, common in the past at the continental scale, persist at present in disjunct (e.g., *Zelkova*, growing in Eastern Asia and South-Western Eurasia) [1] or in restricted distribution areas (e.g., *Hemiptelea*, thriving in several localities of Korea and eastern China) [12]. The oldest fossils of living genera consist of leaves and fruits of *Ulmus* referred to the early Eocene of China (ca. 50 Mya) and the middle to late Eocene of North America [13].

The systematics of elm family has had a very controversial story. It was taxonomically described for the first time in 1815 by de Mirbel [14] and contained at that time only two genera, *Ulmus* and *Celtis*. Subsequently, for nearly 150 years, the family was commonly divided into two subgroups associated with each of these original genera [8]. Until the late 1990s, two subfamilies of Ulmaceae were recognized, the Ulmoideae and the Celtidoideae, often denominated as "ulmoids" and "celtoids," respectively [15], though at the end of the 1960s Grudzinskaya [16] had proposed distinguishing two different families within the family Ulmaceae, the Ulmaceae s.s. and the Celtidaceae. At that time, the number of genera included in the elm family ranged between 15 and 18 [8,9]. The clarification of the taxonomic division of this group came with the molecular phylogenies performed on all closely related urticoid families of the order Rosales, mainly on Cannabaceae. The modern circumscription of Cannabaceae resulted in the integration of the majority of celtoids into this family (e.g., *Celtis*, *Pteroceltis*, *Trema*, *Aphananthe*, and *Lozanella*) and their exclusion from Ulmaceae [15,17]. Furthermore, the genus *Ampelocera*, treated previously as a member of Celtidoideae [8,18], was recognized as an ulmoid taxon [6,15,19].

Modern treatments based on molecular phylogenies therefore clearly separate Cannabaceae and Ulmaceae [15,18–21]. Moreover, Cannabaceae is a sister family to Moraceae and Urticaceae and thus is not the closest taxon of Ulmaceae within the urticalean rosids [17,20,22]. Moreover, recent molecular studies [23,24] divide the Ulmaceae into two distinct taxonomic and biogeographic groups: the temperate clade (including *Ulmus*, *Zelkova*, *Planera*, and *Hemiptelea*) and the tropical clade (including *Ampelocera*, *Phyllostylon*, and *Holoptelea*).

Ulmaceae possess very distinctive fruit structures (and corresponding dispersal mechanisms), which provide interesting elements to outline the evolutionary pathways within the family [16], in addition to being extremely important for the identification of extant and extinct genera and species [8]. The asymmetric akene-type, unwinged fruit of *Zelkova* is suggested as being the most primitive fruit-type within the family. At the opposite, the genera *Ulmus*, *Hemiptelea*, *Holoptelea*, and *Phyllostylon* have the most evolved winged fruits and are thus clearly wind-dispersed [9]. However, the dispersal mechanisms of *Hemiptelea* need more investigation, since its small asymmetric fruits possess a wing-like appendix only on one side of the fruit [3], and for this reason it can be referred to as an intermediate step in the evolutionary pathway of the family. *Planera* have fleshy protuberances, and since they grow mainly along water courses, the tree is probably water dispersed [9]. The members of the neotropical *Ampelocera* possess ellipsoid or even pyriform drupes, which in certain species can be relatively large and colored (e.g., *A. macrocarpa*) and are primarily bird-dispersed [6]. The most sophisticated, however, seems to be the dispersal mechanisms in the genus *Zelkova* due to the drupaceous and unwinged features of its individual fruits. In fact, mature fruits commonly fall with the entire twig, and the dried leaves that are still attached function as a drag-enhancing appendage, carrying the fruits away from the parent tree [25,26].

The members of Ulmaceae show a variety of breeding systems and floral types [27]. Some genera, such as *Zelkova* or *Planera*, exhibit three flower types (staminate, pistillate, and hermaphrodite flowers) on the same individual or even on the same flowering branch [27,28]. Similar to other closely related families belonging to the Urticalean rosids, the flowers of Ulmaceae have only one whorl of 4–8 green or brown perianth lobes (denominated either as sepals or tepals). Stamens usually show the same number as tepals (with the exception of *Holoptelea* and *Ampelocera* with up to 12 or 16 stamens, respectively) [27]. The superior ovary is composed of two fused carpels with two linear styles [4].

Despite the long history of Ulmaceae research, a synthesis of the diversity and biogeography of this family that takes more recent publications and the current state of knowledge into account has yet to be produced. Information dealing with spatial distributions and biodiversity is central to many fundamental questions in biogeography and conservation biology [29,30]. The distribution of different plant taxa (especially families) is basic and essential information fundamental in many studies, but syntheses at global scales are rather scarce [31,32]. Understanding global biogeography is of great importance for the effective conservation of any group of organisms, especially plant families with disjunct distribution patterns [28]. In this paper, we investigate the global diversity, distribution pattern, conservation status, and ecological preferences of the elm family. The aim of the present work is thus to provide an up-to-date synthesis. Our main objectives are to (1) present the actual global distribution of Ulmaceae with the highest possible resolution, (2) contribute to elucidating the diversity hotspots of Ulmaceae at the generic and species levels, (3) elucidate the realized (macro)climatic niche and ecological preferences of all extant Ulmaceae species, and (4) synthesize the conservation status of the elm family.

2. Results

2.1. Taxonomic Division and Species List

Based on the published taxonomic treatments and available literature (Supplementary file S1), the taxonomic division and number of species of the Ulmaceae family is given in Table 1. The elm family consists of 56 species, divided into 2 clades (13 in the tropical clade and 43 in the temperate clade) and 7 genera. In general, the information and published literature on the tropical members of the family are sparser than for the temperate species (Supplementary file S1). The following doubtful *Ulmus* species were not included: *U. chumlia*, treated as a synonym of *U. androssowii* [33,34]; *U. procera*, treated as a synonym of *U. minor*, introduced in North America [34,35]; *U. elliptica* (Caucasus), treated as a synonym of *U. glabra* [33,36]; and *U. densa* (Central Asia), treated as a synonym of *U. minor* [33,37].

Table 1. Summary of the taxonomic division, genera and species number, and general distribution of the elm family (Ulmaceae). # Species: number of species. Parenthesis: number of species by region.

Clade	Genus	# Species	General Distribution
Tropical clade	*Ampelocera*	9	South America and Mesoamerica (8), Caribbean (1)
	Holoptelea	2	Africa (1) Asia (1)
	Phyllostylon	2	South America (1) South America, Mesoamerica and Caribbean (1)
Temperate clade	*Hemiptelea*	1	Eastern Asia
	Planera	1	North America
	Ulmus	35	North America (6) Mesoamerica (2) Europe and Western Asia (3) Asia (24)
	Zelkova	6	Mediterranean Europe and Western Asia (3) Eastern Asia (3)
Total		56	

2.2. Species and Genera Distribution

An up-to-date global distribution map of Ulmaceae was produced, which corresponds to the most actual chorological knowledge of this family with the highest possible resolution (Figure 1). This map combines the individual distribution maps made for each species (see

Methods). In many areas, the distributions of several species overlap. Several regions can be considered hotspots of Ulmaceae diversity, with numerous species co-occurring in the same area. This is especially true for Eastern Asia. China has the highest diversity worldwide, with 12 species and 3 genera. The main Chinese hotspots are in the following provinces: western Henan, Shanxi, Shaanxi, southern Anhui, western Zhejang, northern Jiangxi, and Hubei. At least three species can be found in all provinces of China (except western provinces and Hainan), in Taiwan, in North and South Korea, in Japan and in the Russian Far East. The southeastern United States is another important Ulmaceae hotspot, mainly in Arkansas (six species and two genera), Tennessee, Louisiana, Mississippi, Alabama, western Kentucky, and eastern Texas. In the majority of Central and Eastern Europe, three *Ulmus* species co-occur. In the Sub-Himalayan region, up to three *Ulmus* species can also be found together in India (Kashmir, Himachal Pradesh). In South and Central America, a maximum of three to four *Ampelocera* species occur together, mainly in eastern Colombia (Choco, Antioquia, Cordoba, and Zulia) and marginally in Brazil (Acre).

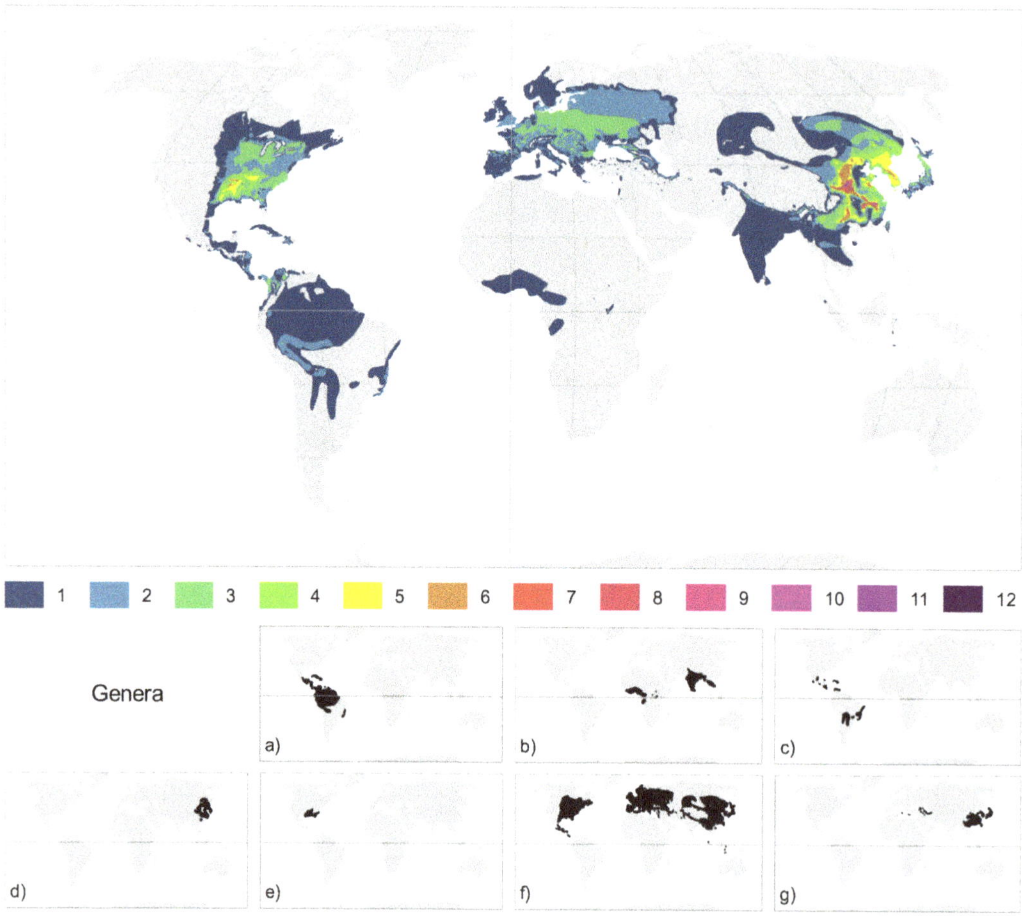

Figure 1. Global distribution of Ulmaceae. The color gradient shows the number of species with overlapping distribution. The small maps below indicate the global distribution of the different genera. Tropical clade: (**a**) *Ampelocera*, (**b**) *Holoptelea*, and (**c**) *Phyllostylon*; Temperate clade: (**d**) *Hemiptelea*, (**e**) *Planera*, (**f**) *Ulmus*, and (**g**) *Zelkova*.

The two clades show very different latitudinal diversity patterns (Figure 2). The tropical clade is nearly entirely confined between the tropical lines, with a peak between 5°

and 12° of N latitude. Only rare species of the temperate clade cross the Tropic of Cancer to the south. The diversity peak of the temperate clade (between 28° N and 38° N) is located in the subtropical zone. Ulmaceae extends to the south to approximately 24° S (*Phyllostylon rhamnoides* in South America) and to the north up to approximately 69° N (*Ulmus glabra* in Europe).

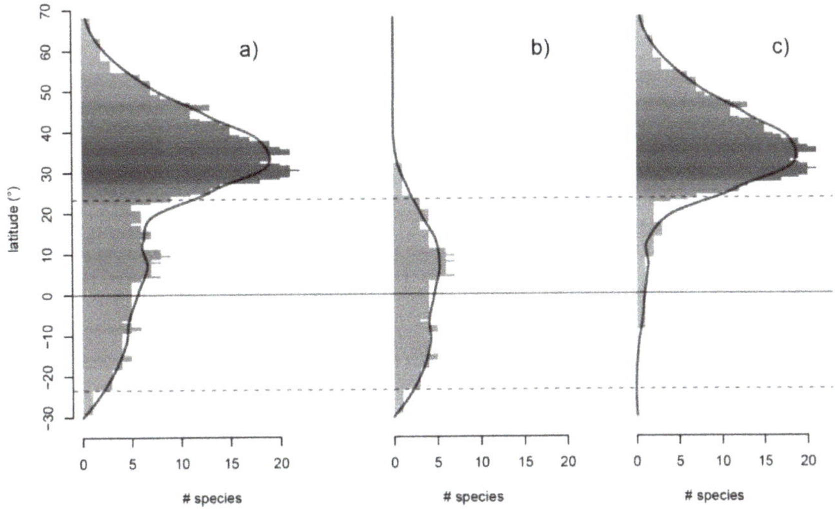

Figure 2. Latitudinal diversity gradient. Species richness by 0.5° latitudinal bin for (**a**) the complete Ulmaceae family, (**b**) the tropical clade and (**c**) the temperate clade. A smooth approximation is shown by the LOESS (locally estimated scatterplot smoothing) curve above the histogram. The equator is represented by the solid horizontal line, and the tropics are represented by the two dashed lines.

2.3. Species Macroclimatic Niche and Ecological Preferences

A detailed overview of the realized macroclimatic niche of all the species of the Ulmaceae family is presented in Figures 3 and 4. Additionally, the ordination plot (Figure 5) allows us to elucidate species that share similar macroclimatic preferences and to highlight outliers. Most of the species of the tropical clade are distributed in areas with tropical climates of (A)f—rainforest, (A)m—monsoon, and (A)w—savannah. The exception is *Ampelocera albertiae*, growing at a high elevation in the mountains in a rather temperate oceanic climate (Cfb), as well as *Phyllostylon rhamnoides* and *Holoptelea integrifolia* found in several climate types, such as BSh (semiarid hot climate), Cwa (dry-winter humid subtropical climate), and Cfa (humid subtropical climate).

The North American members of the temperate clade of Ulmaceae mainly occur in Cfa (humid subtropical climate). Among them, three species also occur in Dfa and Dfb (continental climate without dry season, with warm to hot summer). The European members are mainly in Cfb (oceanic climate), but some species are found in more Mediterranean and/or continental climates. The Asiatic members of the temperate clade are mainly typical elements of Cfa (humid subtropical climate) but are also very common in Cwa (dry-winter humid subtropical climate) and Cwb (dry-winter subtropical highland climate), as well as in Dwa and Dwb (continental climate, with dry winter and warm to hot summer). The niches of some Asiatic species (*Ulmus pumila, U. macrocarpa*, and *U. davidiana*) cover a large gradient of temperatures, including very cold areas, with mean annual temperatures close to or below 0 °C and sometimes with extreme annual temperature variations (Figure 4). Several species are clear outliers among the temperate clade: the Mexican and Mesoamerican *Ulmus* species, the East-Asiatic *U. lanceifolia* and *U. uyematsui* from Taiwan, and the Mediterranean *Zelkova* species.

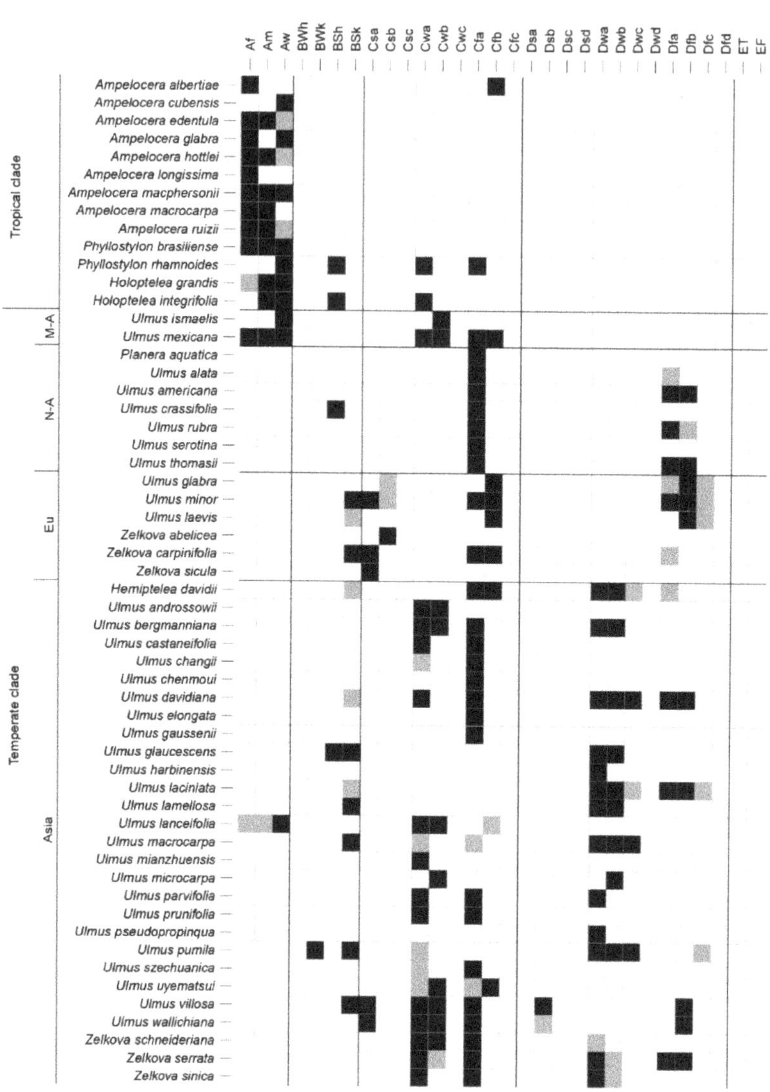

Figure 3. Realized (macro)climatic niches for each species of the Ulmaceae family, according to the Köppen–Geiger climate classification. A black square indicates that a species is largely distributed in the corresponding climate, and a grey square indicates that the species is only marginally distributed in the corresponding climate. The species are presented in the following order from top to bottom: tropical clade and temperate clade, further divided in Mesoamerica (M-A), North America (N-A), Europe (Eu), and Asia. The upper abbreviations indicate the type of climate: tropical rainforest climate (Af), tropical monsoon climate (Am), tropical savanna climate (Aw), arid climate (BW; h—hot, k—cold), semiarid (steppe) climate (BS; h—hot, k—cold), Mediterranean climate (Cs; a—hot summer, b—warm summer, c—cool summer), humid subtropical climate (Cfa), oceanic climate (Cfb), subpolar oceanic climate (Cfc), dry-winter humid subtropical climate (Cwa), dry-winter subtropical highland climate (Cwb), dry-winter subpolar oceanic climate (Cwc), continental climate (D; s—dry summer, w—dry winter, f—no dry season; a—hot summer, b—warm summer, c—cold summer, d—very cold winter), tundra climate (ET), and ice climate (EF).

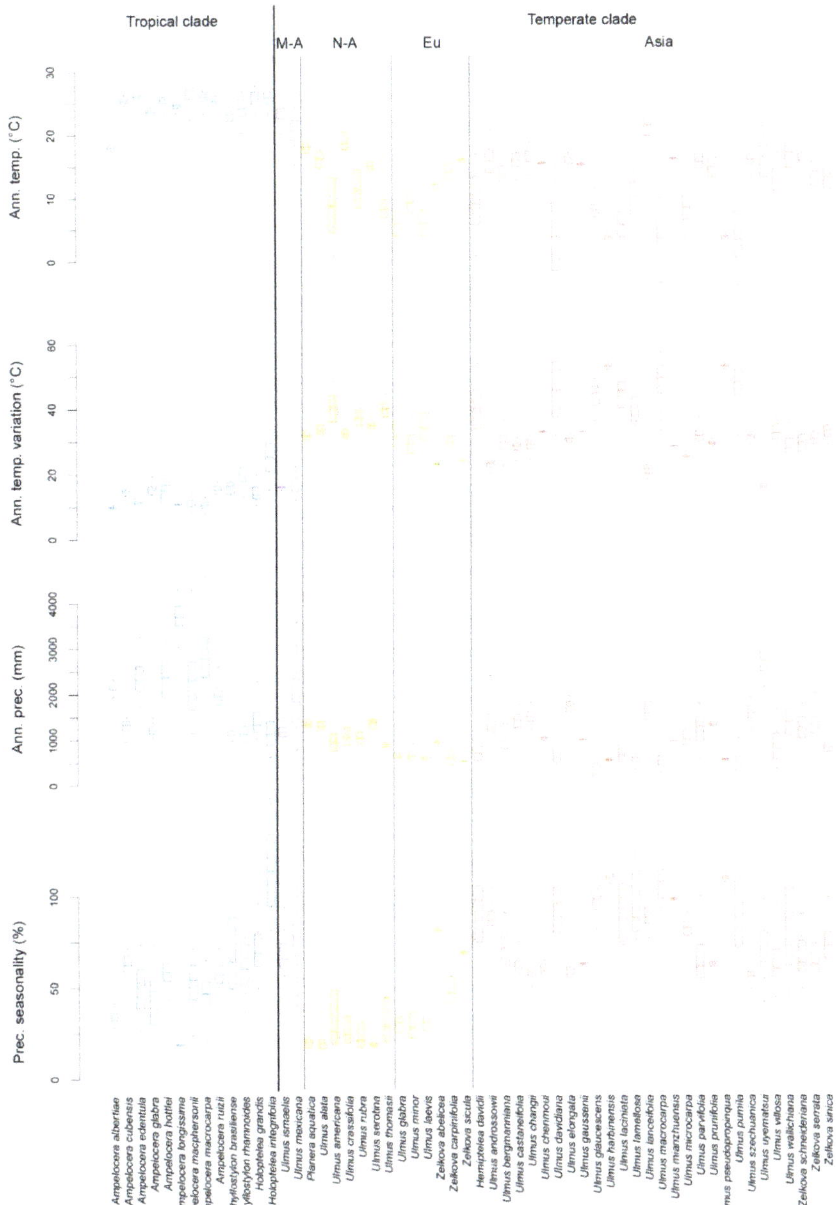

Figure 4. Macroclimatic preferences for each species of the Ulmaceae family according to their natural distribution and for a selection of four climatic variables: mean annual temperature (°C), mean annual temperature variation (maximum temperature of the warmest month—minimum temperature of the coldest month, °C), mean annual precipitation (mm) and precipitation seasonality (variation in monthly precipitation totals over the course of the year, %). The species are presented in the following order from left to right: tropical clade and temperate clade, further divided in Mesoamerica (M-A), North America (N-A), Europe (Eu), and Asia. These five groups are indicated by different colors.

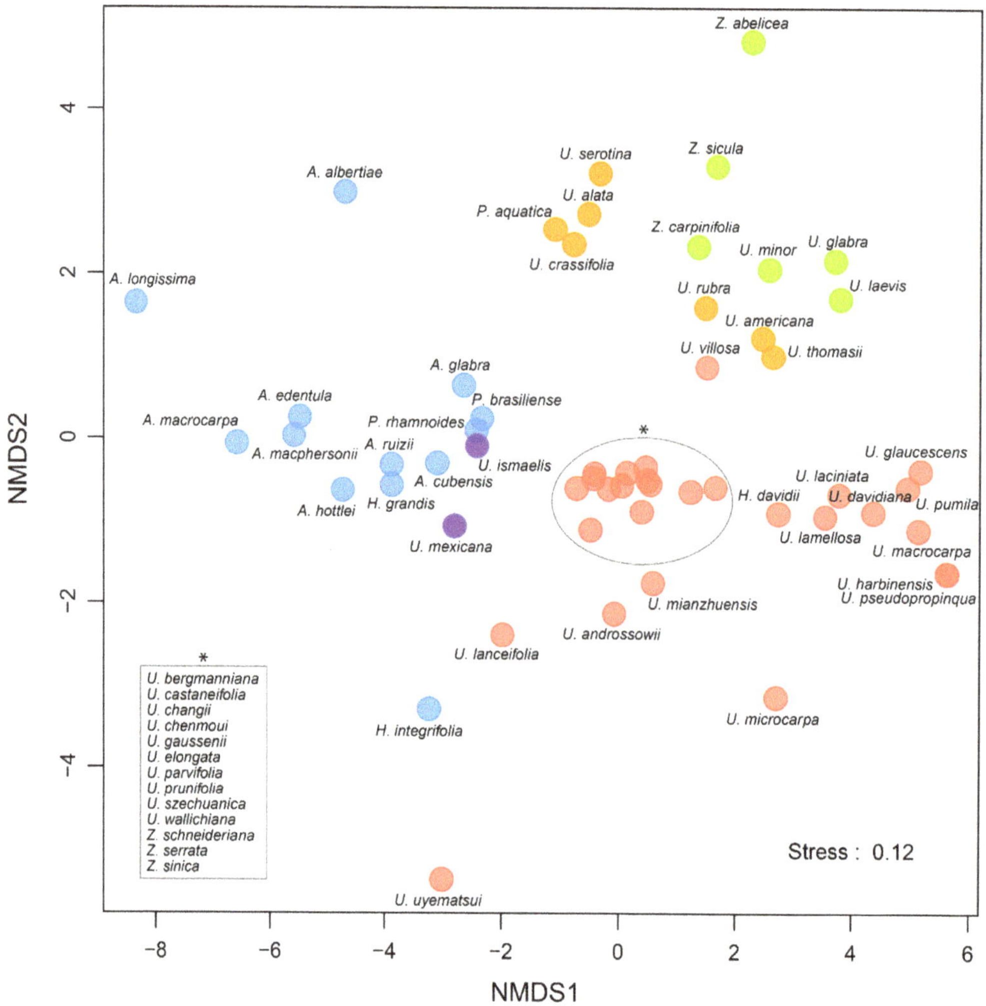

Figure 5. Nonmetric multidimensional scaling (NMDS) ordination plot representing the (macro)climatic similarities between species of the Ulmaceae family. A total of 19 climatic variables were included in the analysis (see Material and Methods). Species appearing close together in the plot share similar macroclimatic preferences. The species are divided into the following groups: tropical clade (blue), temperate clade divided further in Mesoamerica (purple), North America (yellow), Europe (green), and Asia (red).

We collected all available information on the relationship of Ulmaceae species with water and soil conditions (Table 2). Nearly 70% of species belonging to the elm family occur (obligatorily or facultatively) in wet habitats. A majority of species are typical elements of tropical humid forests (15 spp.) or are found exclusively in alluvial and riparian forests (11 spp.). Additionally, there were a relatively large number of Ulmaceae (13 spp.) that occur (not exclusively) in wet microhabitats. Finally, a significant proportion of Ulmaceae species seem to prefer rich, fertile soils.

Table 2. Division of Ulmaceae according to their macro- and microhabitats, with a focus on humid habitats. # Species: number of species.

Category	# Species	%
Species of humid macrohabitats (equatorial and tropical humid rainforests, large amount (>1500 mm) of annual precipitation)	15	26.8
Species of moist microhabitats (alluvial and riparian forests, moist ravines), generally exclusively	11	19.6
Species often occurring in moist microhabitats but also in other habitat types	13	23.2
Species occurring in other habitat types or scarce information available	17	30.4

2.4. Conservation Status

Thirty-eight Ulmaceae species are included to date on the IUCN Red List (i.e., 68% of the total species number). For one-third of all members of the elm family, no global assessment has been made (Table 3). Among the assessed species, two were critically endangered (IUCN category CR), four were endangered (EN), five were vulnerable (VU), and two were near threatened (NT). Thus, 34% of all assessed species and 23% of all Ulmaceae species are considered under threat. This group of threatened species includes all six *Zelkova* species, five *Ulmus* species, and two *Ampelocera* species.

Table 3. Conservation status of the Ulmaceae species. IUCN categories: CR—critically endangered, EN—endangered, VU—vulnerable, NT—nearly threatened, LC—least concerned, DD—data deficiency (IUCN 2020). # Species: number of species.

IUCN Category	# Species	Species
CR	2	*Ulmus gaussenii, Zelkova sicula*
EN	4	*Ampelocera albertiae, Ulmus americana, U. chenmoui, Zelkova abelicea*
VU	5	*Ulmus elongata, U. wallichiana, Zelkova carpinifolia, Z. sinica, Z. schneideriana*
NT	2	*Ampelocera longissima, Zelkova serrata*
LC	22	*Ampelocera edentula, A. hottlei, A. macphersonii, A. macrocarpa, A. ruizii, Hemiptelea davidii, Holoptelea grandis, Phyllostylon rhamnoides, Planera aquatica, Ulmus alata, U. castaneifolia, U. crassifolia, U. davidiana, U. laciniata, U. rubra, U. macrocarpa, U. parvifolia, U. pumila, U. serotina, U. szechuanica, U. thomasii, U. mexicana*
DD	3	*Ulmus glabra, U. laevis, U. minor*
Not assessed	18	*Ampelocera cubensis, A. glabra, Holoptelea integrifolia, Phyllostylon brasiliense, Ulmus androssowii, U. bergmanniana, U. changii, U. glaucescens, U. harbinensis, U. ismaelis, U. lamellosa, U. lanceifolia, U. mianzhuensis, U. microcarpa, U. prunifolia, U. pseudopropinqua, U. uyematsui, U. villosa*

3. Discussion

3.1. Diversity and Distribution

Our global synthesis corroborates recent molecular analyses [23,24] and thus strongly supports the division of Ulmaceae into two taxonomically, biogeographically, and ecologically well-differentiated groups: the so-called temperate clade with 43 species and four genera (*Hemiptelea, Zelkova, Planera,* and *Ulmus*) and the tropical clade with 13 species and three genera (*Holoptelea, Phyllostylon,* and *Ampelocera*). There exists an enormous discrepancy among the number of studies and thus with the exploration and understanding of the biological and evolutionary processes of the temperate species of Ulmaceae in comparison with the tropical species. Therefore, many aspects of the biology, ecology, and phylogenetic relationships of the tropical clade still need much scientific effort [3,6,7]. Furthermore, for many tropical species, in-depth fieldwork is still needed to collect sufficient data about their distribution (e.g., neotropical species with large and scattered distribution: *Phyl-*

lostylon rhamnoides and *Ampelocera macrocarpa*) and to better understand and document their ecology.

Better scientific exploration of northern temperate Ulmaceae is closely linked with their use by humans [38–40]. For millennia, temperate Ulmaceae trees played an important role in rural and forested areas, being part of the traditional landscape as trees with multiple uses [36]. This is especially the case of various species of elm (*Ulmus*) and the keaki tree (*Zelkova serrata*) in Eastern Asia, which is widely used as an ornamental tree in silviculture and timber production [1]. Interestingly, many tropical species are locally commonly used for wood and tool production (e.g., *Phyllostylon brasiliensis*) and/or possess high potential for medicinal purposes (e.g., *Holoptelea*) [4,41] and thus would need an intensification of research efforts.

For similar reasons, the elm family is often described as a nearly exclusively temperate plant group, typical of the Northern Hemisphere [4,9]. Our global synthesis relativizes this assumption. First, the diversity peak of all Ulmaceae is located in the subtropical zone (between 28° and 38° N) and a non-negligible part of the family (ca. 23% of all species) is exclusively distributed in the tropics (the so-called tropical clade). Additionally, several members of the temperate clade are present or even endemic to tropical regions (e.g., *Ulmus ismaelis* and *U. mexicana* in Mexico; and *U. parvifolia*, *U. uyematsui*, or *Zelkova schneideriana* in Taiwan). One member of the temperate clade reaches the Southern Hemisphere (*Ulmus lanceifolia* distributed to the south on the Celebes and Flores islands).

One of the most famous large-scale patterns in biological diversity is the increase in the number of species from the poles to the equator, a trend that has been called the latitudinal diversity gradient (LDG) [29,42]. On a large scale, only a few plant groups do not follow this pattern, such as Poaceae [43] and gymnosperms [31]. Ulmaceae, as a whole, do not exhibit a typical LDG because they have a diversity peak between 28° and 38° of northern latitude (Figure 2). This does not seem to be an unusual pattern in small woody families, with predominantly temperate and relict species (e.g., Juglandaceae, [32]). However, the typical LDG with a peak close to the equator can be observed when only the tropical clade of Ulmaceae is taken into consideration (Figure 2b).

Interestingly, the elm family shows a very similar biogeographical pattern and distribution of diversity centers with other exclusively woody plant families that are rich in relict trees, such as the walnut family (Juglandaceae) [5,32]. Similar to Juglandaceae, the main diversity center of Ulmaceae lies in southeastern Asia (mainly in China) followed by a second center in the southeastern USA and a third in Mesoamerica and northern South America (Figure 1).

Moreover, the fossil record demonstrates prior extirpations and extinction in the history of Ulmaceae, possibly reflecting the response to prior climate change. For example, *Hemiptelea*, which is now endemic to eastern China and the Korean Peninsula, has been confirmed on the basis of its distinctive fossil fruits from the Miocene of Poland and Ukraine (see references in [44]). Although *Ulmus* is no longer native to west-coastal North America, the genus was well established and identifiable from fruits, as well as leaves, in the Eocene to Miocene of California, Oregon, Washington, and British Columbia [8,45]. *Cedrelospermum* was widespread in the Eocene of North America, Europe and Asia and extended well south into the later Cenozoic [46]. The cause of its global extinction remains uncertain.

3.2. Macroclimatic Niche and Ecological Preferences

Our study shows strong differentiation of the macroclimatic niche between the tropical and temperate clades of Ulmaceae (Figures 3–5). Most of the species of the tropical clade are typical elements of tropical and monsoon forests or of savannah. They are thus distributed in tropical climates with a mean annual temperature generally between 20 °C and 28 °C, with very low seasonal variation and annual precipitation that are normally over 1000 mm per year and up to 4000 mm per year for some species in specific regions (e.g., *Ampelocera longissima*, *A. macphersonii*, and *A. macrocarpa*). There are several exceptions, however, with species occupying niches in temperate regions (e.g., in upland areas, *Ampelocera albertiae*) and/or semiarid hot regions (*Phyllostylon rhamnoides* and *Holoptelea integrifolia*). Interestingly, a majority of tropical Ulmaceae possess large leaves (often having entire margins), in contrast to the smaller and usually dentate leaves of temperate genera [6]. This is often interpreted as an adaptation to wet, tropical forest habitats [15]. However, due to the lack of fossil records of the disjunct genera and species of the tropical Ulmaceae clade, their origin and the history of adaptation to tropical climates are less well understood [8].

Similarly, there are also some species of the temperate clade occurring within tropical climates (not exclusively), such as the *Ulmus* species of Mexico and Mesoamerica (*U. mexicana* and *U. ismaelis*) and one species in Eastern Asia (*U. lanceifolia*). These three species are close to the tropical clade in terms of climatic preferences (Figures 4 and 5). North American members of the temperate clade are mainly linked to the humid subtropical climate and in the north to a moderate and humid continental climate. Their climatic niches are generally close to those of European species, but the latter occur in slightly cooler oceanic and continental climates with less precipitation. In Europe, *Zelkova abelicea* and *Z. sicula* are the two notable exceptions among the members of the temperate clade (Figures 4 and 5). They occur within a Mediterranean climate with high precipitation seasonality and dry summers. Recent detailed studies on *Z. sicula* show that this relict and narrow endemic species is isolated and restricted to the Mediterranean climate. The species survived in this region due to suitable microhabitats, but its distribution and dispersal are limited by the Mediterranean climate [47]. Our results highlight that *Zelkova* species found in Mediterranean climates are exceptions among the Ulmaceae.

In Asia, the numerous species of the temperate clade occupy diverse climatic niches with a large gradient of climatic preferences. There is one constant, however, namely, the high precipitation seasonality due to the East Asian monsoon regime, with dry winters. Some notable outliers are *Ulmus uyematsui* (distributed in the mountains of Taiwan, with precipitation between 2000 and 4000 mm per year), as well as *U. pumila*, *U. davidiana*, and *U. macrocarpa*. The last three species possess a broad climatic range, are cold resistant, and occur in monsoon-influenced climates with high seasonality of temperature and precipitation. Our results show that *U. pumila* has a particularly large climatic plasticity among Ulmaceae in terms of temperature and seasonality. It is also one of the most drought-resistant species among the families. These characteristics make this species well adapted for many anthropogenic habitats and probably explain why *U. pumila* is the main species of Ulmaceae reported as invasive [48].

Reliable and precise information about the ecology of the different species of Ulmaceae is scarce, especially for species growing in the tropics or in Asia. Our synthetized compilation (Table 3) demonstrates, however, that a high proportion of Ulmaceae (70%) is linked to humid macro- or microhabitats. This proportion may even be higher, as precise information is missing for many species. Notably, more than 26% of Ulmaceae occur exclusively in tropical humid forests with large amounts of precipitation, higher than 1500 mm per year (e.g., several *Ampelocera* and tropical *Ulmus* species). Moreover, nearly 20% of Ulmaceae are trees exclusive to alluvial and riparian forests or found growing only along river and stream banks. The best examples are North American *Planera aquatica* and *Ulmus americana*, growing exclusively in swamps, along the shores and banks of lakes and rivers and in alluvial flood plains, thus supporting well-waterlogged conditions [49]. In Europe, similar ecological preferences are found for *Ulmus laevis*, which occurs exclusively in riparian forests along large rivers [36]. The two other European *Ulmus* species are also frequently found in alluvial woods, although not exclusively. Furthermore, all six *Zelkova* species are linked to humid habitat conditions to a certain level. East Asiatic *Zelkova* species are particularly common in forests in mountainous regions with high precipitation, as well as in humid ravines and along streams and rivers [1]. This ecological preference is even more pronounced for the Mediterranean members of the genus, with *Z. sicula* growing exclusively along a thalweg, filled in winter with water [47]; and to a lesser extent with *Z. abelicea*, forming large populations only in mountainous areas around winter-moist dolines, summer dry riverbeds or on northern slopes [50].

3.3. Threats and Conservation Status

Nearly one-fourth of all the Ulmaceae species are threatened according to the IUCN Red List (Table 3). Two Ulmaceae species are on the brink of extinction: the Anhui elm (*Ulmus gaussenii*) and the Sicilian zelkova (*Zelkova sicula*) [51,52]. *Ulmus gaussenii* is a narrow endemic species growing in Langya Mountain in Anhui Province of Eastern China [53]. Due to low fertility and habitat destruction, the number of individuals has drastically decreased in recent decades. This is probably the rarest Ulmaceae of the world, since only 26 mature and senescing individual trees are known, growing in a single population and covering an area of less than 10 hectares [54]. *Zelkova sicula*, discovered only in 1991, is also a narrow endemic, occurring only in the Iblei Mountains on the Mediterranean island of Sicily in Italy [52]. The species is known from only two populations, 17 km apart, consisting together of ca. 1860 individuals and covering an extremely small total area of only 0.68 hectares [55]. Furthermore, recent molecular investigations revealed that each population is clonal and considered to be issuing centuries-long sprouting of two single surviving genetic individuals [56,57].

More generally, the genus *Zelkova* is the most endangered group of the elm family, since, in addition to *Z. sicula*, all other members of the genus are endangered (IUCN category EN, *Z. abelicea*), vulnerable (VU, *Z. carpinifolia*, *Z. sinica*, and *Z. schneideriana*) or nearly threatened (NT, *Z. serrata*) according to the IUCN Red List [28,58]. Particularly worrying is the situation of *Z. abelicea*, an endemic species of the Mediterranean island of Crete in Greece [1]. In this species, only well-developed trees can produce fruits. However, the overwhelming majority of individuals are dwarfed and nonfruiting due to overbrowsing by goats [50]. Additionally, the majority of fruit are empty, which is probably due to unfavorable climatic conditions such as pronounced and recurrent droughts [5]. The regeneration of populations via seedlings is nearly impossible due to overgrazing, trampling and soil erosion caused by omnipresent large caprine and ovine flocks [59].

In the genus *Ulmus*, two species are endangered (EN, *U. americana* and *U. chenmoui*), and two species are vulnerable (VU, *U. elongata* and *U. wallichiana*). Although American elm (*U. americana*) still forms large populations in the western parts of the USA and Canada, it is the most susceptible North American elm species to introduced and invasive Dutch elm disease (DED), caused by the fungi *Ophiostoma ulmi* and *O. novo-ulmi* [58,60]. The population decline during the next 100 years is supposed to reach up to 80% due to the continuing threat of DED and due to the destruction of its preferred habitat [60]. European elms were also severely impacted by the DED with a severe mortality [38]. Their conservation status is unclear and should be studied in detail. The other three threatened *Ulmus* species occur in Asia. *Ulmus chenmoui* and *U. elongata* are endemic to China, the number of their populations is very restricted, and their original habitat has been largely destroyed [58,61,62]. *Ulmus wallichiana* is widely distributed in the Himalayan region (Afghanistan, Pakistan, India, and Nepal). However, the species is excessively exploited for fodder and fuel wood, and mature reproducing individuals are extremely rare [63].

Among the neotropical Ulmaceae, *Ampelocera albertiae* (IUCN category endangered, EN, [64]) is the most threatened species. This species is a narrow endemic of Colombia, known from only five populations and is heavily affected by cattle ranching, mining activities, and artificial forest plantations.

Several additional Ulmaceae species that have not yet been assessed and thus are not included on the IUCN Red List [58] have a high probability of being globally threatened. This is probably the case for Mexican/Mesoamerican *Ulmus ismaelis* [65]. The taxon is known only from very few and highly isolated populations in Mexico, Salvador, and Honduras [66]. Similarly, numerous Chinese *Ulmus* species possess very restricted distribution areas (e.g., *U. harbinensis*, *U. lamellosa*, *U. mianzhuensis*, *U. prunifolia*, and *U. pseudopropinqua*), and/or their populations are decreasing due to environmental degradation and habitat loss [67].

Ulmaceae is an evolutionarily ancient family possessing high scientific and conservation value. Much more fieldwork, research, and conservation efforts should be undertaken, especially within the tropical clade and for threatened species with restricted distributions and/or weak biological and ecological knowledge.

4. Materials and Methods

4.1. Taxonomic Division and Species List

The generic division of Ulmaceae was based on recent taxonomic treatments and molecular analyses [10,15,18–21,23,24,68]. The order of the genera in our study follows the phylogenetic trees of Manchester and Tiffney [68] and Jia et al. [10], dividing the elm family into two clades: (1) a tropical clade with *Ampelocera*, *Phyllostylon*, and *Holoptelea* and (2) a temperate clade with *Hemiptelea*, *Zelkova*, *Planera*, and *Ulmus*.

The species numbers of the poorly studied neotropical genera *Ampelocera* and *Phyllostylon* were extracted from two detailed monographies by Todzia [6,7], whereas for the third tropical genus, *Holoptelea*, information was obtained from Boratynska [3] and Todzia [9]. Two genera of the temperate clade, *Planera* and *Hemiptelea*, are monotypic [3,9,49,69]. The species number of the genus *Zelkova* was based on taxonomic and biogeographic compilations and recent molecular studies [1,28,57,70]. The most challenging issue is the taxonomic division of the species-rich genus *Ulmus*. In our study, we extracted taxonomic information from eFloras.org: Flora of China, Flora of North America, Flora of Pakistan, Flora Mesoamericana [49,69,71,72], and several other biogeographic treatments and molecular studies [3,18,36,65,73,74]. We worked at the species level and did not distinguish between subspecies or varieties.

4.2. Data Collection

A literature review was performed to explore the available data about the distribution, ecology, and conservation status of all extant species. Our review protocol was based on Xiao and Watson [75]. We took advantage of open-access online resources, recently published monographs and articles in diverse fields, where useful information could be found. The grey literature was also occasionally consulted. For online research, we used the species names as keyword, alone or accompanied by words like "distribution", "map", "ecology", "habitat", etc. References with inaccurate data (e.g., commercial horticultural websites, personal websites) were excluded in the process. We assessed the quality of the data mainly by crossing them. When the information was concordant and apparently not of the same origin, the references were not excluded. When there was no concordance, we only kept the most reliable sources (known institutions, peer reviewed articles, monographs). All kind of data were taken into account and extracted (maps, tables, texts). The full list of references for each species can be found in Supplementary file S1 in Supplementary material. The authors of all Latin names for Ulmaceae species included in this study, are given in Supplementary file S2. The most important resources are cited here [6,7,12,28,33,36,49,58,69,76–83]. The conservation status assessments and information on threats were taken from the IUCN Red List [58].

4.3. Species Distribution

Distribution maps for each species were georeferenced and produced on GIS (geographic information system) software [84]. The map background comes from different sources [85,86]. The distribution area of each species was determined using literature data, but to improve the reliability and to obtain the most parsimonious results, we also crossed the data with other information such as altitude or climate. The latter was considered by using the Köppen–Geiger climate classification system [87], which was recently made available at a 1 km resolution [88]. The altitudinal data (30 arc-second resolution) were downloaded from WorldClim version 2.1 [89] and are derived from the Shuttle Radar Topography Mission (SRTM) [90]. Finally, for some species, local experts were asked to examine the distribution maps. Only the natural range was considered, although for some species that have been widely planted, this delimitation was not obvious (Supplementary file S2).

4.4. Species Macroclimatic Niche

The distribution of each species was used to assess the realized climatic niche of natural populations. Global data of 19 bioclimatic variables (e.g., mean annual temperature and precipitation, seasonality; the full list of variables is shown in Supplementary file S3) were downloaded from WorldClim at a high resolution (30 arc-seconds) for the 1970–2000 period [91]. For each species, we generated 1000 random points in the area of its distribution, where climatic data were extracted. For each variable, we only kept data between the 5th and 95th percentiles to perform the analyses to remove unwanted outliers due to imprecision in species distributions and for more caution in the evaluation of species climatic preferences. The resolution of the climatic data (30 arc-seconds, ~1 km) does not capture microclimates (e.g., in ravines and slopes), which have been found to be important for some species (e.g., *Zelkova abelicea*, [92]). We therefore must consider that our analyses represent macroclimatic preferences.

The different Köppen–Geiger climates were also recorded for each species, according to its distribution. The Köppen–Geiger climate classification system uses a 2- or 3-letter abbreviation to designate each climate type [87,88]. The first letter indicates the main climate: A—tropical, B—arid, C—temperate, D—cold, and E—polar. The second letter indicates the seasonal precipitation type: m—monsoon, w—savannah, W—desert, S—steppe, s—dry summer, w—dry winter, and f—no dry season. The third letter gives precision regarding the temperature (h—hot, k—cold, a—hot summer, b—warm summer, c—cold summer, and d—very cold winter).

4.5. Statistical Analyses

All data analyses and graphs were performed with R [93]. Ordination was performed with the nonmetric multidimensional scaling procedure (NMDS) using the package vegan [94–96]. The 19 climatic variables available in WorldClim were included in the analysis (see Supplementary file S3). Precipitation data were square-root transformed, and all climatic data were standardized before performing NMDS, with the Euclidean distance as the distance measure.

Supplementary Materials: The following are available online at https://www.mdpi.com/article/10.3390/plants10061111/s1, Supplementary file S1: Full list of references for the distribution of each species, Supplementary file S2: Chorological maps of all Ulmaceae species, Supplementary file S3: List of bioclimatic variables.

Author Contributions: Conceptualization, Y.F. and G.K.; collection and analysis of data, Y.F.; writing, review and editing, Y.F., G.K., Y.-G.S., L.F., G.G. and S.R.M. All authors have read and agreed to the published version of the manuscript.

Funding: Fondation Franklinia (G.K., L.F.).

Institutional Review Board Statement: Not applicable.

Informed Consent Statement: Not applicable.

Data Availability Statement: Data available from the Zenodo open-access repository: https://doi.org/10.5281/zenodo.4600469, (accessed on 1 April 2021).

Acknowledgments: We would like to thank the Fondation Franklinia and the team of the Botanic Garden of the University of Fribourg (Switzerland) for their support and help in preparation of the manuscript.

Conflicts of Interest: The authors declare no conflict of interest.

References

1. Kozlowski, G.; Bétrisey, S.; Song, Y.-G.; Fazan, L.; Garfì, G. *The Red List of Zelkova*; Natural History Museum Fribourg: Fribourg, Switzerland, 2018; ISBN 2-9701096-2-X.
2. Simpson, M.G. *Plant Systematics*, 3rd ed.; Elsevier: Amsterdam, The Netherlands, 2019; ISBN 0-12-812628-0.
3. Boratynska, K. Chorology of the family ulmaceae (sensu stricto). *Arbor Kornikie* **1989**, *34*, 3–29.

4. Christenhusz, M.J.; Fay, M.F.; Chase, M.W. *Plants of the World: An Illustrated Encyclopedia of Vascular Plants*; University of Chicago Press: Chicago, IL, USA, 2017; ISBN 0-226-53670-X.
5. Fazan, L.; Song, Y.-G.; Kozlowski, G. The woody planet: From past triumph to manmade decline. *Plants* **2020**, *9*, 1593. [CrossRef]
6. Todzia, C.A. A Revision of *Ampelocera* (Ulmaceae). *Ann. Mo. Bot. Gard.* **1989**, *76*, 1087–1102. [CrossRef]
7. Todzia, C.A. A Reevaluation of the genus *Phyllostylon* (Ulmaceae). *SIDA Contribut. Bot.* **1992**, *15*, 263–270.
8. Manchester, S.R. Systematics and fossil history of the Ulmaceae. *Evolut. Syst. Foss. Hist. Hamamelidae* **1989**, *2*, 221–251.
9. Todzia, C.A. Ulmaceae. In *Flowering Plants·Dicotyledons*; Springer: Berlin/Heidelberg, Germany, 1993; pp. 603–611.
10. Jia, L.-B.; Manchester, S.R.; Su, T.; Xing, Y.-W.; Chen, W.-Y.; Huang, Y.-J.; Zhou, Z.-K. First occurrence of *Cedrelospermum* (Ulmaceae) in Asia and its biogeographic implications. *J. Plant Res.* **2015**, *128*, 747–761. [CrossRef] [PubMed]
11. Jia, L.-B.; Su, T.; Huang, Y.-J.; Wu, F.-X.; Deng, T.; Zhou, Z.-K. First fossil record of *Cedrelospermum* (Ulmaceae) from the Qinghai–Tibetan plateau: Implications for morphological evolution and biogeography. *J. Syst. Evol.* **2019**, *57*, 94–104. [CrossRef]
12. Fang, J.; Wang, Z.; Tang, Z. *Atlas of Woody Plants in China: Distribution and Climate*; Springer Science & Business Media: Berlin/Heidelberg, Germany, 2011; Volume 1, ISBN 3-642-15017-9.
13. Wang, Q.; Manchester, S.R.; Li, C.; Geng, B. Fruits and leaves of *Ulmus* from the paleogene of fushun, Northeastern China. *Int. J. Plant Sci.* **2010**, *171*, 221–226. [CrossRef]
14. De Mirbel, C.F.B. *Élements de Physiologie Végétale et de Botanique*; Magimel: Paris, France, 1815; Volume 1.
15. Wiegrefe, S.J.; Sytsma, K.J.; Guries, R.P. The Ulmaceae, one family or two? Evidence from chloroplast DNA restriction site mapping. *Plant Syst. Evol.* **1998**, *210*, 249–270. [CrossRef]
16. Grudzinskaya, I.A. Ulmaceae and reasons for distinguishing celtidoideae as a separate family celtidaceae link. *Bot. Zhurn.* **1967**, *52*, 1723–1748.
17. Yang, M.-Q.; van Velzen, R.; Bakker, F.T.; Sattarian, A.; Li, D.-Z.; Yi, T.-S. Molecular phylogenetics and character evolution of Cannabaceae. *Taxon* **2013**, *62*, 473–485. [CrossRef]
18. Zavada, M.S.; Kim, M. Phylogenetic analysis of Ulmaceae. *Plant Syst. Evol.* **1996**, *200*, 13–20. [CrossRef]
19. Ueda, K.; Kosuge, K.; Tobe, H. A molecular phylogeny of Celtidaceae and Ulmaceae (Urticales) based on RbcL nucleotide sequences. *J. Plant Res.* **1997**, *110*, 171–178. [CrossRef]
20. Zhang, S.; Soltis, D.E.; Yang, Y.; Li, D.; Yi, T. Multi-gene analysis provides a well-supported phylogeny of rosales. *Mol. Phylogenet. Evol.* **2011**, *60*, 21–28. [CrossRef]
21. Chase, M.W.; Christenhusz, M.J.M.; Fay, M.F.; Byng, J.W.; Judd, W.S.; Soltis, D.E.; Mabberley, D.J.; Sennikov, A.N.; Soltis, P.S.; Stevens, P.F. An update of the angiosperm phylogeny group classification for the orders and families of flowering plants: APG IV. *Bot. J. Linn. Soc.* **2016**, *181*, 1–20.
22. Sytsma, K.J.; Morawetz, J.; Pires, J.C.; Nepokroeff, M.; Conti, E.; Zjhra, M.; Hall, J.C.; Chase, M.W. Urticalean rosids: Circumscription, rosid ancestry, and phylogenetics based on RbcL, TrnL-F, and NdhF sequences. *Am. J. Bot.* **2002**, *89*, 1531–1546. [CrossRef]
23. Neubig, K.; Herrera, F.; Manchester, S.R.; Abbott, J.R. Fossils, biogeography and dates in an expanded phylogeny of Ulmaceae. In Proceedings of the Botany 2012—Annual Meeting of the Botanical Society of America in Columbus, Columbus, OH, USA, 7–11 July 2012.
24. Zhang, Q.; Deng, M.; Bouchenak-Khelladi, Y.; Zhou, Z.; Hu, G.; Xing, Y. The diversification of the northern temperate woody flora—A case study of the elm family (Ulmaceae) based on phylogenomic and paleobotanical evidence. *J. Syst. Evol.* **2021**. [CrossRef]
25. Oyama, H.; Fuse, O.; Tomimatsu, H.; Seiwa, K. Variable seed behavior increases recruitment success of a hardwood tree, *Zelkova Serrata*, in spatially heterogeneous forest environments. *For. Ecol. Manag.* **2018**, *415*, 1–9. [CrossRef]
26. Certini, D.; Fazan, L.; Nakayama, N.; Viola, I.M.; Kozlowski, G. Velocity of the falling dispersal units in *Zelkova abelicea*: Remarkable evolutionary conservation within the relict tree genus. *Am. J. Bot.* **2020**. [CrossRef]
27. Leme, F.M.; Staedler, Y.M.; Schönenberger, J.; Teixeira, S.P. Ontogeny and vascularization elucidate the atypical floral structure of *Ampelocera Glabra*, a tropical species of Ulmaceae. *Int. J. Plant Sci.* **2018**, *179*, 461–476. [CrossRef]
28. Kozlowski, G.; Gratzfeld, J. *Zelkova—An Ancient Tree: Global Status and Conservation Action*; Natural History Museum Fribourg: Fribourg, Switzerland, 2013.
29. Mutke, J.; Barthlott, W. Patterns of vascular plant diversity at continental to global scales. *Biol. Skr.* **2005**, *55*, 521–531.
30. Collen, B.; Whitton, F.; Dyer, E.E.; Baillie, J.E.; Cumberlidge, N.; Darwall, W.R.; Pollock, C.; Richman, N.I.; Soulsby, A.-M.; Böhm, M. Global patterns of freshwater species diversity, threat and endemism. *Glob. Ecol. Biogeogr.* **2014**, *23*, 40–51. [CrossRef] [PubMed]
31. Fragnière, Y.; Bétrisey, S.; Cardinaux, L.; Stoffel, M.; Kozlowski, G. Fighting their last stand? A global analysis of the distribution and conservation status of gymnosperms. *J. Biogeogr.* **2015**, *42*, 809–820. [CrossRef]
32. Song, Y.-G.; Fragnière, Y.; Meng, H.-H.; Li, Y.; Bétrisey, S.; Corrales, A.; Manchester, S.R.; Deng, M.; Jasińska, A.K.; Văn Sâm, H. Global biogeographic synthesis and priority conservation regions of the relict tree family Juglandaceae. *J. Biogeogr.* **2020**, *47*, 643–657. [CrossRef]
33. Hassler, M. World plants: Synonymic checklists of the vascular plants of the world (Version Nov 2018). In *Species 2000 & ITIS Catalogue of Life, 2020-09-01 Beta*; Roskov, Y., Ower, G., Orrell, T., Nicolson, D., Bailly, N., Kirk, P.M., Bourgoin, T., DeWalt, R.E., Decock, W., van Nieukerken, E., et al., Eds.; Naturalis: Leiden, The Netherlands, 2020.

34. The Plant List A Working List of All Plant Species; Version 1.1. Available online: http://www.theplantlist.org/ (accessed on 15 December 2020).
35. Elbert, L.; Little, J. *Checklist of United States Trees (Native and Naturalized)*; Forest Service, US Department of Agriculture: Washington, DC, USA, 1979.
36. Caudullo, G.; De Rigo, D. *Ulmus*-elms in Europe: Distribution, habitat, usage and threats. In *European Atlas of Forest Tree Species*; Publication Office of the European Union: Luxembourg, 2016; pp. 186–188.
37. Seregin, A.P. Digital herbarium of Moscow State University: The largest Russian biodiversity database. *Biol. Bull.* **2017**, *44*, 584–590. [CrossRef]
38. Dunn, C.P. *The Elms: Breeding, Conservation, and Disease Management*; Springer Science & Business Media: Berlin/Heidelberg, Germany, 2000; ISBN 0-7923-7724-9.
39. Boratyńska, K.; Sękiewicz, M.; Boratyński, A. Morfologia, systematyka i rozmieszczenie geograficzne. In *Wiązy (Ulmus)*; Bugała, W., Boratyński, A., Iszkuło, G., Eds.; Bogucki Wydawnictwo Naukowe: Poznań, Poland, 2015; pp. 24–52.
40. Zhang, Q.; Zhou, Z.; Xing, Y. Phylogeny and biogeographic history of Ulmaceae based on complete chloroplast genome and nuclear sequences. In Proceedings of the 16th National Congress of Botanical Society of China, Kunming, China, 10–13 October 2018.
41. Kumar, V.; Singh, S.; Bhadouria, R.; Singh, R.; Prakash, O. Phytochemical, analytical and medicinal studies of *Holoptelea Integrifolia* roxb: Planch—A Review. *Curr. Tradit. Med.* **2019**, *5*, 270–277. [CrossRef]
42. Hillebrand, H. On the generality of the latitudinal diversity gradient. *Am. Nat.* **2004**, *163*, 192–211. [CrossRef]
43. Visser, V.; Clayton, W.D.; Simpson, D.A.; Freckleton, R.P.; Osborne, C.P. Mechanisms driving an unusual latitudinal diversity gradient for grasses. *Glob. Ecol. Biogeogr.* **2014**, *23*, 61–75. [CrossRef]
44. Manchester, S.R.; Chen, Z.-D.; Lu, A.-M.; Uemura, K. Eastern Asian endemic seed plant genera and their paleogeographic history throughout the northern hemisphere. *J. Syst. Evol.* **2009**, *47*, 1–42. [CrossRef]
45. Denk, T.; Dillhoff, R.M. *Ulmus* leaves and fruits from the early-middle Eocene of Northwestern North America: Systematics and implications for character evolution within Ulmaceae. *Botany* **2005**, *83*, 1663–1681. [CrossRef]
46. Magallón-Puebla, S.; Cevallos-Ferriz, S.R. Latest occurrence of the extinct genus *Cedrelospermum* (Ulmaceae) in North America: *Cedrelospermum Manchesteri* from Mexico. *Rev. Palaeobot. Palynol.* **1994**, *81*, 115–128. [CrossRef]
47. Garfì, G.; Carimi, F.; Fazan, L.; Gristina, A.S.; Kozlowski, G.; Console Livreri, S.; Motisi, P.; Pasta, S. From glacial refugia to hydrological microrefugia: Factors and processes driving the persistence of the climate relict tree *Zelkova sicula*. *Ecol. Evol.* **2021**. [CrossRef]
48. Lykholat, Y.; Khromykh, N.; Didur, O.; Alexeyeva, A.; Lykholat, T.; Davydov, V. Modeling the invasiveness of *Ulmus pumila* in urban ecosystems in conditions of climate change. *Regulat. Mech. Biosyst.* **2018**, *9*, 161–166. [CrossRef]
49. Sherman-Broyles, S.L.; Barker, W.T.; Schulz, L.M. Ulmaceae. *Flora N. Am.* **1997**, *3*, 368–380.
50. Kozlowski, G.; Frey, D.; Fazan, L.; Egli, B.; Bétrisey, S.; Gratzfeld, J.; Garfì, G.; Pirintsos, S. The tertiary relict tree *Zelkova abelicea* (Ulmaceae): Distribution, population structure and conservation status on Crete. *Oryx* **2014**, *48*, 80–87. [CrossRef]
51. World Conservation Monitoring Centre IUCN Red List of Threatened Species: Ulmus Gaussenii. Available online: https://www.iucnredlist.org/en (accessed on 4 November 2020).
52. Garfì, G.; Pasta, S.; Fazan, L.; Kozlowski, G. IUCN Red List of Threatened Species: Zelkova Sicula. Available online: https://www.iucnredlist.org/en (accessed on 6 October 2020).
53. Zhang, Q.; Zhang, H.; Li, Q.; Bai, R.; Ning, E.; Cai, X. Characterization of the complete chloroplast genome sequence of an endangered elm species, *Ulmus gaussenii* (Ulmaceae). *Conserv. Genet. Resour.* **2019**, *11*, 71–74. [CrossRef]
54. Geng, Q.-F.; Yang, J.; He, J.; Wang, D.-B.; Shi, E.; Xu, W.-X.; Jeelani, N.; Wang, Z.-S.; Liu, H. Microsatellite markers for the critically endangered elm species *Ulmus gaussenii* (Ulmaceae). *Genes Genet. Syst.* **2016**, *91*, 11–14. [CrossRef]
55. Garfì, G.; Carimi, F.; Pasta, S.; Rühl, J.; Trigila, S. Additional insights on the ecology of the relic tree *Zelkova sicula* di Pasquale, garfì et quézel (Ulmaceae) after the finding of a new population. *Flora Morphol. Distribut. Funct. Ecol. Plants* **2011**, *206*, 407–417. [CrossRef]
56. Garfì, G.; Buord, S. Relict species and the challenges for conservation: The emblematic Case of *Zelkova sicula* di pasquale, garfì et quézel and the efforts to save it from extinction. *Biodivers. J.* **2012**, *3*, 281–296.
57. Christe, C.; Kozlowski, G.; Frey, D.; Bétrisey, S.; Maharramova, E.; Garfì, G.; Pirintsos, S.; Naciri, Y. Footprints of past intensive diversification and structuring in the genus *Zelkova* (Ulmaceae) in south-western Eurasia. *J. Biogeogr.* **2014**, *41*, 1081–1093. [CrossRef]
58. IUCN. The IUCN Red List of Threatened Species. Available online: https://www.iucnredlist.org (accessed on 11 December 2020).
59. Fazan, L.; Guillet, S.; Corona, C.; Kozlowski, G.; Stoffel, M. Imprisoned in the Cretan mountains: How relict *Zelkova abelicea* (Ulmaceae) trees cope with Mediterranean climate. *Sci. Total Environ.* **2017**, *599*, 797–805. [CrossRef] [PubMed]
60. Stritch, L.; Rivers, M.C.; Barstow, M. IUCN Red List of Threatened Species: Ulmus Americana. Available online: https://www.iucnredlist.org/en (accessed on 20 October 2020).
61. Gao, J.-G.; Wu, Y.-H.; Xu, G.-D.; Li, W.-Q.; Yao, G.-H.; Ma, J.; Liu, P. Phylogeography of *Ulmus elongata* based on fourier transform-infrared spectroscopy (FTIR), thermal gravimetric and differential thermal analyses. *Biochem. Syst. Ecol.* **2012**, *40*, 184–191. [CrossRef]

62. Song, J.; Chen, L.; Chen, F.; Ye, J. Edaphic and host plant factors are linked to the composition of arbuscular mycorrhizal fungal communities in the root zone of endangered *Ulmus chenmoui* Cheng in China. *Ecol. Evol.* **2019**, *9*, 8900–8910. [CrossRef] [PubMed]
63. Mughal, A.H.; Mugloo, J.A. Elm (*Ulmus wallichiana*): A vulnerable lesser known multipurpose tree species of Kashmir valley. *SKUAST J. Res.* **2016**, *18*, 73–79.
64. Lopez-Gallego, C.; Morales, M.P.A. IUCN Red List of Threatened Species: Ampelocera albertiae. Available online: https://www.iucnredlist.org/en (accessed on 9 February 2021).
65. Todzia, C.A.; Panero, J.L. A New species of *Ulmus* (Ulmaceae) from Southern Mexico and a synopsis of the species in Mexico. *Brittonia* **1998**, *50*, 343–347. [CrossRef]
66. Linares, J.L. Primer registro de *Ulmus ismaelis* (Ulmaceae) para Centroamérica. *Rev. Mexicana Biodivers.* **2005**, *76*, 95–96. [CrossRef]
67. Hou, H.; Ye, H.; Wang, Z.; Wu, J.; Gao, Y.; Han, W.; Na, D.; Sun, G.; Wang, Y. Demographic history and genetic differentiation of an endemic and endangered *Ulmus lamellosa* (Ulmus). *BMC Plant Biol.* **2020**, *20*, 526. [CrossRef]
68. Manchester, S.R.; Tiffney, B.H. Integration of paleobotanical and neobotanical data in the assessment of phytogeographic history of holarctic angiosperm clades. *Int. J. Plant Sci.* **2001**, *162*, 19–27. [CrossRef]
69. Wu, Z.; Raven, P.H.; Hong, D. *Flora of China—Ulmaceae through Basellaceae*; Science Press: Beijing, China, 2003; Volume 5.
70. Naciri, Y.; Christe, C.; Bétrisey, S.; Song, Y.-G.; Deng, M.; Garfì, G.; Kozlowski, G. Species delimitation in the East Asian species of the relict tree genus *Zelkova* (Ulmaceae): A complex history of diversification and admixture among species. *Mol. Phylogenet. Evol.* **2019**, *134*, 172–185. [CrossRef]
71. Nee, M. Ulmaceae—Flora Mesoamericana. Available online: http://legacy.tropicos.org/Project/FM (accessed on 11 December 2020).
72. Akhter, R. Ulmaecae in Flora of Pakistan. Available online: www.eFloras.org (accessed on 11 December 2020).
73. Wiegrefe, S.J.; Sytsma, K.J.; Guries, R.P. Phylogeny of elms (*Ulmus*, Ulmaceae): Molecular evidence for a sectional classification. *Syst. Bot.* **1994**, *19*, 590–612. [CrossRef]
74. Buchel, A.S. The species of the genus *Ulmus*. In *The Elms, Breeding, Conservation and Disease Management*; Kluwer Academic Publishers: Berlin, Germany, 2000; pp. 351–358.
75. Xiao, Y.; Watson, M. Guidance on conducting—A systematic literature review. *J. Plann. Educ. Res.* **2019**, *39*, 93–112. [CrossRef]
76. Little, J.; Elbert, L. Atlas of United States Trees. In *Conifers and Important Hardwoods*; USDA Forest Service Miscellaneous Publication: Washington, DC, USA, 1971; Volume 1, p. 1146.
77. Melville, R.; Heybroek, H.M. The elms of the Himalaya. *Kew Bull.* **1971**, *26*, 5–28. [CrossRef]
78. Fu, L.; Xin, Y. Elms of China. In *The Elms*; Springer: Boston, MA, USA, 2000; pp. 21–44.
79. Caudullo, G.; Welk, E.; San-Miguel-Ayanz, J. Chorological maps for the main European woody species. *Data Brief* **2017**, *12*, 662–666. [CrossRef] [PubMed]
80. Pederneiras, L.C.; Machado, A.F.P.; Pederneiras, L.C.; Machado, A.F.P. Flora do estado do Rio de Janeiro: Ulmaceae. *Rodriguésia* **2017**, *68*, 541–543. [CrossRef]
81. GBIF. GBIF: The Global Biodiversity Information Facility. Available online: https://www.gbif.org (accessed on 29 September 2020).
82. Gradstein, S.R. *Catálogo de Plantas y Líquenes de Colombia*; En Bernal, R., Gradstein, S.R., Celis, M., Eds.; Instituto de Ciencias Naturales, Universidad Nacional de Colombia: Bogotá, Colombia, 2015.
83. Royal Botanic Gardens Kew. Plants of the World Online. Available online: http://www.plantsoftheworldonline.org (accessed on 29 September 2020).
84. QGIS. Development Team QGIS Geographic Information System. Available online: http://qgis.org (accessed on 12 November 2020).
85. GADM. GADM Database of Global Administrative Areas. Available online: https://gadm.org/ (accessed on 26 February 2021).
86. Natural Earth Data. Natural Earth—Free Vector and Raster Map Data at 1:10 m, 1:50 m, and 1:110 m Scales. Available online: https://www.naturalearthdata.com/ (accessed on 26 February 2021).
87. Peel, M.C.; Finlayson, B.L.; McMahon, T.A. Updated world map of the Köppen-Geiger climate classification. *Hydrol. Earth Syst. Sci.* **2007**, *11*, 1633–1644. [CrossRef]
88. Beck, H.E.; Zimmermann, N.E.; McVicar, T.R.; Vergopolan, N.; Berg, A.; Wood, E.F. Present and future Köppen-Geiger climate classification maps at 1-km resolution. *Sci. Data* **2018**, *5*, 1–12. [CrossRef]
89. Fick, S.E.; Hijmans, R.J. WorldClim 2: New 1-km spatial resolution climate surfaces for global land areas. *Int. J. Climatol.* **2017**, *37*, 4302–4315. [CrossRef]
90. Earth Resources Observation and Science (EROS). Center shuttle radar topography mission (SRTM) 1 arc-second global. *USGS* **2017**. [CrossRef]
91. O'Donnell, M.S.; Ignizio, D.A. Bioclimatic predictors for supporting ecological applications in the conterminous United States. *US Geol. Surv. Data Ser.* **2012**, *691*, 4–9.
92. Goedecke, F.; Bergmeier, E. Ecology and potential distribution of the Cretan endemic tree species *Zelkova abelicea*. *J. Mediterr. Ecol.* **2018**, *16*, 15–26.
93. R Core Team. *R: A Language and Environment for Statistical Computing*; R Foundation for Statistical Computing: Vienna, Austria, 2018.
94. Kruskal, J.B. Nonmetric multidimensional scaling: A numerical method. *Psychometrika* **1964**, *29*, 115–129. [CrossRef]

95. Oksanen, J.; Kindt, R.; Legendre, P.; O'Hara, B.; Stevens, M.H.H.; Oksanen, M.J.; Suggests, M. The Vegan Package. *Commun. Ecol. Package* **2007**, *10*, 719.
96. Borcard, D.; Gillet, F.; Legendre, P. *Numerical Ecology with R*; Springer: New York, NY, USA, 2018; ISBN 3-319-71404-X.

plants

MDPI

Article

Phylogeny, Taxonomy, and Biogeography of *Pterocarya* (Juglandaceae)

Yi-Gang Song [1,2,3,*,†], Ying Li [1,†], Hong-Hu Meng [4], Yann Fragnière [2], Bin-Jie Ge [1], Hitoshi Sakio [5], Hamed Yousefzadeh [6], Sébastien Bétrisey [2,7] and Gregor Kozlowski [1,2,7,*]

1 Eastern China Conservation Center for Wild Endangered Plant Resources, Shanghai Chenshan Botanical Garden, Chenhua Road No.3888, Songjiang, Shanghai 201602, China; ying909726271@sina.com (Y.L.); gebinjie123@163.com (B.-J.G.)
2 Department of Biology and Botanic Garden, University of Fribourg, Chemin du Musée 10, CH-1700 Fribourg, Switzerland; yann.fragniere@unifr.ch (Y.F.); sebastien.betrisey@unifr.ch (S.B.)
3 Shanghai Chenshan Plant Science Research Center, Chinese Academy of Sciences, Chenhua Road No.3888, Songjiang, Shanghai 201602, China
4 Centre for Integrative Conservation, Xishuangbanna Tropical Botanical Garden, Chinese Academy of Sciences, Xuefu Rd. 88, Wuhua, Kunming 650223, China; menghonghu@xtbg.ac.cn
5 Field Center for Sustainable Agriculture and Forestry, Faculty of Agriculture, Niigata University, Sado-city, Niigata 950-2181, Japan; sakiohit@gmail.com
6 Faculty of Natural Resources, Department of Forestry, Tarbiat Modares University (TMU), Mazandaran 14115-111, Iran; h.yousefzadeh@modares.ac.ir
7 Natural History Museum Fribourg, Chemin du Musée 6, CH-1700 Fribourg, Switzerland
* Correspondence: ygsong@cemps.ac.cn (Y.-G.S.); gregor.kozlowski@unifr.ch (G.K.); Tel.: +86-021-37792288-915 (Y.-G.S.)
† These authors contributed equally to this work.

Received: 5 October 2020; Accepted: 6 November 2020; Published: 9 November 2020

check for updates

Abstract: Relict species play an important role in understanding the biogeography of intercontinental disjunctions. *Pterocarya* (a relict genus) is the valuable model taxon for studying the biogeography of East Asian versus southern European/West Asian disjunct patterns. This disjunction has not been as well studied as others (e.g., between Eastern Asia and North America). Several phylogenetic studies on *Pterocarya* have been conducted, but none have provided a satisfactory phylogenetic resolution. Here, we report the first well-resolved phylogeny of *Pterocarya* using restriction site-associated DNA sequencing data based on the sampling of all taxa across the entire distribution area of the genus. Taxonomic treatments were also clarified by combining morphological traits. Furthermore, fossil-calibrated phylogeny was used to explore the biogeography of *Pterocarya*. Our results support the existence of two sections in *Pterocarya*, which is in accordance with morphological taxonomy. Section *Platyptera* comprises three species: *P. rhoifolia*, *P. macroptera*, and *P. delavayi*. Section *Pterocarya* also comprises three species: *P. fraxinifolia*, *P. hupehensis*, and *P. stenoptera*. The divergence between the two sections took place during the early Miocene (20.5 Ma). The formation of the Gobi Desert and climate cooling of northern Siberia in the Middle Miocene (15.7 Ma) might have caused the split of the continuous distribution of this genus and the formation of the East Asian versus southern European/West Asian disjunct pattern. Lastly, the divergence between *P. hupehensis* and *P. stenoptera* as well as between *P. rhoifolia* and *P. macroptera*/*P. delavayi* (10.0 Ma) supports the late Miocene diversification hypothesis in East Asia.

Keywords: divergence time; East Asia-southern Caucasus disjunction; Late Miocene diversification; phylogenomic relationship; refugia; restriction site-associated DNA sequencing (RAD-seq)

1. Introduction

Understanding geographic patterns of species diversity is one of the central aims of biogeography [1–3]. North temperate disjunctions among East Asia, southern Europe/West Asia, eastern North America, and western North America refugia are certainly the best known and most frequently studied of all the major intercontinental disjunctions [1,2,4–7]. Among the northern temperate disjunctions, many analyses have been conducted between (1) East Asia and eastern North America [8–12], (2) East Asia and western North America [13,14], (3) eastern North America and western North America [15], (4) southern Europe and western North America [3,16], and (5) of the trans-Atlantic disjunction [17,18]. There is a group of genera (e.g., *Acer*, *Aesculus*, *Forsythia*, *Liquidambar*, *Picea*, *Parrotia*, *Pterocarya*, and *Zelkova*) sharing the East Asian versus southern European/West Asian disjunct pattern, including those in North America [19–22]. However, all these studies ignore the biogeography of East Asia versus southern European/West Asian disjunct patterns.

Relicts are species that were abundant and occurred in a large area at an earlier geological time, but now only occur in one or a few small areas (so called refugia) [7]. The Sino-Japanese Floristic Region (SJFR) in East Asia harbors the most diverse temperate flora worldwide and is the most important glacial refugium for Cenozoic relict flora [23]. Many of the previous phylogeographic studies in the SJFR focused on individual regions, such as the Sino-Himalayan Forest [24–26] and the Sino-Japanese Forest [27–29], as well as on a single species. Hence, the expansion of phylogeographic studies to multiple pairs of sister species or groups of closely related taxa has been advocated [23,28,30,31]. More recently, the late Miocene diversification hypothesis was raised, proposing that the Cenozoic relict flora in East Asia split into southern and northern lineages during the late Miocene [1,2,7,30,32]. However, additional relict genera should be studied in detail to test this hypothesis.

Pterocarya Kunth (Juglandaceae) is a small Cenozoic relict genus whose species live in riparian areas, with six to eight species [33–35]. The members of this genus were widely distributed throughout the Northern Hemisphere during the Miocene, while currently they are limited only to the areas of East Asia and the southern Caucasus (part of West Asia) [36–38]. The disjunct distribution between East Asia (five to seven species) and West Asia (one species) makes *Pterocarya* a perfect candidate for the exploration of the evolutionary history of genera with disjunct distribution between East Asia and southwestern Eurasia. Additionally, the main distribution in East Asia provides a chance to test the late Miocene diversification hypothesis.

Molecular phylogeny is an important basic framework for biogeography to study the patterns and processes that shape the distributions of life over a wide range of spatial and temporal scales [12,39–41]. Using chloroplast DNA fragments and low-copy nuclear gene data, previous phylogenetic studies on *Pterocarya* have recovered several provisional frameworks within the genus and identified its position within the Juglandaceae [35,42–45]. The monophyly of *Pterocarya* is strongly supported, whereas phylogenetic relationships among species in the genus are controversial and remain unresolved (Figure 1A–C). In addition, taxon sampling has not adequately addressed issues related to taxonomic treatments for several taxa. Overall, there are three significant conflicts: (1) the traditional division of the genus into two sections (sect. *Platyptera* and sect. *Pterocarya*) is not supported by the current molecular phylogeny; (2) the phylogenetic position of *P. hupehensis* and *P. macroptera* is erratic; and (3) there are still controversies on the taxonomic delimitations among closely related *Pterocarya* species (e.g., three different species mentioned in the Chinese edition of the Flora of China were merged into one taxon (*P. macroptera*), in contrast to the English edition) [33,34] (Figure 1).

In recent years, phylogenomics has provided a more robust phylogenetic framework, and has breathed new life into biogeography [12]. Restriction site-associated DNA sequencing (RAD-seq) produces abundant single-nucleotide polymorphism (SNP) data throughout the genome, which can be used for phylogenetic inference [46–49]. The RAD-seq approach, in particular, has proven useful in reconstructing fine-scale relationships within closely related species, recently diverged species, and species experiencing interspecific gene flow [41,50–52].

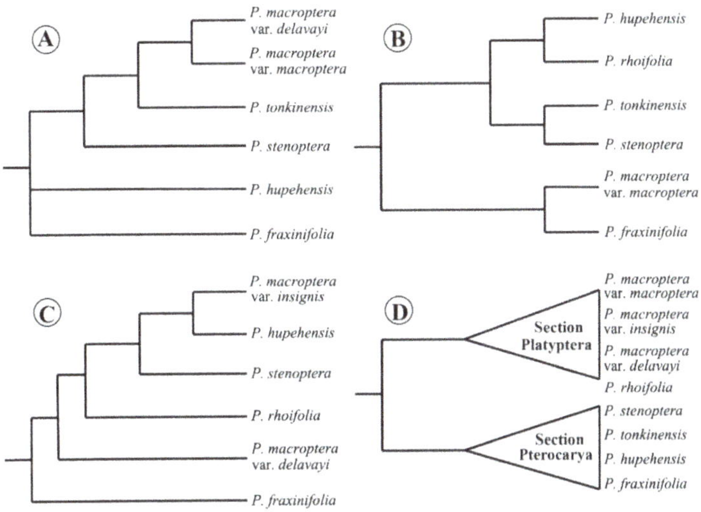

Figure 1. Previous phylogenetic topologies of *Pterocarya* based on different data sets. (**A**) Results based on five chloroplast markers (*rbc*L, *mat*K, *trn*L, *trn*L-F, and *atp*B-*rbc*L) [43]; (**B**) results based on three chloroplast (*rbc*L, *mat*K, and *trn*L-F) and two nuclear loci (ITS and Crabs Claw) [44]; (**C**) results based on nuclear microsatellite and plastid DNA markers [35]; (**D**) two-section classification interpreted as a phylogenetic hypothesis [34].

The present study aimed to answer the following research questions: (1) What are the stable and well-resolved phylogenetic relationships within the genus *Pterocarya*? (2) What are the taxonomic treatments based on molecular and morphological analyses? (3) Which biogeographic and speciation events could be responsible for the disjunct distribution of the genus between the East Asian and southwestern Eurasian refugia? (4) Did the sections *Platyptera* and *Pterocarya* in the East Asia follow the late Miocene diversification? To address these questions, a comprehensive sample collection strategy was used as well as the following analyses: (1) phylogenetic topology was reconstructed based on RAD-seq data; (2) systematic morphometric analysis was used to clarify taxonomic treatments; and (3) divergence times and biogeographic historical events were estimated based on a fossil-calibrated phylogeny.

2. Results

2.1. RAD-seq and Data Matrices for Phylogenetic Inference

The Illumina sequencing yielded an average of 11,055,000 reads (raw reads) per sample, ranging from 4,780,000 to 18,580,000. After quality filtering, the average was reduced to 9,947,083 reads (clean reads) per sample, ranging from 3,790,000 to 17,290,000. The sequencing quality was high because the average Q30 was 91.54% per sample, ranging from 88.55% to 92.39%. The mean GC percentage of all the samples was 47.10%, ranging from 43.03% to 58.57% (Table 1). Detailed information concerning the RAD-seq data processing was given in Table S2.

We recovered an average of 5,502,955 reads (RAD tags) after filtering the data de novo via IPYRAD. We obtained 1,728,343 clusters per sample with a mean depth of 15.67. The consensus loci that passed filtering for paralogs ranged from 38,695 to 204,925, and the average was 102,981. The mean sequencing error (E = 0.0103) was lower than the heterozygosity (H = 0.0413). Lastly, the samples had an average of 9287 (ranging from 4222 to 13,650) unlinked SNP sites in the final data sets for phylogenetic inference (Tables 2 and S2).

Table 1. Summary of RAD-seq data processing (paired-end reads) from 24 samples used in this study.

Summary Statistic	Raw Reads	Clean Reads	Total Length of Clean Reads (Gbp)	Clean Data Percentage (%)	Q30 Percentage (%)	GC Percentage (%)
Average	11,055,000	9,947,083	1.54	84.30	91.54	47.10
Maximum	18,580,000	17,290,000	2.36	90.02	92.39	58.57
Minimum	4,780,000	3,790,000	0.69	76.16	88.55	43.03
SD	3,232,889	3,250,324	0.37	3.64	0.88	3.91

Table 2. Summary statistics of filtering and clustering results of one single end RAD sequences (R1) from 24 samples used for the phylogenetic analysis in this study.

Summary Statistic	RAD Tags (R1)	Total Clusters (R1)	Mean Depth of Clusters	H	E	Consensus Loci	Loci in Final Data Set
Average	5,502,955	1,728,343	15.67	0.0413	0.0103	102,981	9287
Maximum	8,591,043	3,985,579	17.75	0.0526	0.0136	204,925	13,650
Minimum	2,495,755	769,873	13.04	0.0350	0.0075	38,695	4222
SD	1,316,019	733,868	1.11	0.0043	0.0016	39,940	2668

2.2. RAD-seq Phylogenetic Reconstruction

Both ML and BI analyses of the final data set showed that the genus *Pterocarya* is monophyletic with two sections: sect. *Pterocarya* (which includes *P. fraxinifolia*, *P. hupehensis*, *P. stenoptera*, and *P. tonkinensis*), and sect. *Platyptera* (which includes *P. rhoifolia* and three varieties of *P. macroptera*) (Figures 2 and S1).

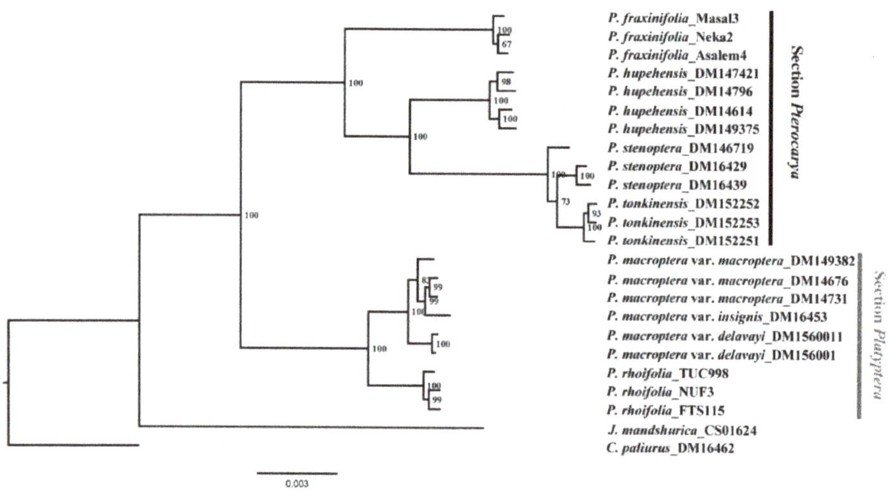

Figure 2. Phylogenetic tree inferred from RAD-seq data for 22 *Pterocarya* individuals and 2 outgroup taxa using RaxML. The numbers next to the nodes of the binary branches are bootstrap values.

Pterocarya fraxinifolia was sister to the other three species in sect. *Pterocarya*. Accessions of *P. stenoptera* were inferred to be paraphyletic, with a population sampled from Zhejiang Province appearing to be more closely related to the accession of *P. tonkinensis* than to the accessions of *P. stenoptera* from Shaanxi Province (Figures 2 and S1). Within sect. *Platyptera*, *P. rhoifolia* was sister to the clades of *P. macroptera*. However, the three varieties of *P. macroptera*, *P. macroptera* var. *delavayi* were inferred to be sister to the other two varieties, with 100% bootstrap (BS) support (Figures 2 and S1).

2.3. Morphological Traits and Taxonomic Conclusions

According to the comparative studies of whole morphologies, together with phylogenetic results of the study, there were four main differences between the two sections: (1) terminal buds, which are either naked (sect. *Pterocarya*) or scaled (sect. *Platyptera*); (2) presence (sect. *Pterocarya*) or absence (sect. *Platyptera*) of lacunae in the walls of the nutlets; (3) presence (sect. *Platyptera*) or absence (sect.

Pterocarya) of bud-scale scars on branchlets; and (4) the position of male spikes on old growth (sect. *Pterocarya*) or new growth (sect. *Platyptera*). *Pterocarya stenoptera* and *P. tonkinensis* were differentiated only by winged and wingless rachises, respectively (Figure 3). With respect to *P. macroptera*, the mature leaves of the variety *delavayi* substantially differed from the other two varieties (var. *macroptera* and *insignis*), especially regarding the microstructures of the trichomes. The mature leaves of variety *delavayi* exclusively had solitary trichomes, whereas the other two varieties have fasciculate trichomes scattered along the main and secondary veins (Figures 3 and 4).

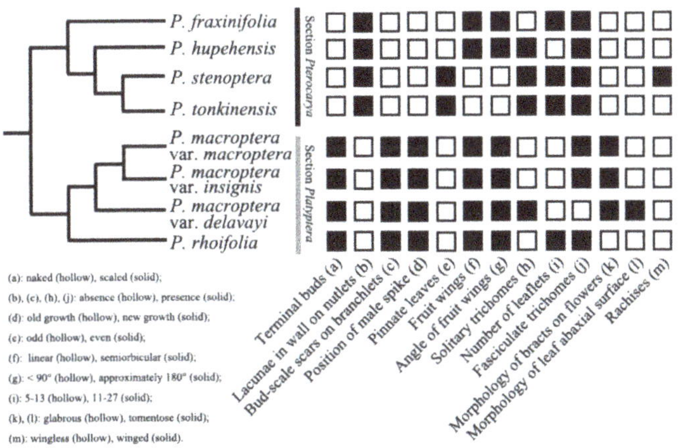

Figure 3. Distribution of taxonomic morphological features in *Pterocarya* based on new phylogenetic tree.

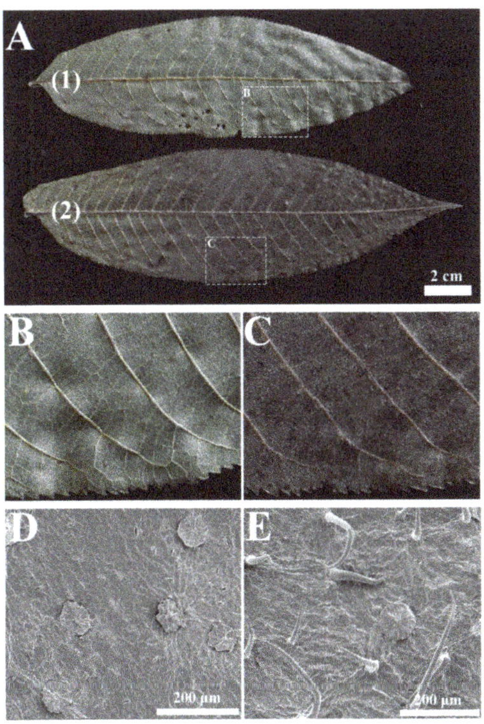

Figure 4. Morphology of leaves of *Pterocarya macroptera* (**A**(**1**),**B**,**D**) and *Pterocarya delavayi* (**A**(**2**),**C**,**E**).

2.4. Estimation of Divergence Times

The estimated divergence of *Pterocarya* from *Juglans* was on the order of 34.73 Ma, with 95% highest posterior density (HPD: 34.0–36.2 Ma). The crown age of *Pterocarya* with the divergence of the two sections was 20.48 Ma (early Miocene, 95% HPD: 15.20–27.96 Ma). The estimated crown age of sect. *Pterocarya* with the divergence of *P. fraxinifolia* from the other species of this section was approximately 15.74 Ma (95% HPD: 14.16–17.37 Ma). The split between *P. hupehensis* and the *P. stenoptera/P. tonkinensis* clade was estimated to have occurred at 9.98 Ma (95% HPD: 7.73–12.73 Ma). The estimated crown age of sect. *Platyptera* with the divergence of *P. rhoifolia* from *P. macroptera* was approximately 10.17 Ma (95% HPD: 5.51–14.30 Ma), and the split of *P. macroptera* var. *delavayi* from the other two varieties was estimated to have occurred in 5.30 Ma (95% HPD: 2.41–8.37 Ma) (Figure 5).

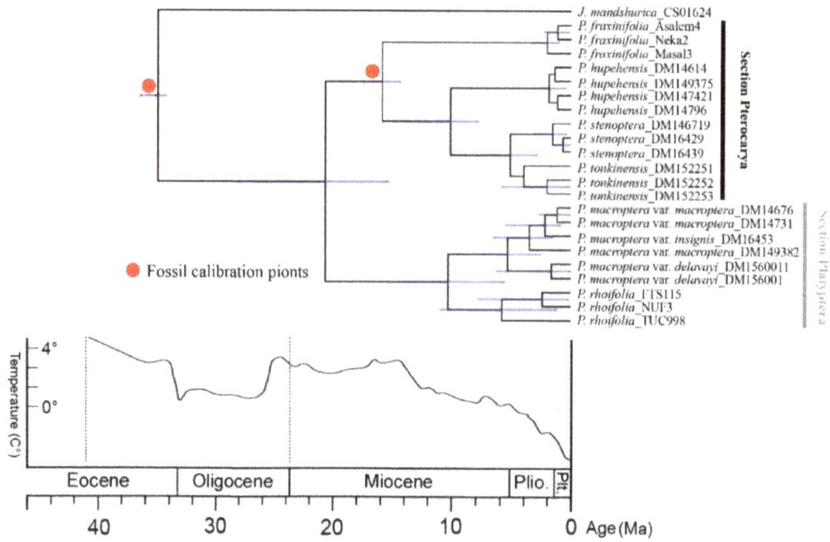

Figure 5. Timing of diversification in *Pterocarya*. Chronogram derived from a MCC tree estimated via the uncorrelated exponential model in BEAST. The blue bars indicate the 95% HPD intervals of the age estimate. Geological time abbreviations: Plio. = Pliocene; Plt. = Pleistocene. The climatic sequence of the major global temperature trends was redrawn from that of [53].

3. Discussion

3.1. Phylogenetic Hypothesis for Pterocarya

We presented a robust phylogenetic reconstruction of *Pterocarya* based on RAD-seq data. Although there have been previous efforts to understand the phylogenetic and biogeographical history of *Pterocarya* [35,43–45], no study to date has included all six species and all three varieties of *P. macroptera* together in a molecular analysis. According to the molecular phylogeny of Xing et al. (2014) [44] in which three chloroplast loci (*rbc*L, *mat*K, and *trn*L-F) and two nuclear loci (internal transcribed spacer [ITS] and Crabs Claw) were used, *Pterocarya* split into two clades: (1) *P. fraxinifolia* and *P. macroptera* var. *macroptera* clustered into one clade, and (2) the other four species clustered into another clade (Figure 1B). Based on the ITS and *trn*H-*psb*A loci, Mostajeran et al. (2016) [45] proposed three clades within *Pterocarya*: (1) *P. fraxinifolia*, (2) *P. hupehensis* and *P. macroptera*, and (3) *P. stenoptera* and *P. tonkinensis*. Xiang et al. (2014) [43] also proposed three clades based on five chloroplast markers (*rbc*L, *mat*K, *trn*L, *trn*L-F, and *atp*B-*rbc*L): (1) *P. fraxinifolia*; (2) *P. hupehensis*; and (3) *P. stenoptera*, *P. tonkinensis*, *P. macroptera* var. *macroptera*, and *P. macroptera* var. *delavayi* (Figure 1A). Maharramova et al. (2018) [35] suggested that *P. fraxinifolia* is the ancestor of the East Asiatic species (Figure 1C) and that *P. macroptera* var. *insignis* was closely related to *P. hupehensis*, and thus, that it is more distantly related to *P. macroptera* var. *delavayi* (Figure 1C).

Compared with traditional methods, RAD-seq can acquire an abundance of polymorphic markers to solve the problems of few identified gene loci and poor representative genomic information [46–49,54]. Unlike previous studies, our molecular phylogenetic topology showed 100% support for the separation of the genus *Pterocarya* into two sections (sect. *Pterocarya* and sect. *Platyptera*), which is consistent with the classical taxonomy based on morphological characteristics summarized in Flora of China (FOC) [33,34]. Section *Pterocarya* showed a disjunct distribution between *P. fraxinifolia* in the Caucasus region and the other three species (*P. hupehensis*, *P. stenoptera*, and *P. tonkinensis*) in East Asia, with 100% support. Section *Platyptera*, also with 100% support, comprises two taxa from East Asia: the Japanese endemic *P. rhoifolia* and the Chinese endemic *P. macroptera*. This RAD-seq tree provides a valuable framework for understanding the phylogeny of all species and varieties within the *Pterocarya* genus.

3.2. Taxonomic Implications and Evolutionary Importance of Morphological Features

Taxonomy requires an integrative approach to effectively define species boundaries [55]. Morphological features provide basic information for species identification. Studies on the micromorphology of the genus *Pterocarya* are lacking, especially concerning the type of trichomes [37]. The present study provides additional identifying characteristics and, for the first time, highlights the importance of trichomes, as well as the morphology of bracts on male and female flowers, for the differentiation of *Pterocarya* species. When the RAD-seq phylogenetic tree data are combined with the morphological characteristics summarized in FOC [34] and Kozlowski et al. (2018) [37], an in-depth speciation analysis and taxonomic treatment for this genus can be performed.

The species of *Pterocarya* have a number of unifying characteristics, such as large two-winged nutlets and a chambered pith [37]. Moreover, our study revealed that all the *Pterocarya* taxa have peltate trichomes (Figure 5). These common features reflect the close affinities among the species and confirm a monophyletic origin of the genus. The presence or absence of terminal buds and lacunae in the nutlet walls provide two main characteristics for differentiating the two sections and thus support the RAD-seq phylogenetic tree. The phylogenetic framework of *Pterocarya* obtained in this study provides an opportunity to analyze of the evolutionary history of related traits used for the delimitations of different sections and species (Figure 5).

On the basis of our results, we hypothesize that the odd-pinnate leaves represent the ancestral character state, whereas even-pinnate leaves represent the derived character states (Figure 4). Additionally, the close relationship between *P. stenoptera* and *P. tonkinensis* is confirmed by only one morphological feature (winged rachises in *P. stenoptera* but wingless rachises in *P. tonkinensis*) that virtually differentiates them (Figure 4).

Two micromorphological features are very important for distinguishing *P. macroptera* from other species, as well as for differentiating among its varieties (Figure 4). First, both female and male flower bracts in all the varieties of this taxon are tomentose, which is exclusive to *P. macroptera* (all the other species have glabrous bracts). The second important feature is the type of trichomes on mature leaves, which, in *P. macroptera* var. *delavayi*, differs from the type of the other two varieties and confirms the phylogenetic resolution within *P. macroptera*. We have concluded that *P. macroptera* var. *delavayi* should be treated morphologically and phylogenetically as a separate species (*Pterocarya delavayi*). In contrast, the lack of resolution of the phylogenetic tree (Figure 3) and the lack of differences in all the morphological features (Figure 4) suggest that the remaining two varieties (*insignis* and *macroptera*) should be merged into one taxon (*P. macroptera*).

On the basis of these morphological and phylogenetic results, we propose that *Pterocarya* should be divided into six species: three (*P. rhoifolia*, *P. macroptera*, and *P. delavayi*) in sect. *Platyptera* and three (*P. fraxinifolia*, *P. hupehensis*, and *P. stenoptera*) in sect. *Pterocarya*. A new identification key for both sections and all species is provided in the Supplementary Material (Doc. S1).

3.3. East Asian versus Southern European/West Asian Disjunctions of Relict Trees: The Importance of the Gobi Desert's Formation and Climatic Cooling after the Middle Miocene Epoch

East Asia and southern Europe/West Asia (or southern Caucasus and the Mediterranean regions) served as the most important refugia of relict trees during previous climatic fluctuations [23,56,57]. There are many Cenozoic relict woody genera that exhibit the pronounced disjunct distribution patterns between East Asia and southern Europe/West Asia, e.g., *Parrotia, Liquidambar, Acer, Albizia, Buxus, Carpinus, Fagus, Diospyros, Hippophae, Sorbus, Taxus,* and *Zelkova* [20,37,58,59]. However, the times and processes leading to the East Asian versus southern European/West Asian disjunct pattern are poorly understood. The results of our study suggest that such a disjunction in the genus *Pterocarya* (sect. *Pterocarya*) occurred during the middle of the Miocene period (15.7 Ma), whereas other studies have suggested different divergence times in other relict genera (e.g., 7.5 Ma for the two species of *Parrotia* in the late Miocene) [20].

The estimated timescale described in our study for the genus *Pterocarya* is in agreement with the known fossil evidence. Fossil records indicate the wide distribution of this genus in Eurasia during the early Neogene period. The absence of fossil data in western Siberia after the Miocene period indicates the disappearance of *Pterocarya* during this period. We hypothesize that the local disappearance of *Pterocarya* in the high latitudes of western Siberia may have been the result of a sharp decrease in global temperatures during the middle Miocene period followed by a major ice sheet expansion from the Arctic [53]. This climatic change may have caused the extinction of *Pterocarya*, along with other relict woody genera, in large parts of western Eurasia and the formation of the isolated refugium in the southern Caucasus and Hyrcanian forests [37].

The second important event was the desertification of the central Asiatic region and, in particular, the formation of the Gobi Desert. The timing and processes leading to the formation of this desert are still debated [60]. However, recent studies indicate that desertification had already started in the early Miocene period [61–64]. The results of our study support this hypothesis by indicating that biological exchanges between eastern and western Eurasia may have been restricted during the early and middle Miocene periods (Figure 6).

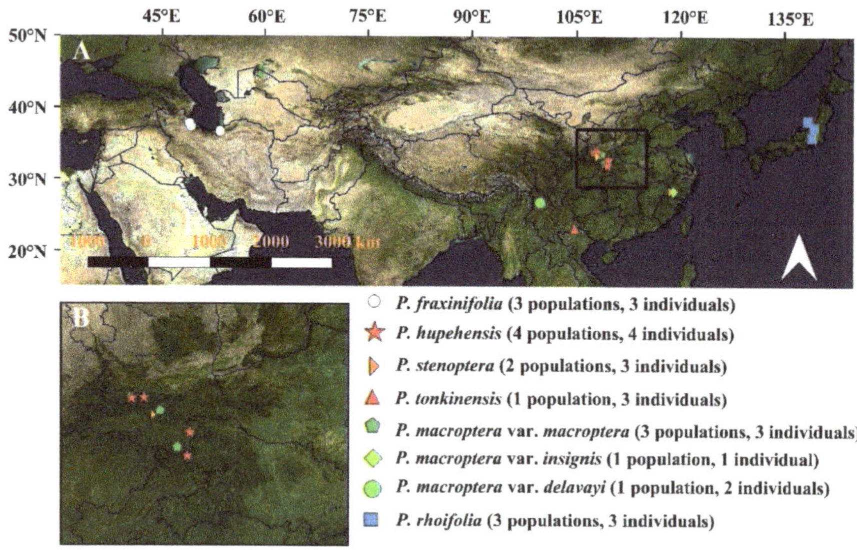

Figure 6. Sampling sites used in this study covering the entire distribution area of *Pterocarya* from the southern Caucasus and East Asia.

In our opinion, these two important climatic and geological events (e.g., cooling of the Siberian region and desertification of Central Asia) could have been responsible for the formation of the current western versus eastern Eurasia disjunct distribution pattern within the *Pterocarya* genus, as well as in many other relict tree genera, since the middle Miocene period.

3.4. Late Miocene Diversification in the East Asian Refugium

With more than 600 endemic genera of the so-called Arcto-Tertiary flora, East Asia is the main refugium for plants, including numerous emblematic relict tree genera and species, such as *Gingko*, *Davidia*, and *Tetracentron* [29]. Toward the late Miocene period, numerous relict tree genera experienced intense diversification and divergence in the region. A prominent example is the split between two species within *Cercidiphyllum* at the Miocene/Pliocene boundary [29] and the divergence between Chinese *Euptelea pleiosperma* and Japanese *E. polyandra* in the late Miocene period (5.5 Ma) [65]. Recently, the same pattern was detected in Asian butternut (*Juglans* section *Cardiocaryon*), the sister genus of *Pterocarya* in Juglandaceae [30]. In addition, our study confirms this biogeographical pattern. In sect. *Platyptera*, the divergence between *P. rhoifolia* (endemic to Japan) and *P. macroptera* (endemic to China) was estimated to have occurred during the late Miocene period (10.17 Ma). These estimated divergence times are very similar to the divergence times of Asian butternuts (10.9 Ma) [30]. The overlapping distributions of *P. rhoifolia* and *Juglans ailantifolia* in Japan and of *P. macroptera* and *J. cathayensis* in China confirm this hypothesis.

Interestingly, in the East Asia, members of the sect. *Pterocarya* in this genus *P. hupehensis* diverged from *P. stenoptera*/*P. tonkinensis* clade at exactly the same time in the late Miocene period (10.0 Ma). *Pterocarya hupehensis* is restricted to the mountainous areas of southern East Asia, whereas members of the *P. stenoptera*/*P. tonkinensis* clade are widely distributed in eastern and southeastern Asia. The evolutionary history of this clade is clearly in need of further population genetic studies. In the future, new emerging molecular methods (e.g., comparative phylogenomics) based on increased numbers of taxa and sampled populations will help to elucidate the detailed evolutionary and population demographic histories of the genus *Pterocarya*, as well as other relict genera of the SJFR.

4. Materials and Methods

4.1. Taxon Sampling and DNA Extraction

In this study, we used twenty-two individuals that represented all species and varieties and covered the entire range of each of the species (n: number of individuals, p: population number): *P. fraxinifolia* ($n = 3, p = 3$), *P. hupehensis* ($n = 4, p = 4$), *P. macroptera* var. *delavayi* ($n = 2, p = 1$), *P. macroptera* var. *insignis* ($n = 1, p = 1$), *P. macroptera* var. *macroptera* ($n = 3, p = 3$), *P. rhoifolia* ($n = 3, p = 3$), *P. stenoptera* ($n = 3, p = 2$), and *P. tonkinensis* ($n = 3, p = 1$) (Figure 6). *Juglans mandshurica* and *Cyclocarya paliurus* were used as outgroups. The voucher specimens are housed in the herbarium of the Shanghai Chenshan Botanical Garden (CSH), at Niigata University, and at Tarbiat Modares University (TMU). None of the field collections of *Pterocarya* species required specific permissions or involved endangered or threatened species.

DNA extraction was performed with a Qiagen DNeasy Plant Tissue Kit from silica-gel dried leaves according to the manufacturer's (Qiagen, Valencia, CA, USA) standard protocol. The DNA extraction quality was checked by 1% agarose gel in conjunction with 1 KB Plus DNA Ladder (Invitrogen) or a New England Biolabs 100 bp DNA ladder marker (Ipswich, MA, USA). The genomic DNA concentrations were subsequently quantified with a dsDNA HS kit on a Qubit 2.0 Fluorometer.

4.2. RAD-seq Library Preparation

Library preparation and sequencing of the RAD markers from genomic DNA were performed by Majorbio (Shanghai, China) using the restriction enzyme *Taqα I*. The Illumina HiSeqTM platform and an Illumina PE150 were used for sequencing, generating 300~500 bp paired-end reads (P1 and P2).

The restriction sites and barcodes were trimmed from each sequence, and bases with FASTQ quality scores below a given value (<20) were replaced with N. Sequences with more than 10% of Ns were discarded. Illumina adapters and sequences smaller than 25 bp were removed, and roughly filtered reads of each individual were obtained.

4.3. Processing and Clustering RAD-seq Data

After receiving the sequencing data, we demultiplexed and processed the roughly filtered reads using the software pipeline IPYRAD v0.7.11 [66]. Nucleotide bases with a Phred quality score (Q) below 33 were replaced with an ambiguous base ("N"), and reads with more than 5% "N"s were discarded. Filtered reads of each individual were first assembled de novo into putative loci. For within-sample clustering, the sequences were clustered at 0.85 similarity by VSEARCH [67]. After clustering, the rates of heterozygosity (H) and sequencing errors (E) were jointly estimated from aligned clusters for each sampled individual [68], and the average parameter values were used when calling consensus bases. Loci containing more than two alleles after error correction were excluded as potential paralogs since *Pterocarya* species are diploid [42]. Consensus sequences were then aligned with Muscle v3.8.31 [69]. A final filtering step excluded any loci containing one or more sites that appeared heterozygous across more than five samples, as such loci may represent a fixed difference among clustered paralogs rather than a true heterozygous site at the broad phylogenetic scale. The remaining clusters representing multiple alignments of putative orthologs were treated as RAD-seq loci and assembled into phylogenetic data matrices.

4.4. Morphological Evaluations and Data Sets

During our fieldwork between 2014 and 2016, we collected all eight taxa of *Pterocarya* and the two outgroups. Afterwards, we evaluated all the morphological characteristics (including the trunks, bark, buds, leaves, flowers, and fruits) as described in *FOC* and other publications [33,34,37]. Due to the absence of leaf epidermal features in the literature, we studied the trichomes of all species and varieties of the mature leaves. The dried materials were directly mounted onto stubs without any treatment. After being sputter coated with gold, the specimens were examined and imaged via scanning electron microscopy (SEM) (Quanta 250). The descriptions and terminology of the trichomes mainly followed those of Deng et al. (2014) [70].

We collected data on 13 total binary morphological characteristics: terminal buds (naked or scaled), lacunae in the walls on nutlets (presence or absence), bud-scale scars on branchlets (presence or absence), position of male spike (old or new growth), pinnate leaves (odd or even), fruit wings (linear or semi- or bi-cular), angle of fruit wings (<90° or approximately 180°), solitary trichomes (presence or absence), number of leaflets (5–13 or 11–27), fasciculate trichomes (presence or absence), morphology of bracts of flowers (glabrous or tomentose), morphology of leaf abaxial surface (glabrous or tomentose), and rachises (wingless or winged). These characteristics were easy to identify and can be treated as binary and were stated on our molecular phylogenetic tree.

4.5. Phylogenetic Reconstruction

The single end of the paired-end reads (P1) of RAD-seq data was used for phylogenetic inference and all the data were submitted to GenBank with information related to taxonomy and GenBank accession numbers (Table S1). Maximum likelihood (ML) and Bayesian Inference (BI) trees were inferred using RAxML v8.2.4 [71] and MrBayes v3.2.6 [72], respectively. An ML tree with random starting trees and a GTR + GAMMA nucleotide substitution model was constructed, and the reliability of the tree topology was determined by 200 nonparametric bootstrapped replicates. The BI analyses were started with random trees, and four parallel Markov Chain Monte Carlo (MCMC) searches were performed for 100 million generations each. The trees were sampled every 100 generations, and the first 20% of each run was discarded as burn-in. Tracer v1.6 [73] was used to check the log-likelihood of sampled trees and determine when stationarity had been reached.

4.6. Fossil Constraints and Estimations of Divergence Times

Two fossils were used as minimum age constraints for two nodes. The first fossil was *Juglans clarnensis*, which was identified as the oldest *Juglans* fossil and dated back to the Eocene epoch (34–55 Ma) in North America [36,37]. The second fossil was *Pterocarya smileyi* from North America. This fossil dated back to the Miocene epoch (5.3–23.0 Ma) and was the first fossil identified as having an affinity with section *Pterocarya* [36]. To infer divergence times, a relaxed clock model was analyzed under a MCMC simulation in BEAST v1.7.5 [74]. A prior Yule tree was used with an uncorrelated lognormal molecular clock. Tree and log files were generated from two runs with different starting seeds. The MCMC length was 100 million generations, with parameter sampling occurring every 1000 generations. Convergence was assessed by Tracer v1.6 [73], and the effective sample sizes (ESSs) of all the parameters were also assessed. A maximum clade credibility (MCC) tree was generated by TreeAnnotator v1.7.4 after the first 20% of the trees had been removed as burn-in [74].

Supplementary Materials: The following are available online at http://www.mdpi.com/2223-7747/9/11/1524/s1, Figure S1: Bayesian Inference (BI) tree of *Pterocarya*. PMM: *P. macroptera* var. *macroptera*, PMI: *P. macroptera* var. *insignis*, and PMD: *P. macroptera* var. *delavayi*, Table S1: List of taxa included in dataset of restriction site-associated DNA sequencing (RAD-seq) for the phylogenetic analysis of *Pterocarya* with information related to taxonomy and GenBank accession numbers, Table S2: Detail information of RAD-seq data processing for *Pterocarya* used in this study, Doc. S1: Identification key for the sections and species in the genus *Pterocarya* (Juglandeceae).

Author Contributions: Conceptualization, G.K. and Y.-G.S.; methodology, Y.L., Y.-G.S., and Y.F.; software, Y.L., B.-J.G, and S.B., and Y.F.; investigation, B.-J.G., H.S., H.Y. S.B., and Y.-G.S.; writing—original draft preparation, Y.-G.S., Y.L., and H.-H.M.; writing—review and editing, G.K., H.S., and H.Y.; supervision, G.K. All authors have read and agreed to the published version of the manuscript.

Funding: This research was funded by (1) Shanghai Municipal Administration of Forestation and City Appearances, grant number G202401; (2) Youth Innovation Promotion Association, CAS in China, grant number 2018432; (3) China Scholarship Council (CSC), grant number 201608310121; and (4) Foundation Franklinia.

Acknowledgments: We would like to thank Duo-Qing Lin and Rui-Bin Liu from Shanghai Chenshan Botanical Garden for their help during the field work.

Conflicts of Interest: The authors declare no conflict of interest.

References

1. Donoghue, M.J.; Bell, C.D.; Li, J.H. Phylogenetic patterns in Northern Hemisphere plant geography. *Int. J. Plant Sci.* **2001**, *162*, S41–S52. [CrossRef]
2. Milne, R.I.; Abbott, R.J. The origin and evolution of tertiary relict floras. *Adv. Bot. Res.* **2002**, *38*, 281–314.
3. Mao, K.S.; Hao, G.; Liu, J.Q.; Adams, R.P.; Milne, R.I. Diversification and biogeography of *Juniperus* (Cupressaceae): Variable diversification rates and multiple intercontinental dispersals. *New Phytol.* **2010**, *188*, 254–272. [CrossRef]
4. Raven, P.H. Plant species disjunctions: A summary. *Ann. Mo. Bot. Gard.* **1972**, *59*, 234–246. [CrossRef]
5. Thorne, R.F. Major disjunctions in the geographic ranges of seed plants. *Q. Rev. Biol.* **1972**, *47*, 365–411. [CrossRef]
6. Tiffney, B.H. The Eocene North Atlantic land bridge: Its importance in Tertiary and modern phytogeography of the Northern Hemisphere. *J. Arnold Arbor.* **1985**, *66*, 243–273. [CrossRef]
7. Milne, R.I. Northern hemisphere plant disjunctions: A window on tertiary land bridges and climate change? *Ann. Bot.* **2006**, *98*, 465–472. [CrossRef]
8. Wen, J. Evolution of eastern Asian and eastern North American disjunct distributions in flowering plants. *Annu. Rev. Ecol. Evol. Syst.* **1999**, *30*, 421–455. [CrossRef]
9. Wen, J. Evolution of eastern Asian-eastern North American biogeographic disjunctions: A few additional issues. *Int. J. Plant Sci.* **2001**, *162*, S117–S122. [CrossRef]
10. Xiang, Q.Y.; Soltis, D.E.; Soltis, P.S.; Manchester, S.R.; Crawford, D.J. Timing the eastern Asian-Eastern North American floristic disjunction: Molecular clock corroborates paleontological estimates. *Mol. Phylogenet. Evol.* **2000**, *15*, 462–472. [CrossRef] [PubMed]

11. Xiang, Q.Y.; Zhang, W.H.; Ricklefs, R.E. Regional differences in rates of plant speciation and molecular evolution: A comparison between eastern Asia and eastern North America. *Evolution* **2004**, *58*, 2175–2184. [CrossRef] [PubMed]
12. Wen, J.; Nie, Z.L.; Ickert-Bond, S.M. Advances in biogeography in the age of a new modern synthesis. *J. Syst. Evol.* **2019**, *57*, 543–546. [CrossRef]
13. Wu, Z. On the significance of Pacific intercontinental discontinuity. *Ann. Mo. Bot. Gard.* **1983**, *70*, 577–590.
14. Wen, J.; Nie, Z.L.; Ickert-Bond, S.M. Intercontinental disjunctions between eastern Asia and western North America in vascular plants highlight the biogeographic importance of the Bering land bridge from late Cretaceous to Neogene. *J. Syst. Evol.* **2016**, *54*, 469–490. [CrossRef]
15. Xiang, Q.Y.; Soltis, D.E.; Soltis, P.S. The eastern Asian and eastern and western North America floristic disjunction: Congruent phylogenetic patterns in seven diverse genera. *Mol. Phylogenet. Evol.* **1998**, *10*, 178–190. [CrossRef]
16. Wen, J.; Ickert-Bond, S.M. Evolution of the Madren-Tethyan disjunctions and the North and South American amphitropical disjunctions in plants. *J. Syst. Evol.* **2009**, *47*, 331–348. [CrossRef]
17. Donoghue, M.J.; Smith, S.A. Patterns in the assembly of temperate forests around the Northern Hemisphere. *Philos. Trans. R. Soc. B Biol. Sci.* **2004**, *359*, 1633–1644. [CrossRef]
18. Manchester, S.R. Biogeographical relationships of North American tertiary floras. *Ann. Mo. Bot. Gard.* **1999**, *82*, 472–522. [CrossRef]
19. Ickert-Bond, S.M.; Wen, J. Phylogeny and biogeography of Altingiaceae: Evidence from combined analysis of five non-coding chloroplast regions. *Mol. Phylogenet. Evol.* **2006**, *39*, 512–528. [CrossRef]
20. Li, J.H.; Tredici, P.D. The Chinese Parrotia: A sibling species of the Persian *Parrotia*. *Arnoldia* **2008**, *66*, 2–9.
21. Li, J.; Stukel, M.; Bussies, P.; Skinner, K.; Lemmon, A.R.; Lemmon, E.M.; Brown, K.; Bekmetjev, A.; Swenson, N.G. Maple phylogeny and biogeography inferred from phylogenomic data. *J. Syst. Evol.* **2019**, *57*, 594–606. [CrossRef]
22. Naciri, Y.; Christe, C.; Betrisey, S.; Song, Y.G.; Deng, M.; Garfi, G.; Kozlowski, G. Species delimation in the East Asian species of the relict tree genus *Zelkova* (Ulmaceae): A complex history of diversification and admixture among species. *Mol. Phylogenet. Evol.* **2019**, *134*, 172–185. [CrossRef] [PubMed]
23. Qiu, Y.X.; Fu, C.X.; Comes, H.P. Plant molecular phylogeography in China and adjacent regions: Tracing the genetic imprints of Quaternary climate and environmental change in the world's most diverse temperate flora. *Mol. Phylogenet. Evol.* **2011**, *59*, 225–244. [CrossRef]
24. Zhou, Z.; Hong, D.Y.; Niu, Y.; Li, G.D.; Nie, Z.L.; Wen, J.; Sun, H. Phylogenetic and biogeographic analyses of the Sino-Himalayan endemic genus *Cyananthus* (Campanulaceae) and implications for the evolution of its sexual system. *Mol. Phylogenet. Evol.* **2013**, *68*, 482–497. [CrossRef] [PubMed]
25. Kou, Y.X.; Cheng, S.M.; Tian, S.; Li, B.; Fan, D.M.; Chen, Y.J.; Soltis, D.E.; Soltis, P.S.; Zhang, Z.Y. The antiquity of *Cyclocarya paliurus* (Juglandaceae) provides new insights into the evolution of relict plants in subtropical China since the late Early Miocene. *J. Biogeogr.* **2016**, *43*, 351–360. [CrossRef]
26. Du, F.K.; Hou, M.; Wang, W.T.; Mao, K.S.; Hampe, A. Phylogeography of *Quercus aquifolioides* provides novel insights into the Neogene history of a major global hotspot of plant diversity in south-west China. *J. Biogeogr.* **2017**, *44*, 294–307. [CrossRef]
27. Qiu, Y.X.; Guan, B.C.; Fu, C.X.; Comes, H.P. Did glacials and/or interglacials promote allopatric incipient speciation in East Asian temperate plants? Phylogeographic and coalescent analyses on refugial isolation and divergence in *Dysosma versipellis*. *Mol. Phylogenet. Evol.* **2009**, *51*, 281–293. [CrossRef]
28. Qiu, Y.X.; Sun, Y.; Zhang, X.P.; Lee, J.; Fu, C.X.; Comes, H.P. Molecular phylogeography of East Asian *Kirengeshoma* (Hydrangeaceae) in relation to Quaternary climate change and landbridge configurations. *New Phytol.* **2009**, *183*, 480–495. [CrossRef] [PubMed]
29. Qi, X.S.; Chen, C.; Comes, H.P.; Sakaguchi, S.; Liu, Y.H.; Tanaka, N.; Sakio, H.; Qiu, Y.X. Molecular data and ecological niche modelling reveal a highly dynamic evolutionary history of the East Asian Tertiary relict *Cercidiphyllum* (Cercidiphyllaceae). *New Phytol.* **2012**, *196*, 617–630. [CrossRef]
30. Bai, W.N.; Wang, W.T.; Zhang, D.Y. Phylogeographic breaks within Asian butternuts indicate the existence of a phytogeographic divide in East Asia. *New Phytol.* **2016**, *209*, 1757–1772. [CrossRef] [PubMed]

31. Yang, L.Q.; Hu, H.Y.; Xie, C.; Lai, S.P.; Yang, M.; He, X.J.; Zhou, S.D. Molecular phylogeny, biogeography and ecological niche modelling of *Cardiocrinum* (Liliaceae): Insights into the evolutionary history of endemic genera distributed across the Sino-Japanese floristic region. *Ann. Bot.* **2017**, *119*, 59–72. [CrossRef]
32. Ye, J.W.; Bai, W.N.; Bao, L.; Wang, T.M.; Wang, H.F.; Ge, J.P. Sharp genetic discontinuity in the aridity-sensitive *Lindera obtusiloba* (Lauraceae): Solid evidence supporting the Tertiary floral subdivision in East Asia. *J. Biogeogr.* **2017**, *44*, 2082–2095. [CrossRef]
33. Kuang, K.R.; Zheng, S.X.; Li, P.Q.; Lu, A.M. Myricaceae, Juglandaceae and Betulaceae. In *Flora of China (Chinese Version)*; Wu, Z.Y., Ed.; Science Press: Beijing, China, 1979; Volume 21, pp. 21–30.
34. Lu, A.M.; Donald, E.S.; Grauke, L.J. Juglandaceae. In *Flora of China*; Wu, Z.Y., Raven, P.H., Eds.; Science Press & Missouri Botanical Garden Press: Beijing, China; St. Louis, MO, USA, 1999; Volume 4, pp. 277–285.
35. Maharramova, E.; Huseynova, I.; Kolbaia, S.; Gruenstaeudl, M.; Borsch, T.; Muller, L.A. Phylogeography and population genetics of the riparian relict tree *Pterocarya fraxinifolia* (Juglandaceae) in the South Caucasus. *Syst. Biodivers.* **2018**, *16*, 14–27. [CrossRef]
36. Manchester, S.R. *The Fossil History of the Juglandaceae. Mongraphs in Systematic Botany from the Missouri Botanical Garden*; Allen Press: Lawrence, KS, USA, 1987; Volume 21, pp. 1–137.
37. Kozlowski, G.; Betrisey, S.; Song, Y.G. *Wingnuts (Pterocarya) and Walnut Family: Relict Trees: Linking the Past, Present and Future*; Natural History Museum Fribourg: Fribourg, Switzerland, 2018; pp. 1–59.
38. Song, Y.G.; Fragnière, Y.; Meng, H.H.; Li, Y.; Bétrisey, S.; Corrales, A.; Manchester, S.; Deng, M.; Jasińska, A.K.; Sâm, H.V.; et al. Global biogeographic synthesis and priority conservation regions of the relict tree family Juglandaceae. *J. Biogeogr.* **2020**, *47*, 643–657. [CrossRef]
39. Tkach, N.V.; Hoffmann, M.H.; Roser, M.; Korobkov, A.A.; Von Hagen, K.B. Parallel evolutionary patterns in multiple lineages of arctic *Artemisia* L. (Asteraceae). *Evolution* **2008**, *62*, 184–198. [CrossRef]
40. Schwery, O.; Onstein, R.E.; Bouchenak-Khelladi, Y.; Xing, Y.W.; Carter, R.J.; Linder, H.P. As old as the mountains: The radiations of the Ericaceae. *New Phytol.* **2015**, *207*, 355–367. [CrossRef]
41. Massatti, R.; Reznicek, A.A.; Knowles, L.L. Utilizing RADseq data for phylogenetic analysis of challenging taxonomic groups: A case study in *Carex* sect. *Racemosae*. *Am. J. Bot.* **2016**, *103*, 337–347. [CrossRef]
42. Manos, P.S.; Stone, D.E. Evolution, phylogeny, and systematics of the Juglandaceae. *Ann. Mo. Bot. Gard.* **2001**, *88*, 231–269. [CrossRef]
43. Xiang, X.G.; Wang, W.; Li, R.Q.; Lin, L.; Liu, Y.; Zhou, Z.K.; Li, Z.Y.; Chen, Z.D. Large-scale phylogenetic analyses reveal fagalean diversification promoted by the interplay of diaspores and environments in the Paleogene. *Perspect. Plant Ecol.* **2014**, *16*, 101–110. [CrossRef]
44. Xing, Y.W.; Onstein, R.E.; Carter, R.J.; Stadler, T.; Linder, H.P. Fossils and a large molecular phylogeny show that the evolution of species richness, generic diversity, and turnover rates are disconnected. *Evolution* **2014**, *68*, 2821–2832. [CrossRef]
45. Mostajeran, F.; Yousefzadeh, H.; Davitashvili, N.; Kozlowski, G.; Akbarinia, M. Phylogenetic relationships of *Pterocarya* (Juglandaceae) with an emphasis on the taxonomic status of Iranian populations using ITS and *trn*H-*psb*A sequence data. *Plant Biosyst.* **2016**, *151*, 1012–1021. [CrossRef]
46. Miller, M.R.; Dunham, J.P.; Amores, A.; Cresko, W.A.; Johnson, E.A. Rapid and cost-effective polymorphism identification and genotyping using restriction site associated DNA (RAD) markers. *Genome Res.* **2007**, *17*, 240–248. [CrossRef]
47. Rowe, H.C.; Renaut, S.; Guggisberg, A. RAD in the realm of next-generation sequencing technologies. *Mol. Ecol.* **2011**, *20*, 3499–3502. [CrossRef]
48. Hipp, A.L.; Eaton, D.A.R.; Cavender-Bares, J.; Fitzek, E.; Nipper, R.; Manos, P.S. A framework phylogeny of the American oak clade based on sequenced RAD data. *PLoS ONE* **2014**, *9*, e93975. [CrossRef]
49. Razkin, O.; Sonet, G.; Breugelmans, K.; Madeira, M.J.; Gomez-Moliner, B.J.; Backeljau, T. Species limits, interspecific hybridization and phylogeny in the cryptic land snail complex *Pyramidula*: The power of RADseq data. *Mol. Phylogenet. Evol.* **2016**, *101*, 267–278. [CrossRef] [PubMed]
50. Eaton, D.A.R.; Ree, R.H. Inferring Phylogeny and Introgression using RADseq Data: An Example from Flowering Plants (*Pedicularis*: Orobanchaceae). *Syst. Biol.* **2013**, *62*, 689–706. [CrossRef]

51. Cruaud, A.; Gautier, M.; Galan, M.; Foucaud, J.; Saune, L.; Genson, G.; Dubois, E.; Nidelet, S.; Deuve, T.; Rasplus, J.Y. Empirical Assessment of RAD Sequencing for Interspecific Phylogeny. *Mol. Biol. Evol.* **2014**, *31*, 1272–1274. [CrossRef]
52. Liu, L.X.; Jin, X.J.; Chen, N.; Li, X.; Li, P.; Fu, C.X. Phylogeny of *Morella rubra* and its relatives (Myricaceae) and genetic resources of Chinese bayberry using RAD sequencing. *PLoS ONE* **2015**, *10*, e0139840. [CrossRef]
53. Zachos, J.; Pagani, M.; Sloan, L.; Thomas, E.; Billups, K. Trends, rhythms, and aberrations in global climate 65 Ma to present. *Science* **2001**, *292*, 686–693. [CrossRef]
54. Mu, X.Y.; Tong, L.; Sun, M.; Zhu, Y.X.; Wen, J.; Lin, Q.W.; Liu, B. Phylogeny and divergence time estimation of the walnut family (Juglandaceae) based on nuclear RAD-Seq and chloroplast genome data. *Mol. Phylogenet. Evol.* **2020**, *147*, 106802. [CrossRef]
55. Zhang, C.Y.; Low, S.L.; Song, Y.G.; Nurainas; Kozlowski, G.; Do, T.V.; Li, L.; Zhou, S.S.; Tan, Y.H.; Cao, G.L.; et al. Shining a light on species delimitation in the tree genus *Engelhardia* Leschenault ex Blume (Juglandaceae). *Mol. Phylogenet. Evol.* **2020**, *152*, 106918. [CrossRef]
56. Wolfe, J.A. Some aspects of plant geography of Northern Hemisphere during late Cretaceous and Tertiary. *Ann. Mo. Bot. Gard.* **1975**, *62*, 264–279. [CrossRef]
57. Kozlowski, G.; Gratzfeld, J. *Zelkova—An Ancient Tree: Global Status and Conservation Action*; Natural History Museum Fribourg: Fribourg, Switzerland, 2013; pp. 1–60.
58. Ickert-Bond, S.M.; Pigg, K.B.; Wen, J. Comparative infructescence morphology in *Liquidambar* (Altingiaceae) and its evolutionary significance. *Am. J. Bot.* **2005**, *92*, 1234–1255. [CrossRef]
59. Jia, D.R.; Bartish, I.V. Climatic changes and orogeneses in the Late Miocene of Eurasia: The main triggers of an expansion at a continental scale? *Front. Plant Sci.* **2018**, *9*, 1400. [CrossRef]
60. Lu, H.Y.; Wang, X.Y.; Wang, X.Y.; Chang, X.; Zhang, H.Z.; Xu, Z.W.; Zhang, W.C.; Wei, H.Z.; Zhang, X.J.; Yi, S.W.; et al. Formation and evolution of Gobi Desert in central and eastern Asia. *Earth Sci. Rev.* **2019**, *194*, 251–263. [CrossRef]
61. Bosboom, R.E.; Abels, H.A.; Hoorn, C.; Van den Berg, B.C.J.; Guo, Z.; Dupont-Nivet, G. Aridification in continental Asia after the Middle Eocene Climatic Optimum (MECO). *Earth Planet. Sci. Lett.* **2014**, *389*, 34–42. [CrossRef]
62. Carrapa, B.; DeCelles, P.G.; Wang, X.; Clementz, M.T.; Mancin, N.; Stoica, M.; Kraatz, B.; Meng, J.; Abdulov, S.; Chen, F.H. Tectono-climatic implications of Eocene Paratethys regression in the Tajik basin of central Asia. *Earth Planet. Sci. Lett.* **2015**, *424*, 168–178. [CrossRef]
63. Zheng, H.B.; Wei, X.C.; Tada, R.J.; Clift, P.D.; Wang, B.; Jourdan, F.; Wang, P.; He, M.Y. Late Oligocene-early Miocene birth of the Taklimakan Desert. *Proc. Natl. Acad. Sci. USA* **2015**, *112*, 7662–7667. [CrossRef]
64. Li, J.X.; Yue, L.P.; Roberts, A.P.; Hirt, A.M.; Pan, F.; Guo, L.; Xu, Y.; Xi, R.G.; Guo, L.; Qiang, X.K.; et al. Global cooling and enhanced Eocene Asian mid-latitude interior aridity. *Nat. Commun.* **2018**, *9*, 3026. [CrossRef]
65. Cao, Y.N.; Comes, H.P.; Sakaguchi, S.; Chen, L.Y.; Qiu, Y.X. Evolution of East Asia's Arcto-Tertiary relict *Euptelea* (Eupteleaceae) shaped by Late Neogene vicariance and Quaternary climate change. *BMC Evol. Biol.* **2016**, *16*, 66. [CrossRef]
66. Eaton, D.A.R. PyRAD: Assembly of de novo RADseq loci for phylogenetic analyses. *Bioinformatics* **2014**, *30*, 1844–1849. [CrossRef]
67. Rognes, T.; Flouri, T.; Nichols, B.; Quince, C.; Mahé, F. VSEARCH: A versatile open source tool for metagenomics. *PeerJ* **2016**, *4*, e2584. [CrossRef] [PubMed]
68. Lynch, M. Estimation of nucleotide diversity, disequilibrium coefficients, and mutation rates from high-coverage genome-sequencing projects. *Mol. Biol. Evol.* **2008**, *25*, 2409–2419. [CrossRef]
69. Edgar, R.C. Search and clustering orders of magnitude faster than BLAST. *Bioinformatics* **2010**, *26*, 2460–2461. [CrossRef]
70. Deng, M.; Hipp, A.; Song, Y.G.; Li, Q.S.; Coombes, A.; Cotton, A. Leaf epidermal features of *Quercus* subgenus *Cyclobalanopsis* (Fagaceae) and their systematic significance. *Biol. J. Linn. Soc.* **2014**, *176*, 224–259. [CrossRef]
71. Stamatakis, A. RAxML version 8: A tool for phylogenetic analysis and post-analysis of large phylogenies. *Bioinformatics* **2014**, *30*, 1312–1313. [CrossRef]
72. Ronquist, F.; Teslenko, M.; Van der Mark, P.; Ayres, D.L.; Darling, A.; Hohna, S.; Larget, B.; Liu, L.; Suchard, M.A.; Huelsenbeck, J.P. MrBayes 3.2: Efficient Bayesian Phylogenetic Inference and Model Choice Across a Large Model Space. *Syst. Biol.* **2012**, *61*, 539–542. [CrossRef]

73. Rambaut, A.; Brummond, A.T. Tracer v1.5. 2007. Available online: http://tree.bio.ed.ac.uk/software/tracer/ (accessed on 11 October 2019).
74. Drummond, A.J.; Rambaut, A. BEAST: Bayesian evolutionary analysis by sampling trees. *BMC Evol. Biol.* **2007**, *7*, 214. [CrossRef]

plants

Article

Species Delimitation and Conservation in Taxonomically Challenging Lineages: The Case of Two Clades of *Capurodendron* (Sapotaceae) in Madagascar

Carlos G. Boluda [1,2,*,†], Camille Christe [1,2,†], Aina Randriarisoa [1,2], Laurent Gautier [1,2] and Yamama Naciri [1,2]

1 Conservatoire et Jardin botaniques de la Ville de Genève, Chemin de l'Impératrice 1, Chambésy, 1292 Geneva, Switzerland; camille.christe@ville-ge.ch (C.C.); aina.randriarisoa@ville-ge.ch (A.R.); laurent.gautier@ville-ge.ch (L.G.); yamama.naciri@ville-ge.ch (Y.N.)
2 Laboratoire de Systématique végétale et Biodiversité, Université de Genève, Chemin de l'Impératrice 1, Chambésy, 1292 Geneva, Switzerland
* Correspondence: carlos.boluda@ville-ge.ch; Tel.: +41-0224185118
† These authors contributed equally to this work.

Abstract: *Capurodendron* is the largest endemic genus of plants from Madagascar, with around 76% of its species threatened by deforestation and illegal logging. However, some species are not well circumscribed and many of them remain undescribed, impeding a confident evaluation of their conservation status. Here we focus on taxa delimitation and conservation of two species complexes within *Capurodendron*: the Arid and Western complexes, each containing undescribed morphologies as well as intermediate specimens alongside well-delimited taxa. To solve these taxonomic issues, we studied 381 specimens morphologically and selected 85 of them to obtain intergenic, intronic, and exonic protein-coding sequences of 794 nuclear genes and 227 microsatellite loci. These data were used to test species limits and putative hybrid patterns using different approaches such as phylogenies, PCA, structure analyses, heterozygosity level, FST, and ABBA-BABA tests. The potential distributions were furthermore estimated for each inferred species. The results show that the *Capurodendron* Western Complex contains three well-delimited species, *C. oblongifolium*, *C. perrieri*, and *C. pervillei*, the first two hybridizing sporadically with the last and producing morphologies similar to, but genetically distinct from *C. pervillei*. The Arid Complex shows a more intricate situation, as it contains three species morphologically well-delimited but genetically intermixed. *Capurodendron mikeorum* nom. prov. is shown to be an undescribed species with a restricted distribution, while *C. androyense* and *C. mandrarense* have wider and mostly sympatric distributions. Each of the latter two species contains two major genetic pools, one showing interspecific admixture in areas where both taxa coexist, and the other being less admixed and comprising allopatric populations having fewer contacts with the other species. Only two specimens out of 172 showed clear genetic and morphological signals of recent hybridization, while all the others were morphologically well-delimited, independent of their degree of genetic admixture. Hybridization between *Capurodendron androyense* and *C. microphyllum*, the sister species of the Arid Complex, was additionally detected in areas where both species coexist, producing intermediate morphologies. Among the two complexes, species are well-defined morphologically with the exception of seven specimens (1.8%) displaying intermediate patterns and genetic signals compatible with a F1 hybridization. A provisional conservation assessment for each species is provided.

Keywords: conservation; current speciation; hybridization; species complex; species delimitation

Citation: Boluda, C.G.; Christe, C.; Randriarisoa, A.; Gautier, L.; Naciri, Y. Species Delimitation and Conservation in Taxonomically Challenging Lineages: The Case of Two Clades of *Capurodendron* (Sapotaceae) in Madagascar. *Plants* **2021**, *10*, 1702. https://doi.org/10.3390/plants10081702

Academic Editor: Igor Bartish

Received: 30 June 2021
Accepted: 11 August 2021
Published: 18 August 2021

1. Introduction

Species conservation assessments, as currently conducted on a wide scale using the International Union for Conservation of Nature (IUCN) criteria, are based on distribution data of clearly defined species and have proved to be a useful pragmatic tool. However, the species concept is sometimes subjective, especially when there is a mismatch between morphospecies (understood as a morphologically delimited group, described or not, considered potentially a valid species and meriting further evaluation) and genetic lineages. Depending on which concept is used, the number of final units to be conserved can vary greatly. With massive DNA sequencing, we can now use unprecedented amounts of genetic information; however, how much this information can help us to establish clear and practical species limits in critical cases is still an open question.

With 33 described species so far and more than a dozen to come, *Capurodendron* (Sapotaceae) is the largest endemic genus of plants in Madagascar [1,2]. It contains trees, rarely shrubs, growing from the most humid to the driest areas of the island, with a great variety of leaf morphology but a highly conserved flower architecture [3]. *Capurodendron*, like most other Sapotaceae, usually produces a reddish hard wood resistant to insect and microorganism damage, and it is therefore highly appreciated locally for furniture and carpentry [4,5]. At the international level, numerous American, Asian and Continental African Sapotaceae species are traded and highly valued. In Madagascar, although trade had essentially developed at the local and national scales so far, signs of illegal logging for overseas exportation have been detected (R. Randrianaivo, pers. comm.). Together with ebonies (*Diospyros* spp.), exportation is thus expected to increase as other precious timbers such as rosewood (*Dalbergia* spp.) become scarcer [6]. Selective logging, together with the massive ongoing deforestation of Madagascar [7,8] has led to 76% of the *Capurodendron* species being threatened according to the IUCN criteria, with even one species out of four being Critically Endangered or possibly Extinct [1,9].

The lack of a robust taxonomy has affected the conservation assessment of many *Capurodendron* species, as for example *C. ludiifolium*, which was considered only as Vulnerable (VU) a few years ago [10]. However, after the revision of Boluda et al. [1], *Capurodendron ludiifolium* was split into five unrelated species (*C. ludiifolium*, *C. naciriae*, *C. sahafariense*, *C. randrianaivoi*, and *C. sakarivorum*), illustrating a case of evolutionary convergence toward a similar leaf venation. Of these five taxa, three are now considered Endangered (EN) and two Critically Endangered (CR). Other examples include two recently described local endemic species [1], both assessed as CR, and which were previously confounded with widespread and genetically distant species: *Capurodendron andrafiamenae* with *C. greveanum* (Least Concern, LC) [11] and *C. birkinshawii* with *C. nodosum*, (Vulnerable, VU) [12]. The latter case additionally highlights the tremendous impact of incorrect taxonomy on extent of occurrence (EOO) calculations. *Capurodendron nodosum* is indeed restricted to the extreme north of Madagascar, while the only known specimen of *C. birkinshawii* was collected in the extreme south. Including the latter in the EOO calculation of the former would have therefore erroneously raised the EOO value from ca. 3000 to 58,000 km^2.

While a 638 gene-based phylogeny showed clear species limits across the major part of the genus *Capurodendron* [1], three species complexes still remain unresolved, impeding the conservation assessment of the taxa they contain. One of them has been named the Eastern Complex as it is found all along the eastern moist evergreen forests of Madagascar. It comprises the morphologically variable species *Capurodendron tampinense*, which according to genetic data, seems to constitute a group of morphologically similar but genetically different species [1]. The resolution of this complex will however require further sampling. In this paper we focus on the resolution of the two other groups, the Western Complex and the Arid Species Complex.

The Western Complex (Table 1) occurs in the deciduous forests of western Madagascar below 300 m elevation and contains three genetically related species, *Capurodendron oblongifolium*, *C. perrieri* and *C. pervillei*, and a fourth undescribed morphospecies similar to *C. pervillei* and here referred to as *C. aff. pervillei*. *Capurodendron oblongifolium* was originally described as a variety of *C. perrieri* [13], then subsumed in the *Flore de Madagascar* Sapotaceae treatment [3]. It has been recently resurrected as a distinct species [1]. Although *Capurodendron oblongifolium* and *C. perrieri* grow in similar habitats, they are allopatric: *C. perrieri* is more widespread and with rare exception is found <50 km from the coast in the regions of Menabe, Melaky and Boeny, while *C. oblongifolium* is always >100 km inland and is restricted to Boeny. *Capurodendron pervillei* grows sympatrically with the two above-mentioned species but is morphologically well differentiated. Finally, the undescribed morphospecies is only known from two specimens embedded in the *Capurodendron pervillei* distribution area, one specimen found among populations of *C. oblongifolium* and the other among populations of *C. perrieri*. Previous phylogenies [14] found the undescribed morphospecies to be polyphyletic, suggesting that more than one taxon may present this morphology.

Table 1. Morphospecies included in each species complex with information related to their delimitation, distribution and phylogenetic status.

	Morphospecies	Details
Western Complex	*C. oblongifolium*	Well delimited, >50 km inland, sympatric with *C. pervillei*.
	C. perrieri	Well delimited, <50 km from the coast, sympatric with *C. pervillei*.
	C. pervillei	Well delimited, widespread and sympatric with both *C. oblongifolium* and *C. perrieri*.
	C. aff. pervillei	Similar to *C. pervillei*, not monophyletic, rare occurrences scattered in the global area of the complex.
Arid Complex	*C. androyense*	Widespread, well delimited morphologically, except three specimens intermediate with *C. microphyllum* and two with *C. mandrarense*.
	C. greveanum-mandrarense	Restricted range N of Toliara, weakly delimited morphologically, characters shared with *C. greveanum* and *C. mandrarense*.
	C. mandrarense	Widespread, variable, weakly differentiated from *C. greveanum-mandrarense* but more hairy and with prominent nerves. Two specimens intermediate with *C. androyense*.
Similar species	*C. greveanum*	Widespread in two disjunct coastal populations. Weakly differentiated from *C. greveanum-mandrarense*, but completely glabrous vegetatively. Phylogenetically far from the Arid Complex.
	C. microphyllum	Restricted range W of Fort-Dauphin, well delimited except three specimens intermediate with *C. androyense*. Sister species to the Arid Complex.

The Arid Complex (Table 1) mainly contains two morphologically well-differentiated species: *Capurodendron androyense* is restricted to the southern and southwestern sub-arid ecosystems, while *C. mandrarense* also extends inland to seasonally dry habitats up to 1000 m altitude. An additional morphospecies occurs in the northwestern edge of the subarid zone which is phenotypically intermediate between, and alternatively identified as, *Capurodendron mandrarense* and *C. greveanum*, the latter being a distantly related widespread species along the western and northern coast. This morphospecies is hereafter called *Capurodendron greveanum-mandrarense*.

The specimens of this Arid Complex form a monophyletic clade sister to *Capurodendron microphyllum*, which has a restricted distribution in sympatry at the extreme southeast of the Arid Complex area. This species is morphologically different from the former ones, although a few specimens exhibit morphologies related to *Capurodendron androyense*, suggesting that hybridization could sporadically occur. The Arid Complex presents two main taxonomical problems: First, the morphospecies *Capurodendron androyense*, *C. mandrarense* and *C. greveanum-mandrarense* might be considered conspecific as they appear intermixed in previously reconstructed phylogenies, showing a mismatch between morphology and detected genetic lineages [1,14]. Second, *Capurodendron greveanum* is a species phylogenetically and morphologically clearly distinct from *C. mandrarense*, and consequently *C. greveanum-mandrarense* is unlikely to represent intermediate morphologies uniting both taxa as a single species. The morphology of *Capurodendron greveanum-mandrarense* could be the result of hybridization events; however, it is absent from the area were the putative parental species coexist.

The goal of this paper is to delimit the taxa of the *Capurodendron* Western and Arid species complexes and explore how the species concept can be applied to lineages in which species are incompletely isolated. For this, we extend the previous use of exonic genetic markers by Christe et al. [14] to intronic and intergenic ones, as well as to microsatellites

(STR), all showing much higher substitution rates than exons alone. We aim to improve the resolution at the population level and address genetic admixture, introgression and hybridization in order to discuss how the IUCN criteria for species conservation ([15] IUCN Species Survival Commission, 2012) can be implemented to ensure the preservation of the genetic diversity of species complexes.

2. Materials and Methods

Taxon sampling—*Capurodendron* herbarium samples stored in G, K, MO, P, TAN and TEF herbaria (ca. 860 gatherings) were morphologically studied and specimens that did not belong to the two target species complexes were discarded, retaining 381 specimens, 43 of which being collected in 2017 during a dedicated field trip in southern Madagascar. Dry specimens were morphologically analyzed using a stereomicroscope (max. 65x), and characteristics of fresh material annotated in the field or deduced from accompanying pictures, when available. Flowers and fruits were boiled 2–10 min to rehydrate them and restore their three-dimensional shape or to isolate the seed. Out of the 381 specimens, 85 (52% from silica-gel preserved specimens and 48% from old herbarium material) representing all the morphological, ecological and geographical variability within species, were selected for DNA extraction: 15 belonged to the Western Complex (*Capurodendron oblongifolium*, *C. perrieri*, *C. pervillei*, and *C. aff. pervillei*) and 57 to the more intricate Arid Complex (*C. androyense*, *C. greveanum*, *C. greveanum-mandrarense*, *C. mandrarense*, *C. microphyllum*). Thirteen specimens belonging to the closest species of both complexes (*Capurodendron gracilifolium*, *C. nanophyllum*, *C. rubrocostatum* and *C.* sp. 20) and three outgroups (*C. birkinshawii*, *C. delphinense* and *Bemangidia lowry*) were added for the phylogenetic study (Table 2).

Table 2. Information on the specimens used. Original identification, followed by collector's name and number, collection year and sample kind.

Lab. Code	Morphospecies	Region	Collector Code	Year	Origin [1]
128	*Capurodendron androyense*	Atsimo-Andrefana	Gautier 6328	2017	Silica gel (G)
141	*C. androyense*	Atsimo-Andrefana	Gautier 6343	2017	Silica gel (G)
139	*C. androyense*	Atsimo-Andrefana	Gautier 6346	2017	Silica gel (G)
140	*C. androyense*	Atsimo-Andrefana	Gautier 6358	2017	Silica gel (G)
138	*C. androyense*	Atsimo-Andrefana	Gautier 6361	2017	Silica gel (G)
143	*C. androyense*	Androy	Gautier 6370	2017	Silica gel (G)
144	*C. androyense*	Androy	Gautier 6371	2017	Silica gel (G)
125	*C. androyense*	Androy	Gautier 6372	2017	Silica gel (G)
145	*C. androyense*	Androy	Gautier 6374	2017	Silica gel (G)
126	*C. androyense*	Androy	Gautier 6376	2017	Silica gel (G)
127	*C. androyense*	Anosy	Gautier 6387	2017	Silica gel (G)
149	*C. androyense*	Anosy	Randrianaivo 2954	2017	Silica gel (G)
29	*C. androyense*	Anosy	Randrianaivo 2959	2017	Silica gel (G)
79	*C. androyense*	Atsimo-Andrefana	Rogers 474	2004	G
70	*C. androyense*	Atsimo-Andrefana	Rakotomalaza 1719	1998	G
150	*C. androyense-mandrarense*	Atsimo-Andrefana	SF 22230	1962	G
161	*C. androyense-mandrarense*	Atsimo-Andrefana	SF 22286	1962	G
56	*C. birkinshawii*	Anosy	Birkinshaw 438	1997	G
98	*C. delphinense*	Anosy	Gautier 5801	2011	Silica gel (G)
151	*C. gracilifolium*	Melaky	Gautier 5736	2011	Silica gel (G)
8	*C. gracilifolium*	Atsimo-Andrefana	Gautier 6318	2017	Silica gel (G)
182	*C. gracilifolium*	Menabe	Randrianaivo 2972	2017	Silica gel (G)
156	*C. gracilifolium*	Atsimo-Andrefana	Messmer 607	1998	G
9	*C. greveanum*	DIANA	Randriarisoa 28	2017	Silica gel (G)
11	*C. greveanum*	Atsimo-Andrefana	Ranaivojaona 267	2000	G
10	*C. greveanum*	Menabe	Randrianaivo 2974	2017	Silica gel (G)
163	*C. mandrarense*	Anosy	Andriamihajarivo 1532	2004	G
20	*C. mandrarense*	Atsimo-Andrefana	Gautier 6349	2017	Silica gel (G)
21	*C. mandrarense*	Atsimo-Andrefana	Gautier 6350	2017	Silica gel (G)
22	*C. mandrarense*	Atsimo-Andrefana	Gautier 6351	2017	Silica gel (G)

Table 2. *Cont.*

Lab. Code	Morphospecies	Region	Collector Code	Year	Origin [1]
23	*C. mandrarense*	Atsimo-Andrefana	Gautier 6356	2017	Silica gel (G)
24	*C. mandrarense*	Androy	Gautier 6366	2017	Silica gel (G)
25	*C. mandrarense*	Androy	Gautier 6378	2017	Silica gel (G)
26	*C. mandrarense*	Androy	Gautier 6379	2017	Silica gel (G)
158	*C. mandrarense*	Anosy	Randrianasolo 204	1991	G
13	*C. mandrarense*	Anosy	Ratovoson 1473	2008	P
159	*C. mandrarense*	Anosy	Randrianaivo 1764	2009	G
110	*C. mandrarense*	Anosy	Randrianaivo 1785	2011	G
27	*C. mandrarense*	Anosy	Randrianaivo 2956	2017	Silica gel (G)
30	*C. mandrarense*	Anosy	Randrianaivo 2960	2017	Silica gel (G)
31	*C. mandrarense*	Anosy	Randrianaivo 2961	2017	Silica gel (G)
32	*C. mandrarense*	Anosy	Randrianaivo 2962	2017	Silica gel (G)
33	*C. mandrarense*	Anosy	Randrianaivo 2964	2017	Silica gel (G)
34	*C. mandrarense*	Ihorombe	Randrianaivo 2966	2017	Silica gel (G)
35	*C. mandrarense*	Ihorombe	Randrianaivo 2967	2017	Silica gel (G)
37	*C. mandrarense*	Menabe	Randrianaivo 2970	2017	Silica gel (G)
38	*C. mandrarense*	Menabe	Randrianaivo 2980	2017	Silica gel (G)
39	*C. mandrarense*	Menabe	Randrianaivo 2981	2017	Silica gel (G)
162	*C. mandrarense*	Ihorombe	SF 6692	1952	G
183	*C. mandrarense-greveanum*	Atsimo-Andrefana	Andrianjafy 1679	2006	P
15	*C. mandrarense-greveanum*	Atsimo-Andrefana	Gautier 6332	2017	G
16	*C. mandrarense-greveanum*	Atsimo-Andrefana	Gautier 6336	2017	G
17	*C. mandrarense-greveanum*	Atsimo-Andrefana	Gautier 6337	2017	G
18	*C. mandrarense-greveanum*	Atsimo-Andrefana	Gautier 6339	2017	G
19	*C. mandrarense-greveanum*	Atsimo-Andrefana	Gautier 6341	2017	G
160	*C. mandrarense-greveanum*	Atsimo-Andrefana	McPherson 17358	1998	G
77	*C. mandrarense-greveanum*	Atsimo-Andrefana	Phillipson 5603	2002	G
12	*C. mandrarense-greveanum*	Atsimo-Andrefana	Razafindraibe 165	2006	G
113	*C. mandrarense-greveanum*	Atsimo-Andrefana	Randrianaivo 1187	2005	G
40	*C. microphyllum*	Anosy	Gautier 6382	2017	Silica gel (G)
186	*C. microphyllum*	Anosy	SF 22411	1963	G
120	*C. microphyllum-androyense*	Anosy	Gautier 5794	2011	Silica gel (G)
41	*C. microphyllum-androyense*	Anosy	Gautier 6393	2017	Silica gel (G)
81	*C. nanophyllum* (Type)	Androy	SF 28521	1968	G
46	*C. perrieri*	Menabe	Noyes 1044	1992	G
47	*C. perrieri*	Atsimo-Andrefana	Razakamalala 5177	2010	G
114	*C. perrieri*	Boeny	Randrianaivo 969	2003	G
36	*C. perrieri*	Menabe	Randrianaivo 2968	2017	Silica gel (G)
45	*C. perrieri*	Menabe	Randrianaivo 2976	2017	Silica gel (G)
190	*C. oblongifolium*	Boeny	PerrierBâthie 1105	1974	P
44	*C. oblongifolium*	Sofia	Rakotonasolo 1601	2015	G
48	*C. oblongifolium*	Sofia	Ramananjanahary 51	2004	G
49	*C. oblongifolium*	Sofia	Razakamalala 1809	2004	G
76	*C. pervillei*	Boeny	Labat 3557	2005	G
164	*C. pervillei*	Sofia	Ramananjanahary 244	2004	G
165	*C. pervillei*	Sofia	Razakamalala 1677	2004	G
50	*C. pervillei*	Sofia	Randrianaivo 2397	2013	G
191	*C. aff. pervillei*	Boeny	Randrianarivelo 307	2005	G
192	*C. aff. pervillei*	Boeny	Randrianaivo 953	2003	G
195	*C. rubrocostatum*	Boeny	Andriamihajarivo 782	2005	G
194	*C. rubrocostatum*	Atsimo-Andrefana	Chauvet 187	1961	G
100	*C. rubrocostatum*	Melaky	Gautier 5936	2012	Silica gel (G)
73	*C. rubrocostatum*	Melaky	Luino 21	2012	G
146	*C. sp.* 20	Boeny	Gautier 6276	2016	Silica gel (G)
74(S26)	*Bemangidia lowry*	Anosy	Gautier 5789	2011	Silica gel (G)

[1] If sampled from a herbarium specimen, then the herbarium code; if sampled in the field, then "Silica gel".

Ordination of Morphological data—In order to objectify identifications in the intricate Arid Complex, morphological data were gathered on an expanded number of specimens. A total of 22 characters (Table 3) were scored across 123 specimens (Table S1). For quantitative characters that displayed variability within a single specimen (e.g., leaf length or number of secondary nerves) an average value of 10 measures was used, while for qualitative variable characters (e.g., type of leaf apex) we selected the dominant state on the specimen. To allow data ordination using qualitative and quantitative variables at once, Factorial Analysis for Mixed Data (FAMD) was conducted and run using the R package FactoMineR [16]; http://factominer.free.fr, accessed on 12 August 2021). This approach was considered unnecessary for the Western Complex.

Table 3. Characters used for the principal coordinate analysis with 123 specimens of the *Capurodendron* Arid Complex.

Character Number	Character	Type	Coding	State
1	Plant height	Continuous	meters	Number of meters
2	Brachyblast	Discrete	0	Absent
			1	Present
3	Prior year's elongating shoots	Discrete	0	Green and glabrous
			1	Brown and hairy
4	Petiole length	Continuous	mm	Number of mm
5	Petiole hairs	Discrete	0	Glabrous
			1	With hairs
			2	Tomentose
6	Leaf base symmetry	Discrete	0	Symmetric
			1	Asymmetric
7	Leaf base	Discrete	0	Decurrent
			1	Cuneate
			2	Obtuse
			3	Subcordate
8	Leaf length	Continuous	mm	Number of mm
9	Leaf width	Continuous	mm	Number of mm
10	Broadest leaf region	Discrete	0	1st third
			1	2nd third
			2	3rd third
11	Leaf apex	Discrete	0	Acute
			1	Obtuse
			2	Rounded
			3	Emarginated
12	Leaf upper side hairs	Discrete	0	Glabrous
			1	With hairs
			2	Tomentose
13	Leaf lower side hairs	Discrete	0	Glabrous
			1	With hairs
			2	Tomentose
14	Midrib on the lower side	Discrete	0	Not prominent
			1	Prominent
15	Secondary nerves on the lower side	Discrete	0	Not prominent
			1	Prominent
16	Midrib hairs on the upper side	Discrete	0	Glabrous
			1	With hairs
			2	Tomentose
17	Midrib hairs on the lower side	Discrete	0	Glabrous
			1	With hairs
			2	Tomentose
18	Pairs of secondary nerves	Continuous	Number	Number of pairs
19	Pedicel length of flower	Continuous	mm	Number of mm
20	Calyx hairs	Discrete	0	Adpressed
			1	Hirsute
21	External sepal length	Continuous	mm	Number of mm
22	Calyx diameter	Continuous	mm	Number of mm

DNA sequencing—DNA was extracted using the CTAB method with chloroform, including sorbitol washes to remove mucilaginous substances [14,17,18]. The sequences were obtained following the methodology explained in Christe et al. [14] combining gene capture with Next Generation Sequencing. For this, a genomic library of each specimen was constructed and labelled with dual indexing barcodes. Specimens were then pooled and 794 protein coding genes and 227 microsatellite loci were captured using a hybridization step with specific biotinylated oligonucleotide probes complementary to the loci of interest. Hybridized sequences were retained by streptavidin-covered magnetic beads while all non-target DNA was washed away. Finally, captured DNA was sequenced using an Illumina HiSeq 4000 machine (2 × 100 bp paired-end).

Capture data processing—The quality of DNA reads was checked with FASTQC (https://www.bioinformatics.babraham.ac.uk/projects/fastqc/, accessed on 12 August 2021) and they were trimmed with Trimmomatic version 0.38 [19]. In order to explore our study question with different type of markers displaying different substitution rates, we extracted our targeted exonic loci as well as the flanking intronic regions when present, as the latter have a higher mutation rate than exons [20]. We also extracted the sequences around the STR loci, which consisted of intergenic non-coding DNA. Four different datasets were gathered: (1) exons, (2) supercontigs (exonic and intronic sequences), (3) STR flanking regions, and (4) STR loci. The aligned sequences and SNPs (single nucleotide polymorphisms) were extracted for the three first datasets.

Aligned sequences—The program HybPiper [21] was run to obtain the 792 nuclear loci and 227 sequences containing STR loci presented in Christe et al. [14], in order to extract the consensus sequence of these loci for all individuals. For the 792 nuclear loci, the same program was run with the intronerate.py script [21] in order to get the supercontig sequences. All these sequences were aligned using the program MAFFT version 7 [22]. Putative paralogs according to Hybpiper were remove, resulting in 638 aligned nuclear loci.

SNPs—For 792 nuclear loci, the longest consensus sequence of each gene was selected as a reference for mapping the reads of each individual in order to extract the SNPs. BWA version 0.7.16 [23] was used for mapping, followed by Picard version 2.21.1 and Samtools version 1.9 [24,25] to sort, remove duplicates, and to index. SNPs and indels were called separately for each individual with HaplotypeCaller from GATK version 4.1.3. The resulting gvcf files were combined and genotyped with the same program. The resulting vcf files were filtered with vcftools version 0.1.16 [26], after removal of putative paralog loci, with the following settings: –minDP 8 –remove-indels –min-alleles 2 –max-alleles 2 –max-missing 0.8). For STR flanking regions, the same strategy was used to extract the SNP, with additionally removing the STR regions with vcftools using a bed file of the concerned positions.

Microsatellites—STR dataset was extracted according to Highnam et al. [27]. Trimmed reads were first mapped to the reference sequence of the STR (STR + flanking region) with Bowtie2 version 2.3.4.2 [28] followed by Picard version 2.21.1 and Samtools version 1.9 to sort, remove duplicates, and to index. Genotyping was accessed with RepeatSeq [27].

In order to exclude polyploidy in some problematic samples, we used the program nQuire [29] to estimate ploidy in each specimen. This method has been used successfully in target capture data for herbarium samples [30].

Phasing—To be able to reconstruct phylogenies using both alleles for each specimen (instead of a consensus sequence) we performed a phasing analysis. For that, supercontigs (containing exons and flanking intronic sequences) for each gene and specimen were obtained using the reads_first.py and intronerate.py scripts of the HybPiper pipeline [14,21]. Then these supercontigs were used as reference sequences for identifying variants for each specimen according to Kates et al. ([31] 2018, pipeline available at https://github.com/mossmatters/phyloscripts/tree/master/alleles_workflow, accessed on 12 August 2021). To assemble the alleles, WhatsHap [32], a Python-based program, combined with Tabix 0.2.6 (https://sourceforge.net/projects/samtools/files/tabix/, accessed on 12 August 2021) were run, and phased sequences were then converted into fasta files using bcftools consensus [33]. As a complete phasing was not expected, especially when using short DNA fragments as here, we retained the biggest phased block of each gene and replaced the remaining sequence by the consensus using haplonerate.py (Kates et al., 2018 [31]; https://github.com/mossmatters/phyloscripts, accessed on 12 August 2021). At the end we obtained two partially or completely phased sequences per gene for each specimen, except for the homozygous loci.

As there is no way to know which allele at a given locus is linked to any other allele at another locus, only gene trees can be estimated, and not species tree. Gene trees of the 638 loci without paralogy signals [13] were performed using RAXML v.8.2.4 [34] with a GTRGAMMA substitution model, discarding nucleotide positions with more than 20% missing data. All the generated trees were manually examined searching for the topological location of each allele for the specimens of interest (e.g., *Capurodendron* aff. *pervillei*).

Heterozygosity—In order to detect individuals with special features such as polyploidy or recent hybridization, we measured the heterozygosity level of each specimen with vcftools version 0.1.16 [26] on each SNP dataset. We calculated the percentage of observed heterozygosity as follows: (total number of sites - homozygotic sites observed)/total number of sites.

Phylogenetic reconstructions.—Out of the 85 specimens, those with more than 20% of loci missing were removed and, for the specimens that were retained, positions missing more than 20% were similarly removed. Phylogenetic reconstructions were performed using three different datasets: A) 600 exonic gene sequences all containing the same 81 specimens, B) 608 genes containing exonic and flanking intronic sequences from 36 to 81 specimens, and C) 195 microsatellite loci flanking regions with 76 to 81 specimens.

A gene tree for each locus was generated using RAXML v.8.2.4 [34] with a GTRGAMMA substitution model. Then Astral-II [35,36], a method based on the multispecies coalescence (MSC), was used to infer the species tree from the gene trees.

We additionally used SplitsTree4 [37] to infer a Neighbor-net network using concatenated sequences and uncorrected P-distances. For phased loci phylogeny see the Phasing section above.

Microsatellites clustering—STRUCTURE v.2.3.4 [38,39] was run on two different datasets. The first one contained the Western Complex, with *Capurodendron oblongifolium*, *C. perrieri*, *C. pervillei* and *C.* aff. *pervillei*, and the second dataset the Arid Complex, with *C. androyense*, *C. mandrarense*, *C. greveanum-mandrarense*, the closely related species *C. microphyllum* and the genetically far but morphologically related *C. greveanum*. Only specimens with less than 18% missing microsatellites and loci with less than 10% missing data were used, leading to the use of 15 specimens and 59 loci in the Western Complex (100% and 26%, respectively), and 52 specimens and 105 loci in the Arid Complex (91% and 46%, respectively).

STRUCTURE was run with 5 million burn-in generations and 5 million iterations, using a *k* value from 1 to 10 with 5 replicates for each *k*. Runs of each *k* value were combined with CLUMMP v.1.1.2 [40]. The ΔK method of STRUCTURE HARVESTER [41] was used to estimate which k value best adjust to our data.

Ordination of genetic data—Principal Coordinate Analyses (PCA) of genetic data were computed with the package smartPCA [42] using plink formatted merged vcf files and the same three datasets as the ones used in the phylogenetic reconstruction, selecting only specimens with less than 20% missing data.

In order to investigate the relationships within the Arid Complex as well as potential internal gene flow, we performed additional analyses using the three morphospecies and clusters based on PCA results from exons and flanking SSR datasets. For accessing the degree of genetic polymorphism, we calculated nucleotide diversity (π), and for genetic differentiation, the weighted pairwise FST. Both analyses were calculated for each site and averaged over all sites using vcftools version 0.1.16. Allele sharing between the putative parental species of *Capurodendron greveanum-mandrarense*, (*C mandrarense* and *C. greaveanum)* as well as within the Arid Complex was accessed with the Patterson's D statistics (ABBA-BABA test) for all possible trios with Dsuite version 0.4 [43] on the exon dataset. Two subgroups were used within *Capurodendron androyense* and *C. mandrarense*, and one in *C. greveanum-mandrarense* (see results). *Capurodendron delphinense* was used as an outgroup. The statistic test Dmin was also used to infer the lower bound of D value for each trio. A significant positive Dmin means that the sharing of derived alleles between the three taxa is inconsistent with a single species-tree relating them, even in presence of incomplete lineage sorting [43,44]. Statistical significance was accessed with the Bonferroni correction and the false discovery rate (FDR) with the Benjamini–Hochberg correction.

Potential species distribution—The potential species distribution for each taxa containing more than three specimens (the minimum required for computation) was calculated with Maxent v.3.3.3a [45]. The 19 environmental variable layers BIO1 to BIO19 from Madagascar, with a spatial resolution of 30 arcsec (about 1 km^2), were obtained from the WordClim database [46], using the raster package in R ([47] R Core Team 2013; https://cran.r-project.org/web/packages/raster/raster.pdf, accessed on 12 August 2021). The BIL layer format was transformed to Esri.asc using DIVA-GIS [48]. Each analysis was run ten times, and the median value of all runs was plotted. Only collections with confident identification were used, with 90 collection points for *Capurodendron androyense*, 79 for *C. greveanum*, 22 for *C. greveanum-mandrarense*, 60 for *C. mandrarense*, 14 for *C. microphyllum*, 6 for *C. oblongifolium*, 46 for *C. perrieri*, and 37 for *C. pervillei*.

3. Results

DNA sequences.—From the 85 analyzed specimens, 72 (85%) provided less than 5% missing data for exon sequences, 10 (12%) between 5–40% missing data, 2 (2%: specimens 162 and 194) between 40–80% and one (1%, specimen 150) more than 80%. Missing data in intronic sequences were usually higher, as our probes were designed specifically to hybridize with exonic loci. Of the 794 protein coding genes, 156 showed putative paralogy signals in one or more specimens and were discarded, thus leaving 638 genes for further analyses.

Morphological ordination—The most important variables contributing to the axes were, in decreasing order of importance: petiole length (dimension 1: 8.8%, dimension 2: 11.1%), presence of hairs in the petiole (8.7%, 11.1%), leaf length (8.7%, 10.5%), presence of hairs in the current year's shoots (8.1%, 9.5%), and leaf width (7.9%, 8.1%). Projections on axes 1 (4.7%) and 2 (3.6%) (Figure 1) show that *Capurodendron greveanum* is clearly different from the Arid Complex specimens. Within the complex, all morphospecies appear well delimited, but *Capurodendron microphyllum* can be divided into two groups, one containing the typical morphotype, the other with specimens displaying character states reminiscent of *C. androyense*. *Capurodendron androyense* and *C. mandrarense* are clearly separated on the plot, which contrasts with their genetic affinities (cf. below). Two specimens with intermediate morphologies between *Capurodendron androyense* and *C. mandrarense* appeared encompassed within the variability of *C. mandrarense* (black dots in Figure 1). The specimens corresponding to the *Capurodendron greveanum-mandrarense* morphotype are grouped together and are clearly separated from *C. greveanum*.

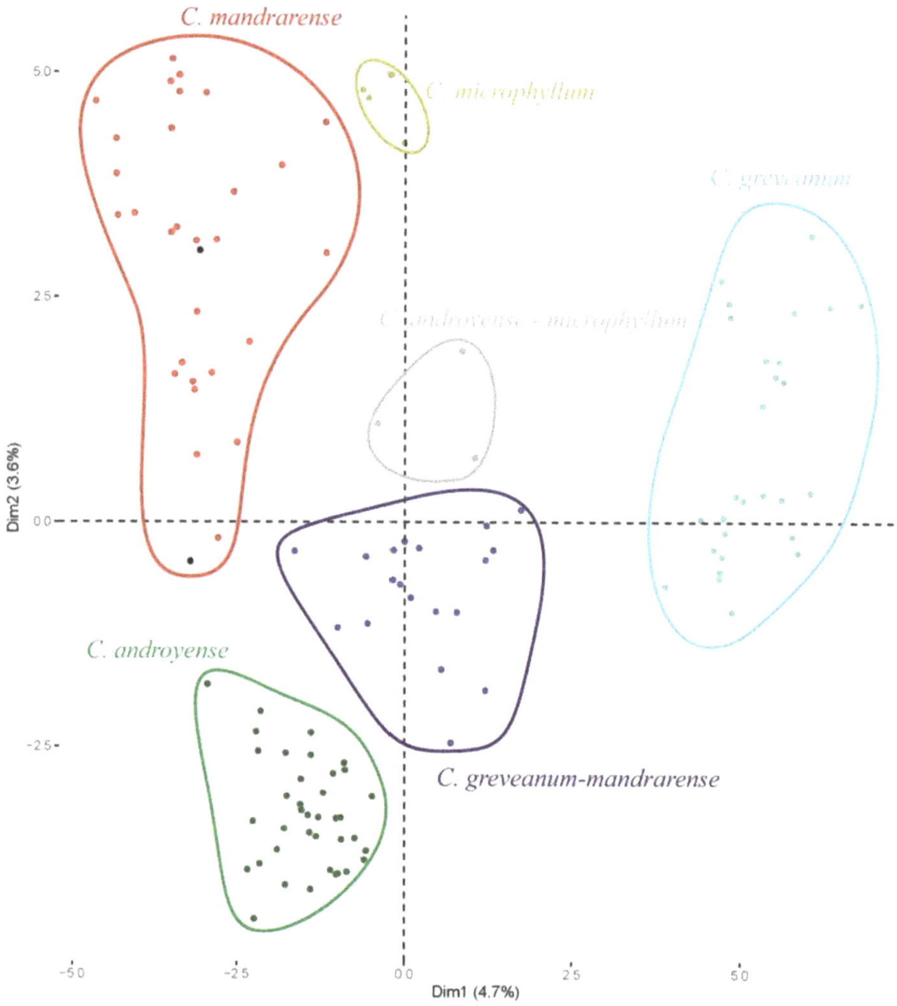

Figure 1. PCA scatter plot of the first two dimensions based on 22 morphological characters and 123 specimens. Black dots within the *Capurodendron mandrarense* cluster represent putative hybrids between that species and *C. androyense* (specimens 150 and 161 in Table 1).

Heterozygosity—This value can theoretically range from 0 for complete homozygotes to 1 for complete heterozygotes. The average heterozygosity for the exonic dataset was 0.051 (standard deviation SD 0.013), while for microsatellite flanking regions it was 0.045 (SD 0.014). Both datasets provided the same pattern of heterozygosity (Figure 2), showing that heterozygous sites are not linked to coding or intergenic regions, but are evenly distributed throughout the whole genome. The lowest heterozygosity levels were found for *Capurodendron greveanum*, *C. gracilifolium*, *C. perrieri* and *C. oblongifolium* ($\leq$0.033), while *C. microphyllum* was the species with the highest value, although with a high standard deviation ($\geq$0.063) (Figure 2). Specimens 191 and 192, both belonging to *Capurodendron* aff. *pervillei*, showed the highest heterozygosity level after *C. microphyllum* specimen 120. The morphospecies *Capurodendron greveanum-mandrarense* did not show a higher heterozygosity than the other two species of the Arid Complex, *C. mandrarense* and *C. androyense*.

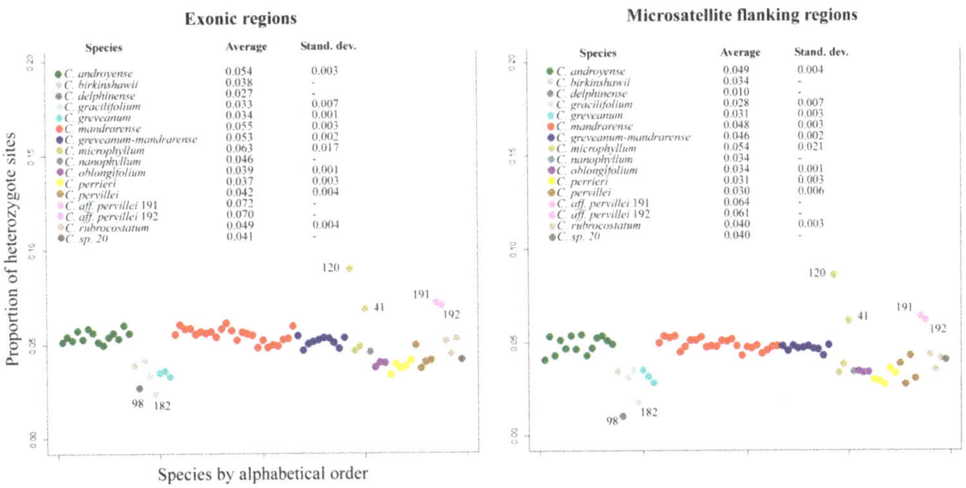

Figure 2. Proportion of heterozygous sites for exons (average 0.051, sd 0.013) and microsatellite flanking regions (0.045, sd 0.014), with average and standard deviation of each taxa indicated.

Phylogenetic reconstructions—The three analyzed datasets (exonic regions, supercontigs, and microsatellite flanking regions) produced trees with a similar topology (Figure 3). The main difference between the three datasets were the positions of *Capurodendron* sp. 20 and *C. rubrocostatum*. *Capurodendron* sp. 20 is sister to the Arid Complex in the supercontig and microsatellite dataset, but sister to (Arid Complex + *C. microphyllum*) when using only exonic sequences. In the case of *Capurodendron rubrocostatum*, it is placed sister to (Western Complex + *C. greveanum*) in microsatellite flanking regions, but sister to *C. greveanum* in the remaining two datasets.

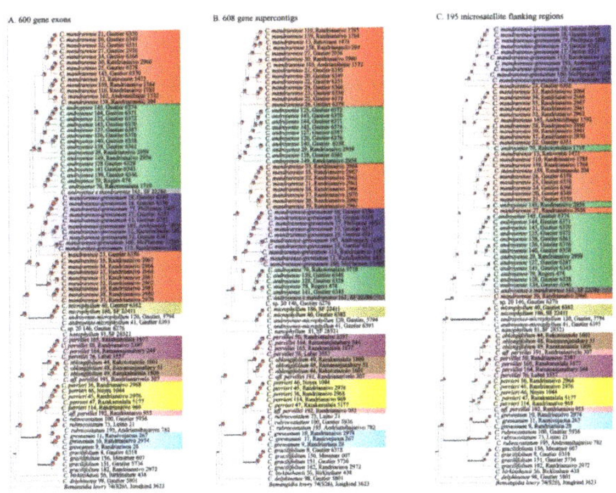

Figure 3. Pseudocoalescent ultrametric phylogenetic tree from Astral inferred from RAxML analyses of (**A**) 600 gene exonic regions, (**B**) 608 gene supercontigs including introns, exons and flanking regions and (**C**) 195 microsatellite flanking regions. All specimens contained less than 20% missing nucleotide positions. Colors refer to morphospecies. Pie charts represent the proportion of gene trees that support the clade of interest (red), support the main alternative bifurcation (blue), or support any other remaining alternative solution (gray). Astral posterior probabilities higher than 0.8 are depicted only for interspecific clades as bold lines. Species names are followed by the specimen number used in this study and the collector code.

Species generally formed supported clades in at least one tree, except for the Arid Complex in which the three morphospecies *Capurodendron androyense*, *C. mandrarense* and *C. greveanum-mandrarense* appeared intermixed. Astral topologies without quartet scores, which are non-ultrametric (Figure S1), showed a radiation-like pattern in the Arid Complex, with all the main clades diverging from a single supported node. In the case of the Western Complex, the three described species appeared separated by long supported branches, but the two specimens of *Capurodendron* aff. *pervillei* are recovered polyphyletic and not sister to *C. pervillei*, but rather one to *C. perrieri* and the other to *C. oblongifolium*. Within *Capurodendron greveanum-mandrarense* morphospecies, specimens LG 6339, LG 6336 and Phillipson 5603, all collected from the same population, appeared tightly clustered at the tip of a long branch (Figure S1).

Split networks produced almost the same clusters for all three datasets (the exon dataset is presented in Figure 4), the only difference being found in the supercontig dataset, which produced the same topology but with longer branches for *Capurodendron pervillei*. *Capurodendron microphyllum* always appeared as a sister species to the Arid Complex, with specimen 120 quite isolated from all remaining ones. Samples from the Arid Complex produced a radiation-like pattern. *Capurodendron androyense* was split into three lineages, *C. mandrarense* into two, and *C. greveanum-mandrarense* was recovered as a single clade. The largest lineage of *Capurodendron androyense* was comprised only of southern specimens (S. Androy and SW Anosy), while the second largest group solely contained the southwestern specimens (surroundings of Toliara and Tsimanampetsotse NP). Within the Western Complex the three described species are monophyletic, but *Capurodendron* aff. *pervillei* appeared polyphyletic, with specimen 191 arising between *C. oblongifolium* and *C. pervillei* and specimen 192 between *C. perrieri* and *C. pervillei*.

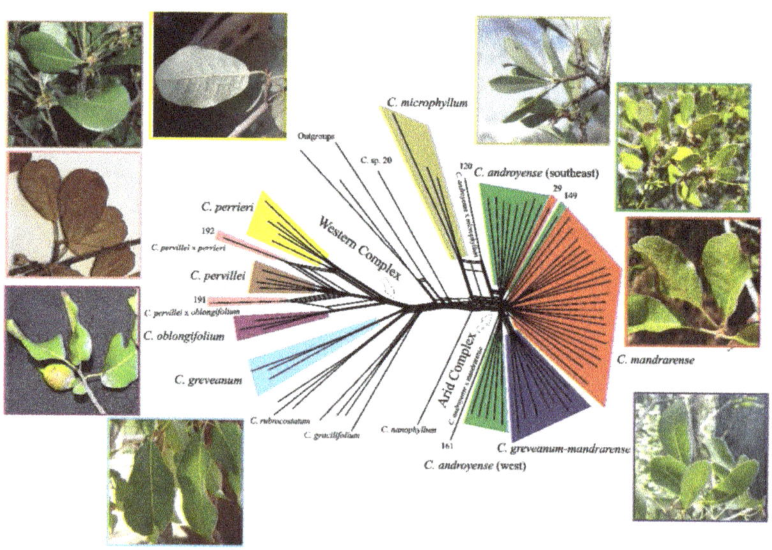

Figure 4. Split network computed from uncorrected P-distances and a concatenated supermatrix of exonic regions from 600 genes and 81 specimens. Splits and picture margin are color-coded according to morphospecies and specimens detected as hybrids are indicated. Numbers refer to specimens in Table 2. (Outgroups: *Bemangidia lowry*, *Capurodendron delphinense* and *C. birkinshawii*).

Phasing—From the 638 gene-trees generated, only those based on alignments with more than 900 bp were taken into account (305 loci in total), as shorter alignments produced unsupported topologies. The trees were manually checked searching for genes displaying alleles clustered in different species, which can be considered a signal of hybridization. Specimens of *Capurodendron greveanum-mandrarense* always displayed alleles clustered within specimens of the Arid Complex, never with *C. greveanum*, and usually only with specimens of their own morphospecies.

In the *Capurodendron* Western Complex, each species displayed homologous alleles clustered together. However, the specimens *Capurodendron* aff. *pervillei* 191 and 192 contained 40% of the informative genes with homologous alleles nested into different species (Table 4). For the remaining 60% genes, both alleles were grouped together but sometimes in one species and sometimes in another: in *C. oblongifolium* and *C. pervillei* for specimen 191, and in *C. pervillei* and *C. perrieri* for 192.

Table 4. Topological positions of the alleles of specimens 191 (Randrianarivelo 307) and 192 (Randrianaivo 953) based on 305 maximum likelihood trees from protein coding genes. Monospecific loci refer to loci in which both alleles appeared nested in a single species, while heterospecific loci point to exons with each allele nesting within distinct species. Percentages are calculated after having discarded any missing or unsupported allele.

	Specimen 191	Specimen 192
Alleles in *C. oblongifolium*	192 (56.8%)	2 (0.7%)
Alleles in *C. perrieri*	2 (0.6%)	160 (56.7%)
Alleles in *C. pervillei*	144 (42.6%)	120 (42.6%)
Missing/unsupported alleles	137	164
Monospecific loci	101 (59.7%)	85 (60.3%)
Heterospecific loci	68 (40.3%)	56 (39.7%)

Microsatellites—STRUCTURE output, using the ΔK method [49], suggested that our data best fit two gene pools for the Western Complex and three for the Arid Complex (Figure S2), however these numbers of clusters do not match well either with the phylogenetic species concept or with the morphological species concept.

In the Western Complex dataset, all specimens appeared completely admixed except the *Capurodendron pervillei* specimens 164 and 165, which are grouped together and without admixture. This pattern is stable from $k = 2$ to $k = 10$.

In the Arid Complex dataset, *Capurodendron greveanum* formed the most clearly isolated group at all k values. However, specimen 9 shared around 60% of its genetic component with the pool composed of *Capurodendron androyense* and *C. mandrarense* clusters, but not with *C. greveanum-mandrarense*. The second-best isolated pool, appearing from $k = 3$ and higher, was composed of the *Capurodendron greveanum-mandrarense* morphospecies, although specimens 19, 113, 160 and 183 displayed admixtures with *C. androyense* and *C. mandrarense*. *Capurodendron microphyllum* never appeared as a single gene pool, even at $k = 10$, nor did *C. androyense* and *C. mandrarense,* both of which displayed a highly intermixed pattern, except for specimens 31, 32, 33 and 34 of *C. mandrarense*, all collected from the same inland area of the Horombe plateau, at ca. 1000 m asl.

Ordination of genetic data—Analyses performed separately on exons, supercontigs and microsatellites flanking regions showed similar outputs and the same groups (Figure 5A). Axes information was always lower than 10%, which is expected when many markers and genetically closely related individuals are used. Three main clusters of dots were detected, one for the outgroup species, another for the Western Complex and related species (*Capurodendron gracilifolium*, *C. greveanum* and *C. rubrocostatum*), and one for the Arid Complex and related species (*C. microphyllum*, *C. nanophyllum* and *C.* sp. 20). Species outside the complexes were well delimited except for *Capurodendron microphyllum*, which showed a wide dot distribution (*C. nanophyllum* and *C.* sp. 20 are both known from a single specimen).

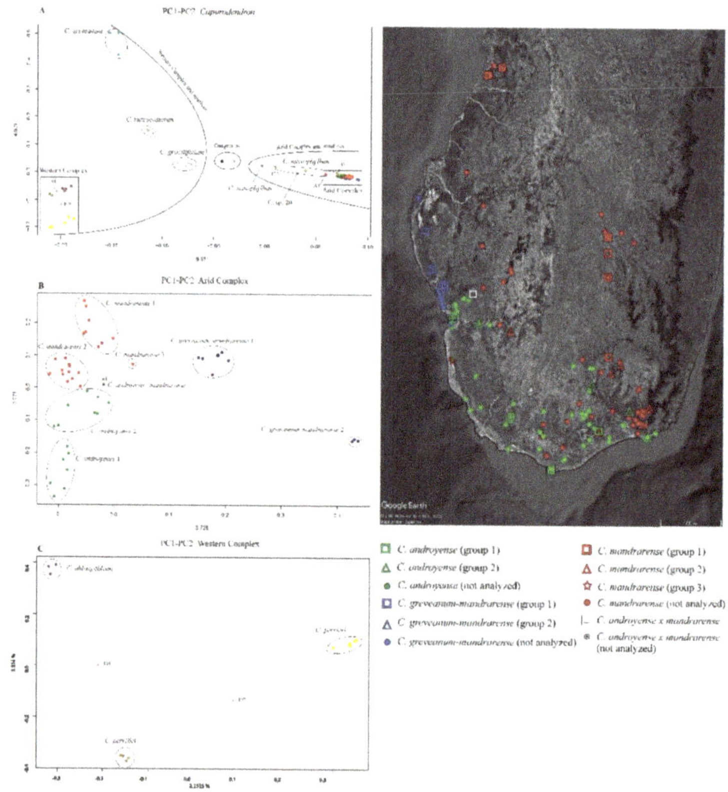

Figure 5. Principal Components Analysis (PCA) of 227 microsatellite flanking regions of (**A**) the complete dataset, (**B**) the Arid Species Complex, and (**C**) the Western Species Complex. The distribution map of the samples of the Arid Complex is shown on the righ side. Specimens 120 and 41 here included under *Capurodendron microphyllum*, share some morphological characters with *C. androyense*.

In the Arid Complex (Figure 5B), the PC1 axis clearly separated *Capurodendron greveanum-mandrarense* specimens from the rest, this morphospecies being further divided into two clusters, with group 2 containing only the three specimens that appeared on a long branch in the non-ultrametric Astral trees. The mutations that supported their differentiation from the other specimens were not caused by genome inversions nor by insertions/deletions, as the variable sites appeared dispersed throughout all loci. As mentioned above, all three specimens were collected from the same population, by two different collectors and in different years, and were processed and sequenced separately, which eliminates contamination as a possible explanation for the pattern.

All the remaining specimens share low values for PC1 and are scattered along the PC2 axis and could thus be considered as forming a large single group. However, the specimens identified as *Capurodendron androyense* were retrieved in the negative values whereas *C. mandrarense* were on the positive coordinates. *Capurodendron androyense* can be divided into two groups: group 1, with the more extreme PC2 values and composed of extreme Southern specimens (Androy and SW Atsimo-Andrefana regions); and group 2, including southwestern (Toliara surroundings and Tsimanampetsotse NP) and southeastern (S Anosy) specimens. *Capurodendron mandrarense* specimens can be split into three subgroups: group 1 containing the specimens from the northern half of the species distribution; group 2 from the southern half; and group 3, with a single specimen, from the extreme southwest in the Tsimanampetsotse NP. Specimen 161 retrieved between *Capurodendron androyense* and *C. mandrarense* on this projection; it was one of the two collected specimens that are morphologically intermediate between these two species, and was the only one that could be analyzed.

In the Western Complex (Figure 5C) the three described species appeared completely isolated, forming three well-differentiated groups. Specimen *Capurodendron* aff. *pervillei* 191 was located halfway between *C. pervillei* and *C. oblongifolium*, while specimen *C.* aff. *pervillei* 192 was equidistant between *C. pervillei* and *C. perrieri*.

To calculate nucleotide diversity, pairwise FST and Patterson's D statistics for the Arid Complex, 5 groups were used based on the results of the PCA, according to genetic affinities but also to geographical proximity of individuals, while keeping the number of specimens in each group as similar as possible (Figure 5). For *Capurodendron androyense*, groups 1 and 2 were analyzed separately, group 1 being the more variable, both genetically and geographically, although not all specimens can be considered as sympatric, and group 2 gathering sympatric specimens with either *C. mandrarense* or *C. greveanum-mandrarense*. For *Capurodendron mandrarense*, two groups were also used with individuals from group 2, from southern areas and growing in sympatry with either *C. androyense* or *C. greveanum-mandrarense*, and group 1 and 3 together, containing specimens from northern areas. For *Capurodendron greveanum-mandrarense* only group 1 was used for pairwise FST, as group 2 shows particularities that might introduce a bias into the analyses.

Nucleotide diversity showed slightly lower values for exons than for flanking SSR but with consistent results among them when the three morphospecies are taken into account. The highest values were those of *Capurodendron androyense* (0.484 $\pm$ 0.1110, 0.506 $\pm$ 0.1145) followed by *C. mandrarense* (0.477 $\pm$ 0.1086, 0.0505 $\pm$ 0.1106), and *C. greveanum-mandrarense* (0.0423 $\pm$ 0.1152, 0.0469 $\pm$ 0.1211). Within the different genetic groups from Figure 5, specimens in sympatry had higher values than specimens in allopatry or partial allopatry (*Capurodendron androyense* group 2, 0.492 $\pm$ 0.117, 0.514 $\pm$ 0.121 *vs* group 1, 0.456 $\pm$ 0.117, 0.481 $\pm$ 0.124 and *C. mandrarense* group 2, 0.473 $\pm$ 0.111, 0.503 $\pm$ 0.114 *vs* group (1+3), 0.455 $\pm$ 0.1180, 0.474 $\pm$ 0.119).

FST comparison (Table 5) showed similar values for exons and flanking STR. Overall the FST values were low, with the highest found between each group of morphospecies (between 0.139 and 0.094). Then the highest value was found between the two genetically most distant groups of *Capurodendron mandrarense* and *C. androyense* (*C. androyense* group 1 and *C. mandrarense* group 1; 0.090 for exons and 0.084 for flanking STR regions). The lowest FST were found between the two groups of *Capurodendron androyense* and *C. mandrarense* that are geographically in contact (*C. androyense* group 2 and *C. mandrarense* group 2; 0.024 and 0.025 for exons and flanking STR respectively).

D-statistics (Supplementary Materials Table S2) did not support any introgression between *Capurodendron greveanum* and *C. greveanum-mandrarense* or any of the other subgroups of the two other morphospecies of the Arid Complex. A low level of introgression (f4 ratio up to 0.08) between *Capurodendron microphyllum* and the two subgroups of *C. androyense* was detected, indicating that the introgression between the two species could predate the split of *C. androyense* in different groups. Introgression between the different subgroups of *Capurodendron androyense* and *C. mandrarense* was also detected, with the highest f4 ratio (0.48) found between the two sympatric populations, *C. androyense* group 2 and *C. mandrarense* group 2. However, Dmin score statistics were never significant for trios including these subgroups. Therefore, the sharing of derived alleles between trios is inconsistent with a single species-tree relating them, even in the presence of incomplete lineage sorting. Significant Dmin score only confirm introgression between *Capurodendron microphylum* and *C. androyense* group 1, group 2 and *C. mandrarense* group 2.

Table 5. FST values (weighted mean and standard deviation) within the Arid complex for the *Capurodendron* groups shown in Figure 5. The groups that were not compared are indicated with the symbol -.

Exons Flanking Regions FST						
	C. androyense	*C. androyense* 1	*C. androyense* 2	*C. mandrarense*	*C. mandrarense* 1	*C. mandrarense* 2
C. androyense 1	-					
C. androyense 2	-	0.037 ± 0.073				
C. mandrarense	0.033 +− 0.052	-	-			
C. mandrarense 1	-	0.090 +− 0.117	-	-		
C. mandrarense 2	-	-	0.024 ± 0.058	-	0.046 ± 0.074	
C. greveanum-mandrarense	0.118 ± 0.114	0.139 ± 0.142	0.095 ± 0.110	0.104 ± 0.111	0.111 ± 0.122	0.111 ± 0.122
STR Flanking Regions FST						
	C. androyense	*C. androyense* 1	*C. androyense* 2	*C. mandrarense*	*C. mandrarense* 1	*C. mandrarense* 2
C. androyense 1	-					
C. androyense 2	-	0.037 ± 0.079				
C. mandrarense	0.033 ± 0.052	-	-			
C. mandrarense 1	-	0.084 ± 0.117	-	-		
C. mandrarense 2	-	-	0.025 ± 0.063	-	0.044 ± 0.081	
C. greveanum-mandrarense	0.112 ± 0.112	0.122 ± 0.140	0.094 ± 0.120	0.108 ± 0.113	0.111 ± 0.139	0.097 ± 0.122

Analyses of potential distribution—The potential distribution predicted by Maxent using 19 bioclimatic variables (Figure 6) showed AUC values of 0.99 for *Capurodendron androyense*, *C. greveanum-mandrarense*, *C. microphyllum*, *C. oblongifolium*, *C. perrieri* and *C. pervillei*, and 0.98 for *C. greveanum* and *C. mandrarense*, indicating a highly supported predicted distribution for all taxa. The three most significant bioclimatic variables contributing to the prediction of each species are shown in Table 6. The predicted distribution for *Capurodendron greveanum* shows two main areas with littoral or sublittoral conditions, one from Morombe (Menabe region) to Besalampy (Melaky region), and the other in the north-east, from Vohemar (SAVA region) to Antsiranana (DIANA region) separated by areas with unsuitable climatic conditions. These disjoint populations match with the species distribution according to specimen collections. *Capurodendron greveanum* cannot develop in the sub-arid regions of southern Madagascar where *C. androyense* grows, but its distribution meets the northwestern populations of *C. mandrarense*. It is only sympatric with *Capurodendron greveanum-mandrarense* in the Mangoky estuary, at the extreme south of its predicted distribution. Within the Arid Complex, each morphospecies shows different habitat preferences, with *Capurodendron androyense* tolerant of the driest habitats in the extreme southwest, and *C. mandrarense* preferring more humid places and extending to medium-altitudes, although overlapping with most of the distribution area of *C. androyense*. *Capurodendron greveanum-mandrarense* is apparently restricted to deciduous-forest habitats near the coast north of the Onilahy estuary, although it is also predicted further south down to Tsimanampetsotse, where it has never been collected. The regions closer to the sharp climatic gradient between dry spiny thicket and moist evergreen forests, just west of Taolagnaro (Fort-Dauphin), showed conditions suitable for *Capurodendron androyense*, *C. mandrarense* and *C. microphyllum*, where all these species have indeed been collected.

Table 6. The most important variables contributing to the potential distribution of each species. AUC = Area below the curve.

Species	Number of Points	Most Important Variables	Jackknife of AUC
Capurodendron androyense	90	Precipitation of wettest month	0.94
		Precipitation of wettest quarter	0.93
		Annual precipitation	0.92
Capurodendron greveanum	79	Annual precipitation	0.94
		Annual mean temperature	0.92
		Mean temperature of warmest quarter	0.92
Capurodendron greveanum-mandrarense	22	Annual precipitation	0.97
		Precipitation of wettest quarter	0.02
		Precipitation of wettest month	0.91

Table 6. *Cont.*

Species	Number of Points	Most Important Variables	Jackknife of AUC
Capurodendron mandrarense	60	Max temperature of warmest Month	0.88
		Temperature seasonality	0.87
		Temperature annual range	0.82
Capurodendron microphyllum	14	Precipitation seasonality	0.94
		Precipitation of wettest month	0.94
		Precipitation of wettest quarter	0.91
Capurodendron oblongifolium	6	Precipitation of wettest month	0.99
		Precipitation of wettest quarter	0.97
		Precipitation seasonality	0.96
Capurodendron perrieri	45	Precipitation seasonality	0.96
		Mean temperature of warmest quarter	0.94
		Mean temperature of wettest quarter	0.94
Capurodendron pervillei	36	Precipitation of wettest month	0.98
		Precipitation of wettest quarter	0.97
		Annual mean temperature	0.97

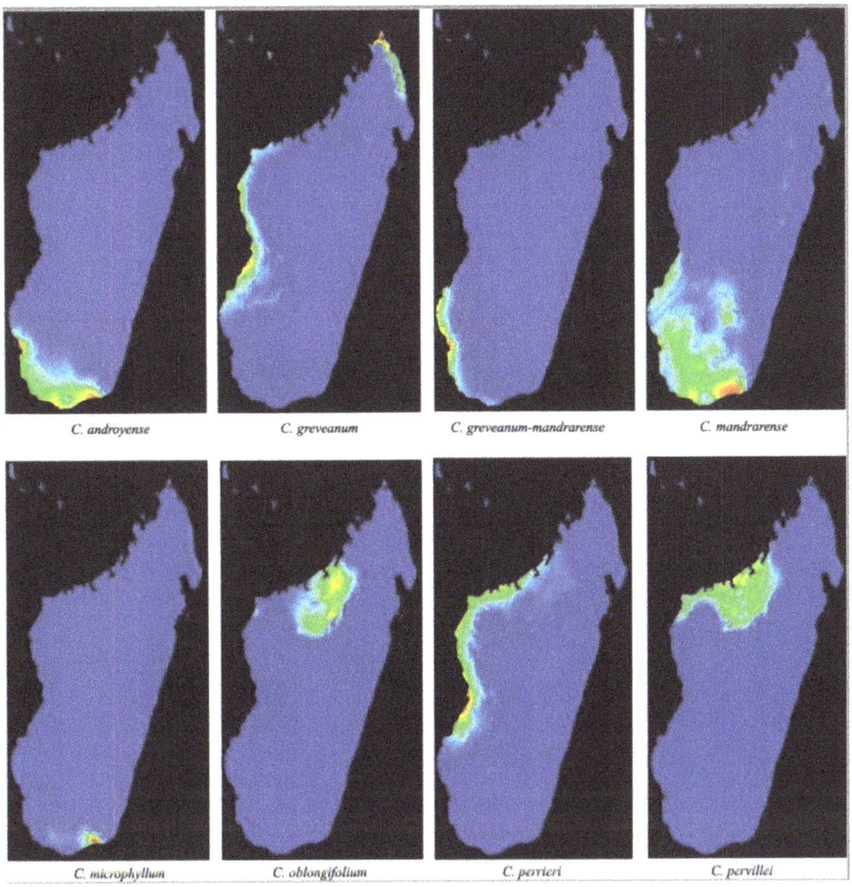

Figure 6. Potential distribution maps predicted by Maxent for *Capurodendron androyense*, *C. greveanum*, *C. greveanum-mandrarense*, *C. mandrarense*, *C. microphyllum*, *C. oblongifolium*, *C. perrieri* and *C. pervillei*.

In the Western Complex, *Capurodendron perrieri* showed the most widespread distribution, occupying the coastal and near-coastal areas of western Madagascar. *Capurodendron oblongifolium* was restricted to inland north-western areas, while *C. pervillei* to inland and coastal north-western areas, partially in sympatry with *C. oblongifolium* and *C. perrieri*.

4. Discussion

Ongoing speciation and the species concept—The species concept is still a pending issue in biology, especially considering that species change across time. Among the 26 or so different species concepts proposed [50,51], those emphasizing species monophyly have been the most common in recent decades, due to the increasing use of genetic data. One of the most popular is that of de Queiroz [50], which defines a species as a lineage composed by a group of populations (a metapopulation) that evolves independently from others. This is also the definition that is most consistent with molecular phylogenies. However, this species concept sometimes fails at establishing the limits within taxonomically challenging groups, leading to mismatches between genetic groups and observed phenotypes [52,53]. This appears to be the case of the *Capurodendron* Arid Complex.

Other proposals rely on a reference-based taxonomy [54], in which well-studied species limits of different living groups (e.g., humans/neanderthals/chimpanzees, in the case of primates) are used as references. Then, in a group under study, the decision that two lineages are different species would be taken if they show a similar or higher genetic differentiation than that of the two closest species in the reference. This could be done comparing, for example, FSTs between the reference and the candidate species under study. In the case of the Arid Complex, a comparison with deeply studied species-complexes such as sweet potato or *Citrus* [55,56] would perpetuate a mismatch between phenotypes and species obtained using such a concept, yielding a single species aggregating different morphologies without transitional forms.

Speciation can give rise to a new species from a parental one that persists unchanged, especially through founder effects, but also through hybridization, or when some subpopulations develop traits that increase their biological fitness and allow a fast adaptation to a different niche [57–59]. In such cases, the new species is monophyletic, but still nested within the parental one, resulting in a paraphyletic taxon. Although ecologically and phenotypically well-defined species can originate from a reduced number of key mutations in regulatory genes, the monophyletic species concepts will never recognize a new species as long as it is nested within the parental taxa [60,61]. Strong or even weak reproductive barriers between the two recently diverging lineages will produce differential accumulation of mutations across time, allowing a monophyletic species concept to eventually recognize two distinct species. However, if the isolation is not strong enough, the diverging lineages will hybridize, resulting in the introgression of genetic material of one species into the other and then leading to the termination of the speciation event.

Rather than the two species concepts mentioned above, each unsatisfactory for lineages in a speciation/introgression process, a population genetics approach in which species can be considered as historically connected populations sharing similar phenotypes and roles in the ecosystem brings together the temporal and phenotypic dimensions of species. This concept is developed by Freudenstein et al. [62] and is compatible with paraphyletic lineages sharing the same phenotype, and accounts for one of the most common problems in species delimitation during speciation processes. This is the concept we will use hereafter for well-characterized morphospecies lacking genetic isolation.

Species delimitation in the *Capurodendron* Western Complex—This species complex forms a well-defined group, sister to *Capurodendron rubrocostatum* and *C. greveanum*. It contains three morphologically and genetically delimited species (*Capurodendron oblongifolium*, *C. perrieri* and *C. pervillei*), and the unique taxonomic problem that we faced was to determine whether specimens 191 (Randrianarivelo 307) and 192 (Randrianaivo 953), both displaying morphologies similar to *C. pervillei*, should be considered a new species.

Observed heterozygosities in both specimens are much higher than statistically expected (Figure 2); they contain the highest proportion of heterozygous sites across all studied *Capurodendron* specimens with the exception of *C. microphyllum* 120 (Gautier 5794). It is well known that hybridization increases the number of heterozygous sites, as each allele comes from a different taxon with a different history of mutations [63]. Phylogenetic networks (Figure 4) show how these specimens arise from lineages belonging to two different species, indicating that a great proportion of the variable sites are not specific to them, but are shared more or less in the same proportion with the parental taxa. The same conclusion holds for PCA analysis (Figure 6), with both specimens located halfway between their putative parental species. It can thus be assumed with confidence that they are of hybrid origin. However, contrary to what could have been expected, STRUCTURE results on microsatellites were unable to show either species isolation or introgression signals on both hybrid specimens. This may be the result of loci selection for gene capture, as phylogenetically distant species were used to design the probes and the resulting microsatellites might not be sufficiently informative. This limitation has also been observed in other groups of *Capurodendron* [14].

The phased phylogenetic gene trees (Table 4) are also key results supporting a hybrid origin: in many genes trees, the two alleles of specimens 191 and 192 are each nested in one of the parental species, a feature never observed with the alleles of the three recognized species of the complex. In a first-generation hybrid, it is expected that half the alleles will appear nested in one species, and the other half nested in the other. However, this proportion can be skewed by events such as shared or uninformative mutations, and can also be biased by phylogenetic reconstruction methods. This balanced pattern furthermore quickly disappears when the hybrid backcrosses with the parental species or reproduces with other hybrids, since introgression from one of the taxa will increase and recombination may recover parental chromosome blocks [64]. Phylogenetic reconstructions might be biased, at least for some genes, especially if parental species are close and if sequences code for proteins. In our case, specimens 191 and 192 may correspond to first generation (F1) hybrids, as the introgression of both parental species is quite balanced. The slightly skewed pattern observed for both samples might be due to the fact that many alleles could not be attributed to a given parental species because they were not informative enough. However, the striking coincidence of the skewness in the two hybrids (42.6% toward *C. pervillei*) is noteworthy. We could not find any reason for this other than the influence of incomplete lineage sorting (ILS) among the parental species.

Although flowers were observed on these hybrid specimens, it is not clear whether they are fertile or sterile, as fruits, even in the first developmental stages, were absent. The parental species are phylogenetically close, which would suggest that hybrids could be fertile. However, if hybridization yielding fertile offspring were a recurrent process in the group, the species of the Western Complex would be expected to be more intermixed genetically, a pattern that was not detected in our analyses (Figures 2–5). From the 91 specimens studied morphologically, just two contained a phenotype different from any described species of the complex, indicating that morphological intermediates are rare and hence probably sterile. However, it is also possible that hybrids are indeed fertile, but that their offspring are not fit in this environment and thus removed by natural selection.

Our data clearly indicate that both specimens correspond to hybrids and should therefore receive the name *Capurodendron pervillei* x *oblongifolium* for 191 (Randrianarivelo 307), and *C. pervillei* x *perrieri* for 192 (Randrianaivo 953). Consistent with this, both specimens were collected in areas where the parental species coexist. The *Capurodendron* Western Complex is therefore composed of three well-differentiated species that can hybridize. Each one needs to be assessed for conservation separately, without inclusion of the hybrids. Hybrids do not require descriptions nor conservation assessment.

Species delimitation in the *Capurodendron* Arid Complex—This species complex forms a well-supported lineage closely related to *Capurodendron microphyllum*. Three morphospecies can be easily distinguished with the naked eye, and they are supported by morphological analysis (Figure 1, red, green, and dark blue). Only two of the 123 specimens studied (1.6%) showed intermediate phenotypes: specimens 150 (SF 22230) and 161 (SF 22286). Genetic analyses, however, show that the three morphospecies are entangled, despite clear morphological discontinuity and different environmental preferences (although with partly sympatric distributions; Figure 6). Since we used loci ranging from low to high substitution rates, even reaching geographical resolution, (Figure 5B) we can conclude that the absence of monophyly is not an artefact produced by non-appropriate loci.

The morphospecies *Capurodendron greveanum-mandrarense* was initially hypothesized as a hybrid between the species referred to in its provisional name [1,14], but no signal supporting this hypothesis has been found so far. High heterozygosity levels (such as the ones observed in the Western Complex) are linked to recent hybridizations, yet in this morphospecies the heterozygosity level is even lower than that found in the other morphospecies of the complex (Figure 2). Phylogenetic networks are suitable diagrams for representing evolutionary relationships in groups that have experienced reticulation [65] and are useful in identifying ambiguous relationships [66]. Applied to our data, hybridization in the Western Complex as well as between *Capurodendron androyense* and *C. microphyllum* was clearly visible (Figure 4). However, no such signal could be observed for *Capurodendron greveanum-mandrarense*, which always constitutes a well-defined clade nested within the Arid Complex (Figures 3 and 4). STRUCTURE on STR separates this morphospecies from *Capurodendron greveanum* even at *k* 2 and shows no admixture within both taxa (Figure S1). PCA (Figure 5B) clearly separates *C. greveanum-mandrarense* from the remaining specimens of the complex and confirms its location far from *C. greveanum*. Finally, phased phylogenetic reconstructions showed no alleles coming from *Capurodendron greveanum*. Thus, *Capurodendron greveanum-mandrarense* is not a hybrid between *C. greveanum* and *C. mandrarense*, but rather an undescribed species displaying morphological convergence with both, and more related to the latter. The differentiation of this species might relate to a local adaptation to the less arid, sandy and coastal conditions found just North of Toliara, limiting its further expansion. It requires formal description and a conservation assessment of its own, and will be referred to hereafter as *Capurodendron mikeorum* nom. prov., as it grows in the forest harboring the Mikea ethnical group.

Conversely, the two morphospecies *Capurodendron androyense* and *C. mandrarense* are not only genetically unresolved, they also display a higher genetic diversity in areas where they grow sympatrically (Figure 5). Two non-exclusive hypotheses can explain this pattern:

i. An ongoing sympatric speciation: Although the morphospecies have a widely overlapping distribution, they show different environmental preferences (Figure 6). *Capurodendron androyense* is the more drought-resistant taxon, extending to the areas with 12 ecologically dry months found along the coast in the extreme south, while *C. mandrarense* prefers relatively more humid habitats, is more cold-tolerant, and has a distribution extending to south-central regions up to 1000 m elevation. Hence, a partially ongoing sympatric speciation mediated by environmental selection might be at work. In such a case, the area with the highest nucleotide diversity, which coincides with the genetic clusters *Capurodendron androyense* 2 and *C. mandrarense* 2 (Table 5; Figure 5) representing the majority of the complex distribution area, could correspond to the diversification center. This is a climatically intermediate area with two less months of dry season than the coastal region, while also escaping the colder night temperatures of the central highlands. From this region, the ancestral species could have undergone a selection pressure towards aridity (enforcing the *Capurodendron androyense* gene pool), and towards more humid and colder habitats (enforcing the *C. mandrarense* gene pool). Then, *Capurodendron androyense* 1 and *C. mandrarense* 1 of Figure 5 would have appeared later, having many fewer introgression signals between them and being therefore genetically 'purer'. This pattern would correspond to a parapatric speciation process driven by ecological adaptation. In such a scenario of recent speciation, it is not surprising to observe such levels of incomplete lineage sorting (ILS), as the two species are additionally found in large areas and are expected to have high population effective sizes [67].

ii. A past allopatric speciation followed by secondary contact: Under this hypothesis the species would have originated in allopatry from a recent common ancestor, adapting to different environments. Posteriorly, *Capurodendron androyense* and *C. mandrarense* distributions would have expanded and come into contact, producing the more admixed *C. androyense* group 2 and *C. mandrarense* group 2. In this scenario, the higher nucleotide diversity of these groups would point to a secondary contact with introgression rather than to an ancient center of diversification. This would parallel a similar situation to the east, where hybrids between *Capurodendron androyense* and *C. microphyllum* (a genetically well differentiated species, sister to the Arid Complex) appear in areas where both taxa coexist. In this case, ILS signal would come from admixture.

Independently of which hypothesis corresponds to what actually occurred (the situation could be more complex still, involving a combination of both), the three morphospecies can coexist in the same forest without forming a population of phenotypically intermediate hybrid specimens. This indicates that selection pressure is keeping each morphospecies separate, and as such, they merit a taxonomical rank. Genetic similarity would tend to support an infraspecific level, such as subspecies. However, a species rank seems more appropriate phenotypically, as more differences are found between these entities than among many other clearly separated *Capurodendron* species. Accordingly, we prefer to use the species concept of Freudenstein et al. [62] and consider each of the three morphospecies of the Arid Complex as valid species: *Capurodendron androyense*, *C. mandrarense* and *C. mikeorum* nom. prov., each deserving a conservation assessment.

Potentials and limitations of genetic data for species delimitation and conservation: lessons from our case study—When implementing the Freudenstein et al. [62] species concept, characterization basically relies on morphology, giving a minor contribution to the genetic analyses we performed. In fact, species identification is simpler by visualization with the naked eye than relying on genetic data. Nonetheless, morphological species delimitation has to be validated by molecular analyses such as the ones we implemented here in order to discard the hypothesis that the morphospecies is the result of recent hybridization.

If a strict morphological species concept had been applied in the Western Complex, a new species (corresponding to *Capurodendron* aff. *pervillei* samples) should have been described. However, genetic data clearly demonstrated that these samples corresponded to sporadic hybrids that do not deserve species recognition. Additionally, since we found a clear parallel between morphology and genetics across the three species of this complex, it seems that the observed hybridization represents sporadic events with little consequence for parental species integrity. Similarly, in the Arid Complex, the genetic analyses conducted on a large number of highly variable loci allowed us to separate the three morphospecies that represent true species (including an undescribed one), from those that should be considered hybrids (*Capurodendron androyense x microphyllum* and *C. androyense x mandrarense*). In the absence of such genetic data, *Capurodendron androyense x microphyllum* specimens indeed could have been erroneously interpreted as representing a plain species. Furthermore, *Capurodendron androyense x mandrarense* specimens might have been considered evolutionary intermediates between *C. androyense* and *C. mandrarense*, an interpretation that would put the latter two species into question. The former hybrids do not merit a species rank even when using the Freudenstein concept, as it appears they need their parental species in order to persist. Conversely, without our molecular analyses excluding a hybrid origin, *Capurodendron mikeorum*, nom. prov. (referred to initially as *C. greveanum-mandrarense*) would have been wrongly interpreted as a nothospecies, instead of a plain species.

Here, four analyses were critical in detecting hybridization: PCA, calculation of heterozygosity levels, reconstruction of phased phylogenies, and reconstruction of phylogenetic networks. From these, the last one provided the best ratio working time/output fidelity, whereas PCA is only conclusive if combined with others. Phasing was the most powerful tool to reveal recent hybrids and their parental species.

The IUCN Red List of Threatened Species ([68,69] IUCN, 2004) is a practical tool allowing the rapid accumulation of results for a broad panel of organisms while providing incentives for additional conservation measures, for example at regional and national levels [70]. However, the underlying assumption is that the organisms assessed are well delimited species with little or no genetic exchange between them. Although this may be the case with the majority of taxa, there nevertheless is a portion for which this is not the case, including the complexes studied here. To accommodate these situations, the IUCN allows conservation assessments on subspecific ranks and even geographically remote subpopulations (https://www.iucnredlist.org/resources/tax-sources, accessed on 12 August 2021). As a consequence, considering each of the morphospecies of the Arid Complex either as species or alternatively as subspecific taxa should have no impact on their conservation.

If we aim at preserving the genetic diversity within the species as accepted here, we should address the subpopulations separately. The Arid Complex contains lineages genetically 'purer' than others in both *Capurodendron androyense* and *C. mandrarense*. Genetically pure populations may deserve more attention compared to admixed ones. However, the impact of admixture on species fitness and genetic diversity is globally not well understood, and certainly not for *Capurodendron*. As long as we are unsure whether the observed admixture is ancestral or caused by secondary contact (resolving this would imply other methodologies involving population genetics modeling [71], we cannot decide if this phenomenon is increasing or curtailing genetic diversity. Furthermore, these populations, despite being partially delimited geographically, can only be circumscribed using genomic tools. As a consequence, genetic assessment of subpopulations is impracticable and their management almost impossible in the field.

As assessments of hybrids are not recommended for the IUCN Red list (except plant hybrids treated as species), the morphospecies *Capurodendron androyense* x *mandrarense*, *C. androyense* x *microphyllum*, *C. pervillei* x *oblongifolium*, and *C. pervillei* x *perrieri* do not need a conservation status. From a conservation point of view, the role of hybrids in population dynamics is unclear [72–74]. On the one hand, they could favor species extinction due to the introgression of one taxon in another, leading, for example, to a reduction of the populations of *Capurodendron microphyllum*, which would be partially replaced by *C. androyense x microphyllum*. On the other hand, hybridization may produce recurrent introgression of genes able to increase and diversify population richness, allowing greater resilience of a species to climatic change or other biological factors.

Conservation assessments—*Western complex*—species conservation assessments for *Capurodendron perrieri* (Near Threatened, NT; [75]) and *C. pervillei* (NT; [76]) have been already published and are not affected by the results obtained here. *Capurodendron oblongifolium* is a recently described species [1] for which we propose the following provisional conservation assessment:

Capurodendron oblongifolium: The Extent of Occurrence (EOO) is estimated to be 2024 km^2 and the Area of Occupancy (AOO) 24 km^2; the species is documented from five locations with respect to the most plausible threat which is habitat destruction due to uncontrolled forest fires, one location being outside the protected area network. With low values in AOO, EOO and one location outside the protected area network in a region regularly impacted by forest fires, continuing decline is projected and the species is preliminarily assessed as Endangered (EN B1ab(i,ii,iii,iv)+2ab(i,ii,iii,iv)).

Arid Complex—The conservation assessment of *Capurodendron androyense* was previously assessed as Least Concern (LC; [77]), however its evaluation included a subpopulation recognized here as *C. mandrarense* as well as the specimens identified here as *C. androyense x microphyllum*, which should be excluded. Their exclusion does not alter the Least Concern status of this species, but EOO and AOO values have been recalculated and are now 44,052 and 244 km^2, respectively. In the case of *Capurodendron mandrarense*, the difficulty differentiating it from the *C. greveanum-mandrarense* specimens (= *C. mikeorum* nom. prov.) complicated its evaluation, leaving this taxon as data deficient. Finally, *Capurodendron mikeorum* nom. prov. should be considered a valid species. Preliminary conservation assessments for these two taxa are proposed below.

Capurodendron mandrarense: This species occurs in southern and southwestern Madagascar. The estimated extent of occurrence (EOO) calculated with all available collections is 120,210 km^2, and the minimum area of occupancy (AOO) is 244 km^2 (qualifying for Endangered under criterion B2). It is known from 78 collections, from 22 locations, 16 outside the protected areas network. Threats include agriculture expansion, selective logging, charcoal production and uncontrolled forest fires. Despite a projected continuing decline, at least outside protected areas, this species cannot be considered severely fragmented and is here assessed as Least Concern.

Capurodendron mikeorum nom. prov: This species is restricted to sandy soils from the south-west of Madagascar. The estimated extent of occurrence (EOO) calculated with all available herbarium specimen data is 1676 km^2, and the minimum area of occupancy (AOO) is 72 km^2 (both qualifying for EN under criterion B). The species is known from 24 herbarium collections from merely five locations, and faces threats from large-scale agriculture, uncontrolled forest fires, and selective logging. Despite the five locations being in or near protected areas, the pressures facing dry forests in the southwestern part of Madagascar, even in protected areas, point toward a continued decline, justifying assigning this species to the category Endangered (B1ab(i,ii,iii,iv,v)+2ab(i,ii,iii,iv,v)).

Supplementary Materials: The following are available online at https://www.mdpi.com/article/10.3390/plants10081702/s1, Figure S1: Pseudocoalescent phylogenetic tree from Astral inferred from RAxML analyses of A. 600 gene exonic regions, B. 608 gene supercontigs including introns, exons and flanking regions and C. 195 microsatellite flanking regions. All specimens contained less than 20% missing nucleotide positions. Figure S2: STRUCTURE output from k = 2 to k = 6 for the Arid and Western Species Complexes ordered by morphospecies, with specimen numbers given according to Table 2 and indicated at the end of each dataset. The probability of each k according to the ΔK method ([48] Evanno et al., 2005) is given, Table S1: Specimen information used in the morphological PCA. Character number and states are given in Table 2. Table S2: Results for the ABBA BABA tests. Numbers in taxa names correspond to the groups shown in Figure 5. *p*-value threshold for Z-scores after Benjamini and Bonferroni corrections are indicated in bold and bold underlined, respectively.

Author Contributions: Conceptualization, C.G.B., C.C., L.G. and Y.N.; methodology, C.G.B., C.C., L.G. and Y.N.; software, C.G.B., C.C. and A.R.; validation, C.G.B., C.C., L.G. and Y.N.; formal analysis, C.G.B., C.C., L.G. and Y.N.; investigation, C.G.B., C.C., L.G. and Y.N.; resources, L.G. and Y.N.; data curation, C.G.B., C.C. and L.G.; writing—original draft preparation, C.G.B.; writing—review and editing, C.G.B., C.C., L.G. and Y.N; visualization, C.G.B.; supervision, L.G. and Y.N.; project administration, L.G. and Y.N.; funding acquisition, L.G. and Y.N. All authors have read and agreed to the published version of the manuscript.

Funding: This research was funded by the Swiss National Science Foundation, grant number 31003A-166349, 2016–2019, the Schmidheiny Foundation, and the Franklinia foundation, grant number 2019-20.

Institutional Review Board Statement: Not applicable.

Informed Consent Statement: Not applicable.

Data Availability Statement: All data are indicated in the manuscript or available in the Supplementary Materials.

Acknowledgments: This work was supported by a grant (N° 31003A-166349, 2016–2019) from the Swiss National Science Foundation attributed to YN and LG and by two grants from the Schmidheiny Foundation attributed to YN in 2016 and 2018. Since 2019, this work is supported by a grant from the Franklinia foundation to LG. We thank people from the iGE3 platform for their help in the sequencing process, and Richard Randrianaivo, Lucie Garnier as well as the Malagasy Government and the

Malagasy local people for their help in sampling Sapotaceae specimens. We furthermore thank Daniel Hoffman for English Language editing and review, the University of Geneva High Performance Computing cluster Baobab for computer resources. We are grateful to the herbaria MO, P, TAN and TEF for allowing us to study herbarium specimens and to perform destructive sampling.

Conflicts of Interest: The authors declare no conflict of interest.

References

1. Boluda, C.G.; Christe, C.; Naciri, Y.; Gautier, L. A 638-gene phylogeny supports the recognition of twice as many species in the Malagasy endemic genus *Capurodendron* (*Sapotaceae*). *Taxon* **2021**. Accepted for publication.
2. Buerki, S.; Devey, D.S.; Callmander, M.W.; Phillipson, P.B.; Forest, F.; Garden, M.B.; Box, P.O.; Louis, S. Spatio-temporal history of the endemic genera of Madagascar. *Bot. J. Linn. Soc.* **2013**, *171*, 304–329. [CrossRef]
3. Aubréville, A. *Sapotaceae*. In *Flore de Madagascar et des Comores: Plantes Vasculaires*; Humbert, H., Leroy, J.-F., Eds.; Imprimerie Officielle: Tananarive, Madagascar, 1974; Volume 164.
4. Rakotovao, G.; Rabevohitra, A.R.; Collas de Chatelperron, P.; Guibal, D.; Gérard, J. *Atlas des Bois de Madagascar*; Edition Quae: Versailles, France, 2012; 418p, ISBN 978-2-7592-1871-4.
5. Gautier, L.; Boluda, C.G.; Randriarisoa, A.; Randrianaivo, R.; Naciri, Y. The new natural history of Madagascar. In *Sapotaceae*; Goodman, S.M., Ed.; Princeton University Press: Princeton, NJ, USA,, 2021.
6. Schuurman, D.; Lowry, P.P. The Madagascar rosewood massacre. *Madag. Conserv. Dev.* **2009**, *4*, 98–102. [CrossRef]
7. Patel, E.R. Logging of Rare Rosewood and Palisandre (*Dalbergia* spp.) within Marojejy National Park, Madagascar. *Madag. Conserv. Dev.* **2007**, *2*, 11–16. [CrossRef]
8. Hassold, S.; Lowry, P.P.; Bauert, M.R.; Razafintsalama, A.; Ramamonjisoa, L.; Widmer, A. DNA Barcoding of Malagasy Rosewoods: Towards a Molecular Identification of CITES-Listed Dalbergia Species. *PLoS ONE* **2016**, *11*. [CrossRef] [PubMed]
9. Gautier, L.; Naciri, Y. Three Critically Endangered new species of Capurodendron (*Sapotaceae*) from Madagascar. *Candollea* **2018**, *73*, 121–129. [CrossRef]
10. Faranirina, L.; Rabarimanarivo, M.; Rivers, M. The IUCN Red List of Threatened Species: *Capurodendron ludiifolium*. Available online: https://dx.doi.org/10.2305/IUCN.UK.2019-2.RLTS.T128658609A128660112.en (accessed on 12 August 2021).
11. Faranirina, L.; Rabarimanarivo, M. *Capurodendron greveanum* The IUCN Red List of Threatened Species: *Capurodendron greveanum*. 2019. Available online: https://dx.doi.org/10.2305/IUCN.UK.2019-2.RLTS.T128658599A128660107.en (accessed on 12 August 2021).
12. Faranirina, L.; Rabarimanarivo, M. *Capurodendron nodosum* The IUCN Red List of Threatened Species: *Capurodendron nodosum*. 2019. Available online: https://dx.doi.org/10.2305/IUCN.UK.2019-3.RLTS.T79062312A79062323.en (accessed on 12 August 2021).
13. Lecomte, M.H. Sapotacées recueilles à Madagascar par M. Perrier de la Bathie. *Bull. Muséum Natl. D'histoire Nat.* **1919**, *25*.
14. Christe, C.; Boluda, C.G.; Koubínová, D.; Gautier, L.; Naciri, Y. New genetic markers for *Sapotaceae* phylogenomics: More than 600 nuclear genes applicable from family to population levels. *Mol. Phylogenet. Evol.* **2021**, *160*, 107123. [CrossRef] [PubMed]
15. Commission, I.S.S. *IUCN Red List Categories and Criteria*, 2nd ed.; version 3.1; IUCN: Gland, Switzerland; Cambridge, UK, 2012; ISBN 9782831714356.
16. Lê, S.; Josse, J.; Husson, F. FactoMineR: An R Package for Multivariate Analysis. *J. Stat. Softw.* **2008**, *1*, 1–18. [CrossRef]
17. Russell, A.; Samuel, R.; Rupp, B.; Barfuss, M.H.J. Phylogenetics and cytology of a pantropical orchid genus Polystachya (*Polystachyinae*, *Vandeae*, *Orchidaceae*): Evidence from plastid DNA sequence data. *Taxon* **2010**, *59*, 389–404. [CrossRef]
18. Souza, H.A.V.; Muller, L.A.C.; Brandão, R.L.; Lovato, M.B. Isolation of high quality and polysaccharide-free DNA from leaves of Dimorphandra mollis (*Leguminosae*), a tree from the Brazilian Cerrado. *Genet. Mol. Res.* **2012**, *11*, 756–764. [CrossRef] [PubMed]
19. Bolger, A.M.; Lohse, M.; Usadel, B. Trimmomatic: A flexible trimmer for Illumina sequence data. *Bioinformatics* **2014**, *30*, 2114–2120. [CrossRef]
20. Weitemier, K.; Straub, S.C.K.; Cronn, R.C.; Fishbein, M.; Schmickl, R.; McDonnell, A.; Liston, A. Hyb-Seq: Combining target enrichment and genome skimming for plant phylogenomics. *Appl. Plant. Sci.* **2014**, *2*. [CrossRef]
21. Johnson, M.G.; Gardner, E.M.; Liu, Y.; Medina, R.; Goffinet, B.; Shaw, A.J.; Zerega, N.J.C.; Wickett, N.J. HybPiper: Extracting coding sequence and introns for phylogenetics from high-throughput sequencing reads using target enrichment. *Appl. Plant. Sci.* **2016**, *4*. [CrossRef]
22. Katoh, K.; Standley, D.M. MAFFT Multiple Sequence Alignment Software Version 7: Improvements in Performance and Usability. *Mol. Biol. Evol.* **2013**, *30*, 772–780. [CrossRef]
23. Li, H.; Durbin, R. Fast and accurate long-read alignment with Burrows–Wheeler transform. *Bioinformatics* **2010**, *26*, 589–595. [CrossRef]
24. Broad Institute Picard Toolkit. Available online: http://broadinstitute.github.io/picard/ (accessed on 12 August 2021).
25. Li, H.; Handsaker, B.; Wysoker, A.; Fennell, T.; Ruan, J.; Homer, N.; Marth, G.; Abecasis, G.; Durbin, R. The sequence alignment/map format and SAMtools. *Bioinformatics* **2009**, *25*, 2078–2079. [CrossRef] [PubMed]
26. Danecek, P.; Auton, A.; Abecasis, G.; Albers, C.A.; Banks, E.; DePristo, M.A.; Handsaker, R.; Lunter, G.; Marth, G.; Sherry, S.T.; et al. The Variant Call Format and VCFtools. *Bioinformatics* **2011**, *27*, 2156–2158. [CrossRef] [PubMed]

27. Highnam, G.; Franck, C.; Martin, A.; Stephens, C.; Puthige, A.; Mittelman, D. Accurate human microsatellite genotypes from high-throughput resequencing data using informed error profiles. *Nucleic Acids Res.* **2013**, *41*, 1–7. [CrossRef] [PubMed]
28. Langmead, B.; Salzberg, S.L. Fast gapped-read alignment with Bowtie 2. *Nat. Methods* **2012**, *9*, 357–359. [CrossRef]
29. Weiß, C.L.; Pais, M.; Cano, L.M.; Kamoun, S.; Burbano, H.A. nQuire: A statistical framework for ploidy estimation using next generation sequencing. *BMC Bioinform.* **2018**, *19*. [CrossRef]
30. Viruel, J.; Conejero, M.; Hidalgo, O.; Pokorny, L.; Powell, R.F.; Forest, F.; Kantar, M.B.; Soto Gomez, M.; Graham, S.W.; Gravendeel, B.; et al. A Target Capture-Based Method to Estimate Ploidy from Herbarium Specimens. *Front. Plant. Sci.* **2019**, *10*. [CrossRef]
31. Kates, H.R.; Johnson, M.G.; Gardner, E.M.; Zerega, N.J.C.; Wickett, N.J. Allele phasing has minimal impact on phylogenetic reconstruction from targeted nuclear gene sequences in a case study of Artocarpus. *Am. J. Bot.* **2018**, *105*, 404–416. [CrossRef] [PubMed]
32. Patterson, M.; Marschall, T.; Pisanti, N.; van Iersel, L.; Stougie, L.; Klau, G.W.; Schönhuth, A. WhatsHap: Weighted Haplotype Assembly for Future-Generation Sequencing Reads. *J. Comput. Biol.* **2015**, *22*, 498–509. [CrossRef] [PubMed]
33. Li, H. A statistical framework for SNP calling, mutation discovery, association mapping and population genetical parameter estimation from sequencing data. *Bioinformatics* **2011**, *27*, 2987–2993. [CrossRef]
34. Stamatakis, A. RAxML version 8: A tool for phylogenetic analysis and post-analysis of large phylogenies. *Bioinformatics* **2014**, *30*, 1312–1313. [CrossRef] [PubMed]
35. Mirarab, S.; Reaz, R.; Bayzid, M.S.; Zimmermann, T.; Swenson, M.S.; Warnow, T. ASTRAL: Genome-scale coalescent-based species tree estimation. *Bioinformatics* **2014**, *30*, i541–i548. [CrossRef]
36. Mirarab, S.; Warnow, T. ASTRAL-II: Coalescent-based species tree estimation with many hundreds of taxa and thousands of genes. *Bioinformatics* **2015**, *31*, i44–i52. [CrossRef] [PubMed]
37. Huson, D.H. SplitsTree: Analyzing and visualizing evolutionary data. *Bioinformatics* **1998**, *14*, 68–73. [CrossRef] [PubMed]
38. Pritchard, J.K.; Stephens, M.; Donnelly, P. Inference of population structure using multilocus genotype data. *Genetics* **2000**, *155*, 945–959. [CrossRef]
39. Falush, D.; Stephens, M.; Pritchard, J. Inference of population structure using multilocus genotype data: Linked loci and correlated allele frequencies. *Genetics* **2003**, *155*, 945–959.
40. Jakobsson, M.; Rosenberg, N.A. CLUMMP: A cluster matching and permutation program for dealing with label switching and multimodality in analysis of population structure. *Bioinformatics* **2007**, *23*, 1801–1806. [CrossRef] [PubMed]
41. Earl, D.A.; von Holdt, B.M. STRUCTURE HARVESTER: A website and program for visualizing STRUCTURE output and implementing the Evanno method. *Conserv. Genet. Resour.* **2012**, *4*, 359–361. [CrossRef]
42. Patterson, N.; Price, A.L.; Reich, D. Population structure and eigenanalysis. *PLoS Genet.* **2006**, *2*, e190. [CrossRef]
43. Malinsky, M.; Matschiner, M.; Svardal, H. Dsuite—Fast D-statistics and related admixture evidence from VCF files. *Mol. Ecol. Resour.* **2021**, *21*, 584–595. [CrossRef]
44. Malinsky, M.; Svardal, H.; Tyers, A.M.; Miska, E.A.; Genner, M.J.; Turner, G.F.; Durbin, R. Whole-genome sequences of Malawi cichlids reveal multiple radiations interconnected by gene flow. *Nat. Ecol. Evol.* **2018**, *2*, 1940–1955. [CrossRef]
45. Phillips, S.J.; Anderson, R.P.; Schapire, R.E. Maximum entropy modeling of species geographic distributions. *Ecol. Modell.* **2006**, *190*, 231–259. [CrossRef]
46. Hijmans, R.J.; Cameron, S.E.; Parra, J.L.; Jones, P.G.; Jarvis, A. Very high resolution interpolated climate surfaces for global land areas. *Int. J. Climatol.* **2005**, *25*, 1965–1978. [CrossRef]
47. R Core Team. *R: A Language and Environment for Statistical Computing*; R Foundation for Statistical Computing: Vienna, Austria, 2014.
48. Hijmans, R.; Cruz, M.; Rojas, E.; Guarino, L. *DIVA-GIS Version 1.4: A Geographic Information System for the Analysis of Biodiversity Data*; Manual; International Potato Center: Lima, Peru, 2001.
49. Evanno, G.; Regnaut, S.; Goudet, J. Detecting the number of clusters of individuals using the software STRUCTURE: A simulation study. *Mol. Ecol.* **2005**, *14*, 2611–2620. [CrossRef] [PubMed]
50. De Queiroz, K. Species Concepts and Species Delimitation. *Syst. Biol.* **2007**, *56*, 879–886. [CrossRef]
51. Wilkins, J.S. Philosophically speaking, how many species concepts are there? *Zootaxa* **2011**, *2765*, 58–60. [CrossRef]
52. Boluda, C.G.; Rico, V.J.; Divakar, P.K.; Nadyeina, O.; Myllys, L.; McMullin, R.T.; Zamora, J.C.; Scheidegger, C.; Hawksworth, D.L. Evaluating methodologies for species delimitation: The mismatch between phenotypes and genotypes in lichenized fungi (*Bryoria sect. Implexae, Parmeliaceae*). *Persoonia* **2019**, *42*, 75–100. [CrossRef]
53. Naciri, Y.; Christe, C.; Bétrisey, S.; Song, Y.-G.; Deng, M.; Garfí, G.; Kozlowski, G. Species delimitation in the East Asian species of the relict tree genus Zelkova (*Ulmaceae*): A complex history of diversification and admixture among species. *Mol. Phylogenetics Evol.* **2019**, *134*, 172–185. [CrossRef] [PubMed]
54. Galtier, N. Delineating species in the speciation continuum: A proposal. *Evol. Appl.* **2019**, *12*, 657–663. [CrossRef]
55. Muñoz-Rodríguez, P.; Carruthers, T.; Wood, J.R.I.; Williams, B.R.M.; Weitemier, K.; Kronmiller, B.; Ellis, D.; Anglin, N.L.; Longway, L.; Harris, S.A.; et al. Reconciling Conflicting Phylogenies in the Origin of Sweet Potato and Dispersal to Polynesia. *Curr. Biol.* **2018**, *28*, 1246.e12–1256.e12. [CrossRef]
56. Wu, G.A.; Terol, J.; Ibanez, V.; López-García, A.; Pérez-Román, E.; Borredá, C.; Domingo, C.; Tadeo, F.R.; Carbonell-Caballero, J.; Alonso, R.; et al. Genomics of the origin and evolution of Citrus. *Nature* **2018**, *554*, 311–316. [CrossRef] [PubMed]

57. Kumar, V.; Lammers, F.; Bidon, T.; Pfenninger, M.; Kolter, L.; Nilsson, M.A.; Janke, A. The evolutionary history of bears is characterized by gene flow across species. *Sci. Rep.* **2017**, *7*, 46487. [CrossRef]
58. Chang, J.-T.; Huang, B.-H.; Liao, P.-C. Genetic evidence of the southward founder speciation of *Cycas taitungensis* from ancestral C. revoluta along the Ryukyu Archipelagos. *Conserv. Genet.* **2019**, *20*, 1045–1056. [CrossRef]
59. Grünig, S.; Fischer, M.; Parisod, C. Recent hybrid speciation at the origin of the narrow endemic *Pulmonaria helvetica*. *Ann. Bot.* **2021**, *127*, 21–31. [CrossRef] [PubMed]
60. Hörandl, E.; Stuessy, T.-F. Paraphyletic groups as natural units of biological classification. *Taxon* **2010**, *59*, 1641–1653. [CrossRef]
61. Kato, M.; Werukamkul, P.; Won, H.; Koi, S. Paraphyletic Species of *Podostemaceae*: *Cladopus fallax* and *Polypleurum wallichii*. *Phytotaxa* **2019**, *401*, 33–48. [CrossRef]
62. Freudenstein, J.V.; Broe, M.B.; Folk, R.A.; Sinn, B.T. Biodiversity and the Species Concept—Lineages are not Enough. *Syst. Biol.* **2017**, *66*, 644–656. [CrossRef] [PubMed]
63. Milne, R.I.; Abbott, R.J. Reproductive isolation among two interfertile Rhododendron species: Low frequency of post-F1 hybrid genotypes in alpine hybrid zones. *Mol. Ecol.* **2008**, *17*, 1108–1121. [CrossRef]
64. Buerkle, C.A.; Lexer, C. Admixture as the basis for genetic mapping. *Trends Ecol. Evol.* **2008**, *23*, 686–694. [CrossRef] [PubMed]
65. Rutherford, S.; Rossetto, M.; Bragg, J.G.; McPherson, H.; Benson, D.; Bonser, S.P.; Wilson, P.G. Speciation in the presence of gene flow: Population genomics of closely related and diverging Eucalyptus species. *Heredity* **2018**, *121*, 126–141. [CrossRef]
66. Solís-Lemus, C.; Bastide, P.; Ané, C. PhyloNetworks: A Package for Phylogenetic Networks. *Mol. Biol. Evol.* **2017**, *34*, 3292–3298. [CrossRef] [PubMed]
67. Naciri, Y.; Linder, P. Species identification and delimitation: The dance of the seven veils. *Taxon* **2015**, *64*, 3–16. [CrossRef]
68. Hilton-Taylor, C.; Brackett, D. *2000 IUCN Red List of Threatened Species*; IUCN: Gland, Switzerland; Cambridge, UK, 2000; ISBN 9782831705644.
69. IUCN. *2004 IUCN Red List of Threatened Species*; IUCN: Gland, Switzerland; Cambridge, UK, 2004.
70. Beech, E.; Rivers, M.; Rabarimanarivo, M.; Ravololomanana, N.; Manjato, N.; Lantoarisoa, F.; Andriambololonera, S.; Ramandimbisoa, B.; Ralimanana, H.; Rakotoarisoa, S.; et al. *Jeannoda, Red List of Trees of Madagascar*; BGCI: Richmond, UK, 2021.
71. Rougeux, C.; Bernatchez, L.; Gagnaire, P.-A. Modeling the multiple facets of speciation-with-gene-flow toward Inferring the divergence history of lake whitefish species pairs (*Coregonus clupeaformis*). *Genome Biol. Evol.* **2017**, *9*, 2057–2074. [CrossRef]
72. Grant, P.R.; Grant, B.R. Hybridization increases population variation during adaptive radiation. *Proc. Natl. Acad. Sci. USA* **2019**, *116*, 23216–23224. [CrossRef]
73. Chan, W.Y.; Hoffmann, A.A.; van Oppen, M.J.H. Hybridization as a conservation management tool. *Conserv. Lett.* **2019**, *12*, e12652. [CrossRef]
74. Hirashiki, C.; Kareiva, P.; Marvier, M. Concern over hybridization risks should not preclude conservation interventions. *Conserv. Sci. Pract.* **2021**, *3*, e424. [CrossRef]
75. Rakotoarisoa, A.A.; Faranirina, L.; Rabarimanarivo, M.; Gautier, L. *Capurodendron perrieri* The IUCN Red List of Threatened Species: *Capurodendron perrieri*. 2020. Available online: https://dx.doi.org/10.2305/IUCN.UK.2020-1.RLTS.T128658666A128660127.en (accessed on 12 August 2021).
76. Rakotoarisoa, A.A.; Gautier, L. *Capurodendron pervillei* The IUCN Red List of Threatened Species: *Capurodendron pervillei*. 2020. Available online: https://dx.doi.org/10.2305/IUCN.UK.2020-1.RLTS.T58388774A58393285.en (accessed on 12 August 2021).
77. Faranirina, L.; Rabarimanarivo, M. *Capurodendron androyense* The IUCN Red List of Threatened Species: *Capurodendron androyense*. 2019. Available online: https://dx.doi.org/10.2305/IUCN.UK.2019-2.RLTS.T128658510A128660087.en (accessed on 12 August 2021).

plants

MDPI

Article

Genetic Diversity and Structure of Rear Edge Populations of *Sorbus aucuparia* (Rosaceae) in the Hyrcanian Forest

Hamed Yousefzadeh [1,*], Shahla Raeisi [2], Omid Esmailzadeh [2], Gholamali Jalali [2], Malek Nasiri [3], Łukasz Walas [4] and Gregor Kozlowski [5,6,7]

1 Department of Environmental Science, Faculty of Natural Resources, Tarbiat Modares University (TMU), Mazandaran 14115-111, Iran
2 Department of Forest Science and Engineering, Faculty of Natural Resources, Tarbiat Modares University (TMU), Mazandaran 14115-111, Iran; reisi.shahla@yahoo.com (S.R.); oesmailzadeh@modares.ac.ir (O.E.); jalali_g@modares.ac.ir (G.J.)
3 Department of Forestry, Faculty of Natural Resources, Tehran University (TU), Tehran 31587-77871, Iran; nasiri.malek@gmail.com
4 Department of Biogeography and Systematics, Institute of Dendrology, Polish Academy of Sciences, Parkowa 5, PL-62-035 Kornik, Poland; lukaswalas@man.poznan.pl
5 Department of Biology and Botanic Garden, University of Fribourg, Chemin du Musée 10, CH-1700 Fribourg, Switzerland; gregor.kozlowski@unifr.ch
6 Natural History Museum Fribourg, Chemin du Musée 6, CH-1700 Fribourg, Switzerland
7 Eastern China Conservation Centre for Wild Endangered Plant Resources, Shanghai Chenshan Botanical Garden, 3888 Chenhua Road, Songjiang, Shanghai 201602, China
* Correspondence: h.yousefzadeh@modares.ac.ir; Tel.: +98-445-531-013; Fax: +98-4455-3499

Citation: Yousefzadeh, H.; Raeisi, S.; Esmailzadeh, O.; Jalali, G.; Nasiri, M.; Walas, Ł.; Kozlowski, G. Genetic Diversity and Structure of Rear Edge Populations of *Sorbus aucuparia* (Rosaceae) in the Hyrcanian Forest. *Plants* **2021**, *10*, 1471. https://doi.org/10.3390/plants10071471

Academic Editor: Calvin O. Qualset

Received: 27 May 2021
Accepted: 23 June 2021
Published: 19 July 2021

Abstract: *Sorbus aucuparia* (Rosaceae) is a small tree species widely distributed in Eurasia. The Hyrcanian forest is the southernmost distribution limit of this species. Severe habitat degradation and inadequate human interventions have endangered the long-term survival of this species in this region, and it is necessary to develop and apply appropriate management methods to prevent the loss of its genetic diversity. In this study, we used 10 SSR markers in order to evaluate the genetic diversity of this taxon. Leaf samples were collected from five known populations of *S. aucuparia* throughout its distribution area in the Hyrcanian forest. Expected heterozygosity ranged from 0.61 (ASH) to 0.73, and according to the M-ratio, all populations showed a significant reduction in effective population size, indicating a genetic bottleneck. Global F_{ST} was not statistically significant and attained the same values with and without excluding null alleles (ENA) correction (F_{ST} = 0.12). Bayesian analysis performed with STRUCTURE defined two genetic clusters among the five known populations, while the results of discriminant analysis of principal components (DAPC) identified three distinct groups. The average proportion of migrants was 22. In general, the gene flow was asymmetrical, with the biggest differences between immigration and emigration in Barzekoh and Asbehriseh. The Mantel test showed that there was no significant correlation between genetic distance (F_{ST}) and geographic distance in *S. aucuparia*. The best pathway for theoretical gene flow is located across the coast of the Caspian Sea and significant spatial autocorrelation was observed in only one population. In order to reduce the extinction risk of very small and scattered populations of *S. aucuparia* in the Hyrcanian forest, it is very important to establish and/or enhance the connectivity through habitat restoration or genetic exchange.

Keywords: conservation genetics; inbreeding depression; range-edge populations; rowan tree; Hyrcanian forest

1. Introduction

The Hyrcanian forest, located along the southern coast of the Caspian Sea in Iran and Azerbaijan, is one of the most important biodiversity centers on our planet [1]. The area possesses a remarkable amount of nearly 150 woody species, among them numerous relict trees [2,3]. The main reason of this impressive tree and shrub diversity lies in the fact that this region was never covered by glaciers during the Pleistocene [4,5].

The rowan tree (*Sorbus aucuparia* L.) is one of the most important species of the genus *Sorbus*, which has a medicinal value, and has a wide natural range in areas with low and high altitudes from the Atlantic coasts of Europe to the Kamchatka Peninsula and East China in Asia [6–8]. The Hyrcanian forest is the southernmost distribution limit of *S. aucuparia*, with only a few and small populations of this species remaining in this region. The rowan tree is distributed in the Hyrcanian forest at higher altitudes in mountainous regions, reaching the upper forest limit (1800–2800 m a.s.l.) and often growing on rocky slopes [9]. Iranian occurrences of this species are typical rear-edge populations, isolated from each other, and occurring in a scattered distribution. Hence, this species in the Hyrcanian region is vulnerable and highly sensitive to climate change.

Future climate change may alter the genetic diversity within species [10] through the reduction of species distribution and increase of habitat fragmentation [11]. It has been reported by many researchers that rising temperatures and drought stress over the last half-century, increased mortality and decreased the growth of plants. This effect is especially strong in the case of edge populations [12].

Marginal populations are potentially important for conservation, since they may preserve rare alleles and gene combinations important for adaptation to extreme environmental conditions [13,14]. However, the assessment of genetic diversity and evolution of peripheral populations is still insufficient [15]. Hoffmann et al. [16] showed that decreasing adaptation potential to severe conditions is often encountered at range edges [17]. This is connected with increased genetic drift, which leads to a reduction in gene diversity. On the other hand, Sáenz-Romero et al. [18] mentioned that in some cases, the migratory fluxes from core populations may improve genetic diversity in peripheral populations [18].

Severe habitat degradation and inadequate human interventions have endangered the survival of many plant species in the Hyrcanian forest [19] and this is even more worrying for marginal species with very low density and abundance. Additionally, the severe habitat conditions (rocky sites with shallow soil and harsh habitat conditions) of *S. aucuparia* in the Hyrcanian forest are responsible for weak regeneration, as there is a strong relationship between habitat quality and genetic diversity [20]. Thus, the long-term survival of this species in the Hyrcanian forest is uncertain. Moreover, due to the rapid degradation of the Hyrcanian forest, it is necessary to apply appropriate management methods to prevent the decline of plant populations, and consequently, the genetic richness of many species, especially those present in the upper forest border due to their higher vulnerability [14].

Knowledge about the levels and patterns of genetic diversity within and between populations is crucial to adopt a good conservation strategy for potentially threatened species [21]. Several molecular techniques have been used as efficient methods for considering the genetic diversity of the genus *Sorbus* [22–27]. Simple sequence repeat marker (SSR) is a cross-selective marker and a powerful tool in evaluating diversity levels, phylogenetic relationships, and genetic structure of the genus *Sorbus* [23,28]. This type of marker has been frequently used in the last years due to its co-dominant character and abundance in the plant's genome [29], and due to the high transferability to the closely related species [29,30].

This study was designated to investigate the genetic diversity and population genetic structure of *S. aucuparia* in its southernmost distribution area, using a set of 10 SSR markers and using plant material covering the whole known natural distribution of this species in Iran. More specifically, we aimed to answer the following specific questions: (1) What is the genetic diversity of *S. aucuparia* within and between its natural populations in Iran? (2) What is the spatial genetic structure of natural populations of *S. aucuparia* in the study area? (3) What is the migration rate and gene flow between the populations of this species? Finally, based on our results, we are discussing the conservation implications and measures needed for the long-term conservation of this species in the Hyrcanian forest.

2. Materials and Methods

2.1. Sampling, DNA Extraction, and SSR Amplification

Leaf samples were collected from five known populations of *S. aucuparia* throughout its distribution area in the Hyrcanian forest (Table 1).

Table 1. Geographical characteristics of the studied populations.

Population Name	Sample Code	Longitude	Latitude	Altitude	Sample Size	Associate Species
Khalkhal	KH	373929.7	483540.9	2100–2500	28	*Quercus macranthera-Sorbus graeca*
Olsehposht	NAV	373936.3	484015.2	1650–2100	12	*Fagus orientalis, Acer* spp.
Asbehriseh	ASH	373741.4	484416.2	1500–1900	12	*Carpinus betulus, Acer* spp.
Barzekoh	LOM	373231.5	484634.4	1450–1850	11	*Fagus orientalis, Carpinus orientalis, Acer mazandaranicum*
Sangedeh	BAN	360601.6	531232.5	1900–2400	15	*Betula pendula-Acer hyrcanum*

Hoebee et al. [23] concluded that trees are very unlikely to be clones if the minimum distance between trees is 30 m. Hence, depending on the population size, 10–28 mature trees were chosen from each population with at least a 50–100 m distance between trees to avoid recurring genotypes [23]. In total, 78 trees were sampled.

DNA was extracted from the leaf tissue using the CTAB methods [31,32] with some modifications [33]. The quantity and quality of the extracted DNA were determined by loading the samples on agarose 1% gel and using spectrophotometry, respectively. In total, a set of 10 polymorphic SSR markers from 15 initially screened SSR markers were selected to detect the genetic variation among populations (Table S1). Markers were amplified using a DNA Engine Thermal Cycler (Bio-rad, Hercules, CA, USA). The reaction mixtures of 10 μL contained 1× buffer, 0.2 mM dNTPs, 2.5 mM $MgCl_2$, 0.2 μM each SSR forward and reverse primer, 30 ng of genomic DNA, and 1 U of Taq polymerase (Thermo Scientific). The PCR program involved an initial denaturation step of 5 min at 94 °C, followed by 30 cycles at 94 °C for 30 s, the appropriate annealing temperature for 30 s, 72 °C for 40 s, and an extension cycle of 1 min at 72 °C. PCR product was run on 8% polyacrylamide gel and dyed with silver nitrate protocol [34]. The multimode bands were coded in the Gel-Pro analyzer 32 software.

2.2. Genetic Diversity

The null allele frequencies of each locus were assessed using Microchecker 2.2.3 software [35]. The average number of alleles (A), number of private alleles [36], and the effective number of alleles (Ae) were calculated using the GENEALEX 6.501 software [37], INEst v. 2.0 [38] was used to estimate the expected heterozygosity (He), observed heterozygosity (Ho), and the inbreeding coefficient (FIS), as well as for a bottleneck test [39]. FSTAT was used to estimate allelic richness (Ar). Global and pairwise F_{ST} were estimated using FREENA and tested with bootstrapping over loci [36]. The significance of a deviation from the Hardy–Weinberg equilibrium, including a Bonferroni correction and the estimated frequency of null alleles, were estimated using CERVUS software.

2.3. Genetic Structure

Analysis of molecular variance (AMOVA) among and within populations was performed using GenAlex [37,40]. From AMOVA, the fixation index (F_{ST}) and Nm (haploid number of migrants) within the population were obtained. The Bayesian algorithm implemented in STRUCTURE [41] was used to clustering individuals, whereas discriminant analysis of principal components (DAPC; [42] provided an independent, non-Bayesian method. STRUCTURE procedure included 10^5 MCMC iterations, 10^4 burn-in, and 10 independent runs with the maximum number of clusters set to K = 6. Evanno's delta K method from CLUMPAK software was used to choose the best K. Function 'find.cluster,' implemented in the adegenet package in R, was used to estimate the optimal number of clusters for the DAPC. Next, the 'dapc' function was used to perform this analysis. To estimate the contemporary dispersal patterns and determining the degree of connectivity in populations under study, assignment analysis was done by GENEALEX 6.501 software. In order to infer historical gene flow (Nm) patterns, MIGRATE-N v3.6 [43] was used to estimate the effective population sizes (θ) and mutation-scaled immigration (M) among the stands [44,45]. Four independent runs with different initial seeds were performed and the Bezier approximation for the marginal likelihood was used to test which run has the best fit for the data. Each run consisted of 50,000 sampled parameter values and 5000 recorded steps after a burn-in of 1000 steps. A static heating scheme was used (chains set at 1, 1.5, 3, 10^5). The software CIRCUITSAPE [46,47] was used for testing how topography could shape the gene flow between populations. The altitude raster was a resistance surface with analyzed populations as nodes.

2.4. Mantel Test

Patterns of isolation by distance (IBD; [48]) were investigated using function 'Mantel test' with 9999 iterations implemented in R. The matrix of the genetic distance (pairwise F_{ST} with ENA correction) was tested against the matrix of spatial distance between populations created using the program QGIS.

2.5. Spatial Autocorrelation

Spatial autocorrelation analysis [49] was performed in GenAlEx [37]. The spatial autocorrelation coefficient (r) was computed using the multilocus genetic distance and the Euclidean distance between individuals.

3. Results

3.1. Genetic Diversity

Analysis of 10 microsatellite loci in 78 individuals (genets) showed 41 different alleles. The number of different alleles per locus ranged from 3 (MSS5) to 5.83 (MSS9). The values of He and Ho per locus varied from 0.47 (MSS5) to 0.73 (MSS9, SA08) and from 0.23 (MSS9) to 0.82 (MSS1), respectively. The highest and lowest frequency values of null alleles were in MSS9 (0.28) and MSS16 (0.003), respectively. The mean null allele frequency for all examined populations was 0.10 (Table S2).

Genetic diversity estimates obtained for each population at the genet level are summarized in Table 2. The expected heterozygosity ranged from 0.61 (ASH) to 0.73 (KH), while Ho ranged from 0.45 (NAV) to 0.56 (BAN and KH). The highest Ar value was in KH (4.56) and the lowest in ASH (3.48). Private alleles were observed in the eastern and western populations (BAN and KH).

Table 2. Parameters of the genetic diversity of the studied populations. For abbreviations of the populations, see Table 1.

Pop	Lat	Long	n	A	Ae	Ar	Ap	Null	Ho	He	Fis	M-Ratio
BAN	36.06	53.124	15	4.56	3.29	4.12	2.00	0.102	0.56	0.70	0.16	0.006
KH	37.393	48.354	28	5.78	3.64	4.56	5.00	0.125	0.56	0.73	0.21	0.050
NAV	37.394	48.401	12	4.56	3.12	4.08	0.00	0.134	0.45	0.69	0.31	0.037
ASH	37.374	48.442	12	3.67	2.79	3.48	0.00	0.082	0.49	0.61	0.17	0.004
LOM	37.323	48.463	11	3.89	2.89	3.61	0.00	0.123	0.46	0.63	0.29	0.000

n—total number of individuals, A—the average number of alleles, Ae—effective number of alleles, Ar—allelic richness, Ap—number of private alleles, Null—frequency of null alleles, Ho—observed heterozygosity, He—expected heterozygosity, F_{Is}—fixation index, M-ratio—*p*-value of Wilcox sign-rank test after 10,000 permutations.

According to the *M*-ratio, all populations showed a significant reduction in the effective population size, indicating a genetic bottleneck (Table 2). The spatial pattern of He and Ar is demonstrated in Figure 1; the highest values were observed in the eastern population.

F_{IS} ranged from 0.16 to 0.31; according to the DIC in all populations under study, inbreeding was not the likely factor of the deviation in the Hardy–Weinberg equilibrium (Table 2).

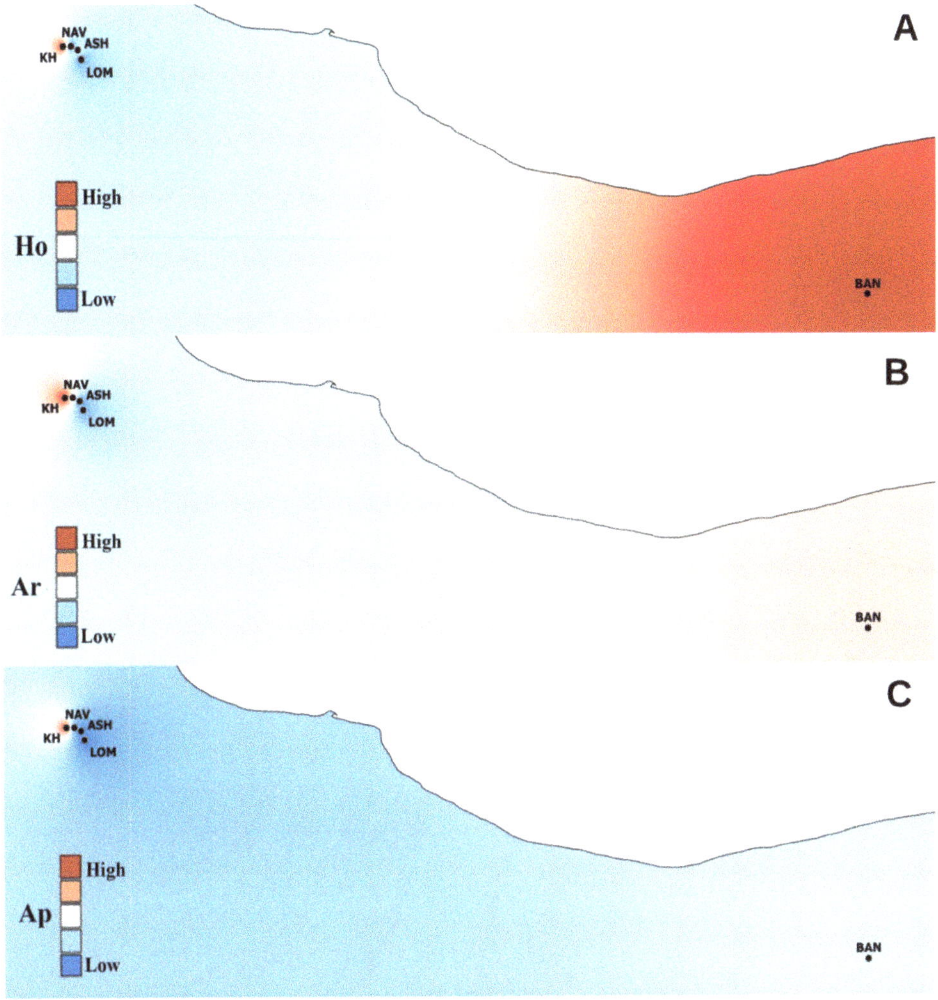

Figure 1. Maps of the genetic diversity of *Sorbus aucuparia* populations visualized by QGIS software. (**A**): expected heterozygosity (Ho), (**B**): allelic richness (Ar), (**C**): number of private alleles (Ap).

Global F_{ST} was statistically insignificant and attained the same values with and without ENA correction (F_{ST} = 0.12). This result suggests that the presence of null alleles does not influence the level of differentiation. The pairwise F_{ST} ranged from 0.005 (between BAN and LOM) to 0.08 (between BAN and ASH), indicating a varied level of differentiation among populations (Table 3).

3.2. Spatial Genetic Structure and Gene Flow

Bayesian analysis of a genetic structure, performed in STRUCTURE, defined two genetic clusters among the five analyzed populations (Figure 2a). Three populations of BAN, KH, and NAV were grouped as a single cluster, whereas two populations of ASH and LOM were assigned to the second cluster. We used DAPC analysis to increase the validation and support the output of Bayesian clustering. The results of DAPC for K = 3—best K for

STRUCTURE—were relatively dissimilar to those obtained with STRUCTURE. The results of DAPC for K = 3 revealed that the BAN population from the eastern part and KH from the western part of the Hyrcanian forest comprised a separated group. However, three populations (NAV, ASH, and LOM) were not assigned to either of the two detected clusters and presented a relatively high admixture (Figure 2b). This result was also confirmed by the population assignment test (Figure S1).

Table 3. Matrix of the genetic distance between populations. For abbreviations of the populations, see Table 1.

	BAN	KH	NAV	ASH	LOM
BAN		0.0337	0.0326	0.0831	0.0057
KH	0.0330		0.0162	0.0655	0.0411
NAV	0.0332	0.0126		0.0779	0.0532
ASH	0.0777	0.0597	0.0732		0.0078
LOM	0.0064	0.0366	0.0527	0.0108	

F_{ST} with ENA correction above the diagonal, F_{ST} without ENA correction below the diagonal.

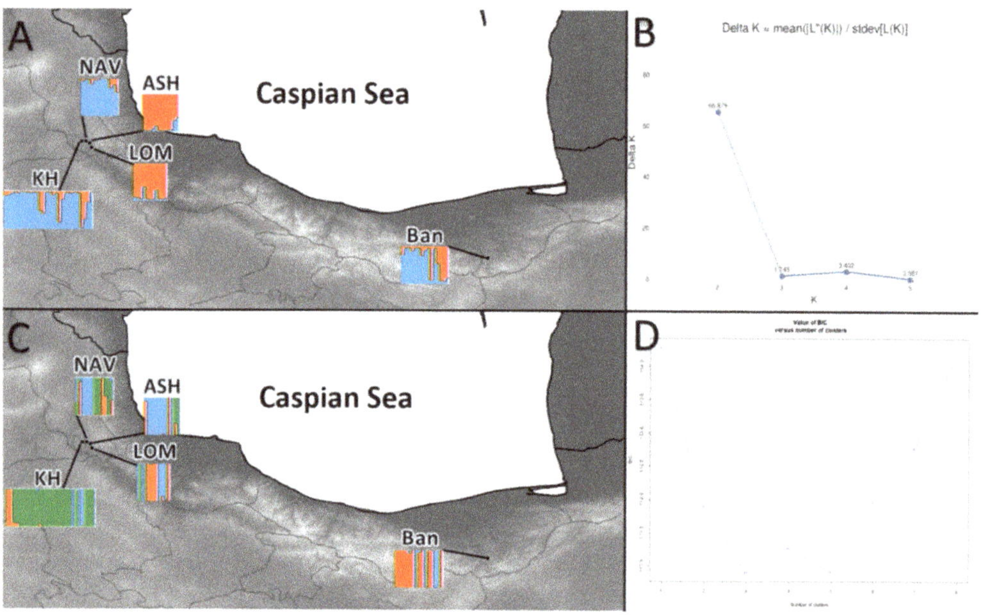

Figure 2. (**A**): Results from STRUCTURE for K = 2 for populations of *Sorbus aucuparia*, (**B**): the optimal number of clusters (K) for STRUCTURE estimated by method from Evanno et al. (2005) [50], (**C**): results from DAPC for K = 3, (**D**): best number of cluster determined by find cluster in R. For abbreviations of the populations, see Table 1.

The results of recent migration rates are shown in Figure 3 and Table S3. The average proportion of migrants was 22.02, suggesting that more individuals than 20 per population may be migrants. However, differences between populations are very strong—in LOM, the number of migrants was 34.9, whereas in BAN and ASH, this value was lower than 20 (15.9 and 14.0, respectively). In general, the gene flow was asymmetrical, with the biggest differences between immigration and emigration in LOM (strong immigration) and ASH (strong emigration). The intensity of gene flow between populations from the western and eastern parts of the Hyrcanian forests was rather low.

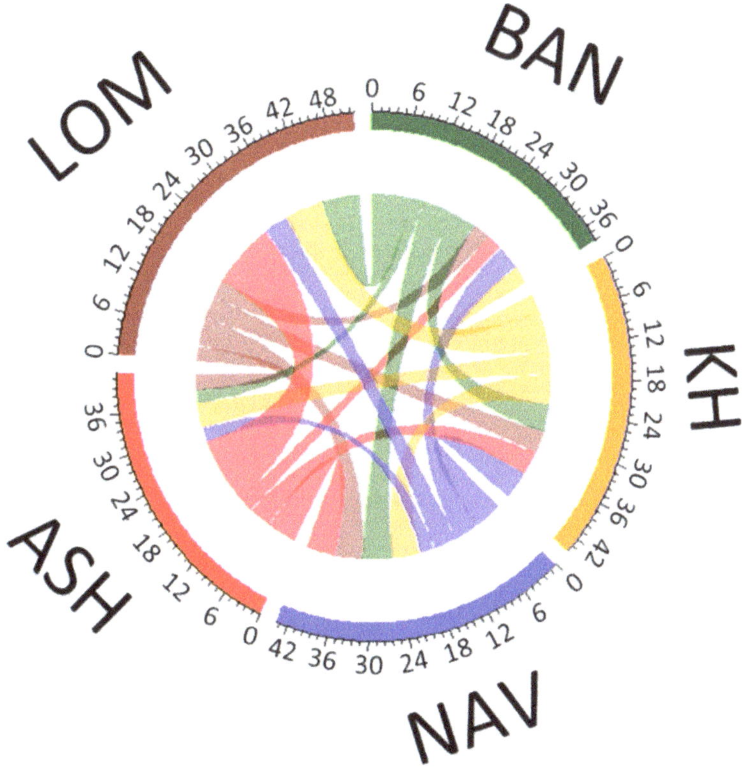

Figure 3. Theoretical gene flow between populations of *Sorbus aucuparia* estimated with MIGRATE-N. For abbreviations of the populations, see Table 1.

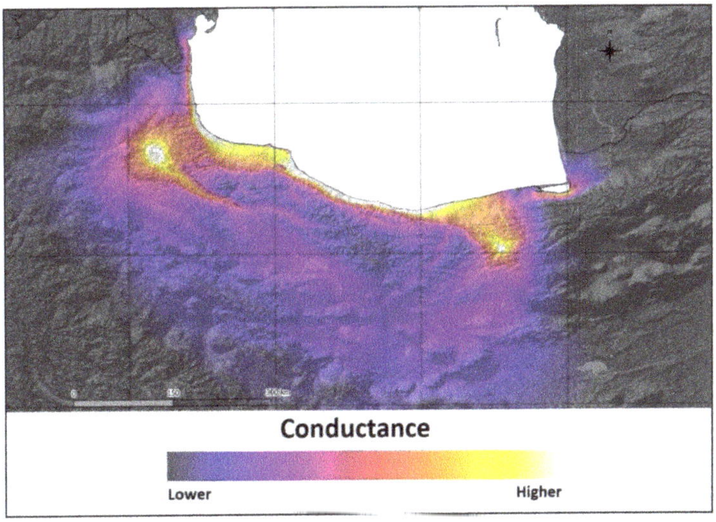

Figure 4. Theoretical gene flow between populations of *Sorbus aucuparia* in relation to the topography determined using CIRCUITSCAPE.

The Mantel test showed that there was no significant correlation between genetic distance (F_{ST}) and geographic distance in *S. aucuparia* ($r = -0.282$, $p = 0.7$). However, a resistance analysis made in CIRCUITSCAPE, with elevation as a matrix of resistance against the F_{ST} matrix, indicated that elevation could be a significant barrier for gene flow across the eastern and western range of the species. CIRCUITSCAPE models the connectivity between stands as a landscape resistance distance (isolation-by-resistance). In our analysis, altitude was used as a resistance raster, and paths without topographical barriers were estimated as best ways for a gene flow. The map generated by CIRCUITSCAPE showed the path of high conductance that represents possible pathways of gene flow among populations; theoretical conductance is presented in Figure 4.

The best pathway for theoretical gene flow is located across the coast of the Caspian Sea. Theoretical southern path across the mountains is less probable, because of topographic complexity. Significant spatial autocorrelation was observed only in population KH (Figure 5). Lack of spatial autocorrelation confirms the Mantel test result and suggests that IBD is irrelevant in the studied populations.

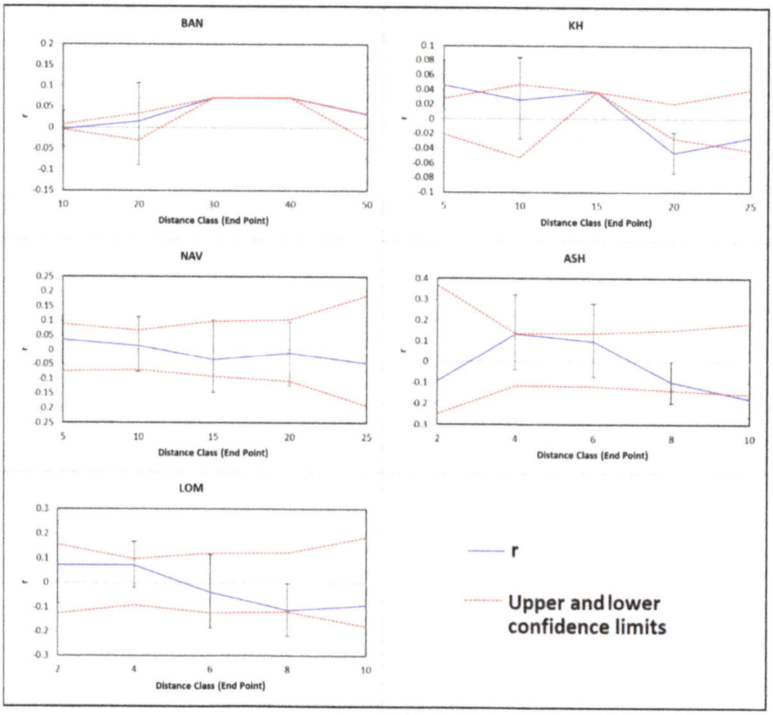

Figure 5. Correlograms illustrating spatial autocorrelation for all analyzed populations. Upper and lower error bars bound the 95% confidence interval to r, as determined by bootstrap resampling.

4. Discussion

Our study revealed that the Hyrcanian populations of *S. aucuparia* in the southern Caspian Sea have higher genetic diversity compared with reported results for other species of the genus *Sorbus* [23,51,52] and even compared with the populations of. *S. aucuparia* in refugial regions of Europe [52]. This is not surprising because the Hyrcanian region was a potential refugium during the last glacial maximum for a wide range of woody taxa [2,53]. An interesting result is the increase in Ho, Ar, and Ap (except the KH population) from the western to the eastern limit distribution of the *S. aucuparia* in Iran. The western population of *S. aucuparia* in the Hyrcanian forest (KH) with a small area and high tree density that is separated about 100 km from the main region of the Hyrcanian forest, showed the highest heterozygosity, allelic richness, and number of private alleles. An appropriate interpretation is that this population is the nearest to the European populations, and it may has acted as a receiver of genes from western and south-eastern Europe, especially from countries with access to the Black Sea (Turkey and Georgia). On the other hand, the BAN population, as the easternmost population of *S. aucuparia* in the Hyrcanian forest, showed high heterozygosity and private alleles. There are several examples where genetic diversity within populations showed an increase towards species distribution margins [54,55]. Kucerova et al. [56] found higher differentiation over central-European populations than those located in southern locations for *S. torminalis*. Additionally, Jankowska-Wroblewska et al. demonstrated that peripheral populations of *S. torminalis* have relatively high levels of genetic diversity [26].

On a global scale, the *S. aucuparia* populations in the Hyrcanian forest are considered as a range-edge population of this species in the Northern Hemisphere. These range-edge populations are isolated from the populations in Europe and are vulnerable to genetic drift [57]. Inbreeding depression, genetic drift, and differentiation of peripheral populations are all exacerbated by persistent reductions in gene flow among small isolated and less dense populations [57,58]. These interpretations contrast strongly with the high levels of individual heterozygosity, suggesting a heavy selection against selfed offspring [59]. All five stands studied had higher expected heterozygosity than observed heterozygosity, resulting in positive inbreeding coefficients. This is contrary to the gametophytic self-incompatibility system of woody Rosaceae [23]. Given that the size and/or density of a population can influence the outcrossing rate of self-compatible plants [60], it seems that harsh habitat conditions, small size, and low tree density, as well as a severe human intervention, has caused, contrary to expectations for the genus *Sorbus* [26], positive inbreeding in *S. aucuparia* populations in the Hyrcanian forest. Additionally, because of their often high levels of heterozygosity, outcrossing trees such as *S. aucuparia* can be disproportionately vulnerable to a reduction in pollen-mediated gene flow, which can mask deleterious recessive alleles that, if expressed, can lead to a reduction in population's fitness [61].

A bottleneck was detected in the Hyrcanian populations of *S. aucuparia* using the *M*-ratio with positive inbreeding. Genetic drift is inversely related to the effective population size (1/2Ne; [62]) and typically occurs in small populations, where rare and private alleles face a greater chance of being lost. The current populations of *S. aucuparia* in the north of Iran may be the remnants of a large population from the past, which over time, due to low competitiveness with other species, has decreased their density and nested in harsh sites with steep slopes and rocky outcrops. Additionally, habitat disturbance, such as forest fires or logging, could lead to fragmented habitat and influence genetic patterns and structures, local extinctions, and subsequent colonization.

In the periphery of a species range, abiotic and biotic environments may differ from those in the center, and there are likely less suitable habitats [59]. Habitat suitability, the historical colonization, migration pattern, and geographical distance among populations shaped the genetic structure of a species [63]. In this study, the habitat conditions of KH population were completely different from the other sites. *Sorbus aucuparia* is usually found above the timberline and in the rocky and steep habitat of the Hyrcanian forest, while the KH habitat is a dune forest with relatively suitable soil and a much smaller habitat slope than other habitats. This could be the reason for the higher genetic diversity and completely different genetic structure of *S. aucuparia* in this habitat. Due to the distance of at least 500 km of the population of BAN from the other four Hyrcanian populations and possibly the complete cessation of gene flow over time, it has been differentiated from other populations.

5. Conclusions

The results of our study demonstrate a positive inbreeding in *S. aucuparia* populations in the Hyrcanian forest, showing evidence of a past bottleneck. To reduce the extinction risk of very small and isolated populations of *S. aucuparia* in this region, there is a need to establish and/or enhance the connectivity between isolated populations through habitat restoration or genetic exchange. In fact, gene movement via seedling could provide a pathway for dispersal and, as a result, greater genetic diversity retention through increased effective population size, reducing the effects of drift [62,64]. To achieve the above-mentioned goals, suitable new areas for afforestation with *S. aucuparia* should be identified to reduce the geographical distance for gene flow among the main populations. Additionally, improving the habitat quality and increasing the density of trees by planting additional seedlings should be used as another alternative to increase connectivity among neighboring trees and reduce the inbreeding depression within the populations.

Supplementary Materials: The following are available online at https://www.mdpi.com/article/10.3390/plants10071471/s1, Table S1: Repeat motif, primer sequence, fragment size and Tm information for 15 under study microsatellite loci, Table S2: Null allele's analysis results by Freena; Table S3. Migration with correction on the sink population (theta*M)/4.

Author Contributions: Plant material collection and preparation, H.Y., S.R., O.E. and G.J.; experiments and data analysis, H.Y., M.N. and Ł.W.; writing—original draft preparation, H.Y; writing—review and editing, H.Y., Ł.W. and G.K.; supervision, H.Y. and G.K. All authors agreed to be accountable for all aspects of the work. All authors have read and agreed to the published version of the manuscript.

Funding: This research was funded by the Tarbiat Modares University, the Iran National Science Foundation (INSF, Grant Number: 96000370), the Fondation Franklinia and University of Fribourg, Switzerland.

Institutional Review Board Statement: Not applicable.

Informed Consent Statement: Not applicable.

Data Availability Statement: Data is attached in Supplementary Materials.

Acknowledgments: We thank Mansour Pouramin and Mohsen Yousefzadeh for the assistance during field data collection. We also wish to thank the two anonymous reviewers for the useful comments.

Conflicts of Interest: The authors declare no conflict of interest.

References

1. Bayat, M.; Burkhart, H.; Namiranian, M.; Hamidi, S.K.; Heidari, S.; Hassani, M. Assessing biotic and abiotic effects on biodiversity index using machine learning. *Forests* **2021**, *12*, 461. [CrossRef]
2. Leestmans, R. Le refuge caspiens et son importance en biogéographie. *Linneana Belg.* **2005**, *20*, 97–102.
3. Zohary, M. *Geobotanical Foundations of the Middle East*; Gustav Fischer Verlag Press: Stuttgart, Germany; Swets & Zeitlinger: Amsterdam, The Netherlands, 1973; Volume 1.
4. Kozlowski, G.; Gibbs, D.; Huan, F.; Frey, D.; Gratzfeld, J. Conservation of threatened relict trees through living ex situ collections: Lessons from the global survey of the genus *Zelkova* (Ulmaceae). *Biodivers. Conserv.* **2012**, *21*, 671–685. [CrossRef]
5. Kozlowski, G.; Song, Y.-G.; Bétrisey, S.; Alvarado, E.V.; Bétrisey, S. *Wingnuts (Pterocarya) & Walnut Family: Relict Trees: Linking the Past, Present and Future*; Natural History Museum: Fribourg, Switzerland, 2018.
6. Hultén, E.; Fries, M. *Atlas of North European Vascular Plants. North of the Tropic of Cancer*; Koeltz Scientific Books: Königstein, Germany, 1986.
7. Raspé, O.; Findlay, C.; Jacquemart, A. Sorbus aucuparia L.: Biological flora of the British Isles, list br. Vasc. Pl. 232, 1. *J. Ecol.* **2000**, *88*, 910–930. [CrossRef]
8. Żywiec, M.; Ledwoń, M. Spatial and temporal patterns of rowan (*Sorbus aucuparia* L.) regeneration in west Carpathian subalpine spruce forest. *Plant Ecol.* **2008**, *194*, 283–291. [CrossRef]
9. Sabeti, H. *Forest, Trees and Shrubs of Iran*; Yazd University Publishers: Yazd, Iran, 1997; 810p.
10. McInerny, G.; Turner, J.; Wong, H.; Travis, J.; Benton, T. How range shifts induced by climate change affect neutral evolution. *Proc. R. Soc. B Biol. Sci.* **2009**, *276*, 1527–1534. [CrossRef]
11. Wróblewska, A.; Mirski, P. From past to future: Impact of climate change on range shifts and genetic diversity patterns of circumboreal plants. *Reg. Environ. Change* **2018**, *18*, 409–424. [CrossRef]
12. Jump, A.S.; Hunt, J.M.; Penuelas, J. Rapid climate change-related growth decline at the southern range edge of *Fagus sylvatica*. *Glob. Change Biol.* **2006**, *12*, 2163–2174. [CrossRef]
13. Lesica, P.; Allendorf, F.W. When are peripheral populations valuable for conservation? *Conserv. Biol.* **1995**, *9*, 753–760. [CrossRef]

14. Hampe, A.; Petit, R.J. Conserving biodiversity under climate change: The rear edge matters. *Ecol. Lett.* **2005**, *8*, 461–467. [CrossRef]
15. Arnaud-Haond, S.; Teixeira, S.; Massa, S.I.; Billot, C.; Saenger, P.; Coupland, G.; Duarte, C.M.; Serrao, E. Genetic structure at range edge: Low diversity and high inbreeding in southeast Asian mangrove (*Avicennia marina*) populations. *Mol. Ecol.* **2006**, *15*, 3515–3525. [CrossRef] [PubMed]
16. Hoffmann, A.A.; Blows, M.W. Species borders: Ecological and evolutionary perspectives. *Trends Ecol. Evol.* **1994**, *9*, 223–227. [CrossRef]
17. Bradshaw, A.D. The croonian lecture, 1991. Genostasis and the limits to evolution. *Philos. Trans. R. Soc. Lond. Ser. B Biol. Sci.* **1991**, *333*, 289–305.
18. Sáenz-Romero, C.; O'Neill, G.; Aitken, S.N.; Lindig-Cisneros, R. Assisted migration field tests in Canada and Mexico: Lessons, limitations, and challenges. *Forests* **2021**, *12*, 9. [CrossRef]
19. Yosefzadeh, H.; Tabari, M.; Akbarinia, M.; Akbarian, M.R.; Bussotti, F. Morphological plasticity of *Parrotia persica* leaves in eastern Hyrcanian forests (Iran) is related to altitude. *Nord. J. Bot.* **2010**, *28*, 344–349. [CrossRef]
20. Gaitán-Espitia, J.D.; Hobday, A.J. Evolutionary principles and genetic considerations for guiding conservation interventions under climate change. *Glob. Change Biol.* **2021**, *27*, 475–488. [CrossRef]
21. Rischkowsky, B.; Pilling, D. *The State of the World's Animal Genetic Resources for Food and Agriculture*; Food & Agriculture Org.: Rome, Italy, 2007.
22. Robertson, A.; Newton, A.; Ennos, R. Breeding systems and continuing evolution in the endemic *Sorbus taxa* on Arran. *Heredity* **2004**, *93*, 487–495. [CrossRef]
23. Hoebee, S.E.; Menn, C.; Rotach, P.; Finkeldey, R.; Holderegger, R. Spatial genetic structure of *Sorbus torminalis*: The extent of clonal reproduction in natural stands of a rare tree species with a scattered distribution. *For. Ecol. Manag.* **2006**, *226*, 1–8. [CrossRef]
24. Angelone, S.; Hilfiker, K.; Holderegger, R.; Bergamini, A.; Hoebee, S. Regional population dynamics define the local genetic structure in *Sorbus torminalis*. *Mol. Ecol.* **2007**, *16*, 1291–1301. [CrossRef]
25. Rasmussen, K.K.; Kollmann, J. Low genetic diversity in small peripheral populations of a rare European tree (*Sorbus torminalis*) dominated by clonal reproduction. *Conserv. Genet.* **2008**, *9*, 1533–1539. [CrossRef]
26. Jankowska-Wroblewska, S.; Meyza, K.; Sztupecka, E.; Kubera, L.; Burczyk, J. Clonal structure and high genetic diversity at peripheral populations of *Sorbus torminalis* (L.) Crantz. *iForest-Biogeosciences For.* **2016**, *9*, 892. [CrossRef]
27. Kavaliauskas, D.; Šeho, M.; Baier, R.; Fussi, B. Genetic variability to assist in the delineation of provenance regions and selection of seed stands and gene conservation units of wild service tree (*Sorbus torminalis* (L.) Crantz) in southern Germany. *Eur. J. For. Res.* **2021**, *140*, 1–15. [CrossRef]
28. Liu, C.; Dou, Y.; Guan, X.; Fu, Q.; Zhang, Z.; Hu, Z.; Zheng, J.; Lu, Y.; Li, W. De novo transcriptomic analysis and development of est-ssrs for *Sorbus pohuashanensis* (Hance) hedl. *PLoS ONE* **2017**, *12*. [CrossRef]
29. Abid, M.; Scheffran, J.; Schneider, U.A.; Ashfaq, M. Farmers' perceptions of and adaptation strategies to climate change and their determinants: The case of Punjab province, Pakistan. *Earth Syst. Dyn.* **2015**, *6*, 225–243. [CrossRef]
30. Ellis, J.; Burke, J. Est-SSRs as a resource for population genetic analyses. *Heredity* **2007**, *99*, 125–132. [CrossRef]
31. Doyle, J.J.; Doyle, J.L. Isolation of plant DNA from fresh tissue. *Focus* **1990**, *12*, 39–40.
32. Dellaporta, S.L.; Wood, J.; Hicks, J.B. A plant DNA minipreparation: Version ii. *Plant Mol. Biol. Rep.* **1983**, *1*, 19–21. [CrossRef]
33. Janfaza, S.; Nasr, S.M.H.; Yousefzadeh, H.; Botta, R. Phylogenetic relationships of the genus *Castanea* based on chloroplast rbcl with focusing Iranian chestnut. *JBES* **2015**, *5*, 312–323.
34. Bassam, B.J.; Caetano-Anollés, G.; Gresshoff, P.M. Fast and sensitive silver staining of DNA in polyacrylamide gels. *Anal. Biochem.* **1991**, *196*, 80–83. [CrossRef]
35. Van Oosterhout, C.; Hutchinson, W.F.; Wills, D.P.; Shipley, P. Micro-checker: Software for identifying and correcting genotyping errors in microsatellite data. *Mol. Ecol. Notes* **2004**, *4*, 535–538. [CrossRef]
36. Chapuis, M.-P.; Estoup, A. Microsatellite null alleles and estimation of population differentiation. *Mol. Biol. Evol.* **2007**, *24*, 621–631. [CrossRef]
37. Peakall, R.; Smouse, P.E. Genalex 6: Genetic analysis in excel. Population genetic software for teaching and research. *Mol. Ecol. Notes* **2006**, *6*, 288–295. [CrossRef]
38. Chybicki, I.J.; INEST 2.2. The User Manual. 2017. Available online: https://www.ukw.edu.pl/pracownicy/strona/igor_chybicki/software_ukw (accessed on 12 October 2018).
39. Garza, J.; Williamson, E. Detection of reduction in population size using data from microsatellite loci. *Mol. Ecol.* **2001**, *10*, 305–318. [CrossRef]
40. Excoffier, L.; Smouse, P.E.; Quattro, J.M. Analysis of molecular variance inferred from metric distances among DNA haplotypes: Application to human mitochondrial DNA restriction data. *Genetics* **1992**, *131*, 479–491. [CrossRef] [PubMed]
41. Wilson, G.A.; Rannala, B. Bayesian inference of recent migration rates using multilocus genotypes. *Genetics* **2003**, *163*, 1177–1191. [CrossRef] [PubMed]
42. Jombart, T.; Devillard, S.; Balloux, F. Discriminant analysis of principal components: A new method for the analysis of genetically structured populations. *BMC Genet.* **2010**, *11*, 1–15. [CrossRef] [PubMed]
43. Beerli, P. Comparison of bayesian and maximum-likelihood inference of population genetic parameters. *Bioinformatics* **2006**, *22*, 341–345. [CrossRef]

44. Beerli, P.; Felsenstein, J. Maximum-likelihood estimation of migration rates and effective population numbers in two populations using a coalescent approach. *Genetics* **1999**, *152*, 763–773. [CrossRef]
45. Beerli, P.; Palczewski, M. Unified framework to evaluate panmixia and migration direction among multiple sampling locations. *Genetics* **2010**, *185*, 313–326. [CrossRef]
46. McRae, B.H. Isolation by resistance. *Evolution* **2006**, *60*, 1551–1561. [CrossRef]
47. McRae, B.H.; Beier, P. Circuit theory predicts gene flow in plant and animal populations. *Proc. Natl. Acad. Sci. USA* **2007**, *104*, 19885–19890. [CrossRef]
48. Wright, S. Isolation by distance. *Genetics* **1943**, *28*, 114. [CrossRef]
49. Smouse, P.E.; Peakall, R. Spatial autocorrelation analysis of individual multiallele and multilocus genetic structure. *Heredity* **1999**, *82*, 561–573. [CrossRef]
50. Evanno, G.; Regnaut, S.; Goudet, J. Detecting the number of clusters of individuals using the software STRUCTURE: A simulation study. *Mol. Ecol.* **2005**, *14*, 2611–2620. [CrossRef]
51. Bednorz, L.; Myczko, L.; Kosinski, P. Genetic variability and structure of the wild service tree (*Sorbus torminalis* (L.) Crantz) in poland. *Silvae Genet.* **2006**, *55*, 197–201. [CrossRef]
52. George, J.-P.; Konrad, H.; Collin, E.; Thevenet, J.; Ballian, D.; Idzojtic, M.; Kamm, U.; Zhelev, P.; Geburek, T. High molecular diversity in the true service tree (*Sorbus domestica*) despite rareness: Data from Europe with special reference to the Austrian occurrence. *Ann. Bot.* **2015**, *115*, 1105–1115. [CrossRef]
53. Alavi, S.J.; Veiskarami, R.; Esmailzadeh, O.; Gadow, K.V. Analyzing the biological and structural diversity of Hyrcanian forests dominated by *Taxus baccata* L. *Forests* **2020**, *11*, 701. [CrossRef]
54. Yakimowski, S.B.; Eckert, C.G. Populations do not become less genetically diverse or more differentiated towards the northern limit of the geographical range in clonal *Vaccinium stamineum* (Ericaceae). *New Phytol.* **2008**, *180*, 534–544. [CrossRef] [PubMed]
55. Assis, J.; Coelho, N.C.; Alberto, F.; Valero, M.; Raimondi, P.; Reed, D.; Serrão, E.A. High and distinct range-edge genetic diversity despite local bottlenecks. *PLoS ONE* **2013**, *8*, e68646. [CrossRef]
56. Kucerova, E.; Clifton, S.W.; Xia, X.-Q.; Long, F.; Porwollik, S.; Fulton, L.; Fronick, C.; Minx, P.; Kyung, K.; Warren, W. Genome sequence of *Cronobacter sakazakii* baa-894 and comparative genomic hybridization analysis with other *Cronobacter* species. *PLoS ONE* **2010**, *5*, e9556. [CrossRef] [PubMed]
57. Kottler, E.J.; Dickman, E.E.; Sexton, J.P.; Emery, N.C.; Franks, S.J. Draining the Swamping Hypothesis: Little Evidence that Gene Flow Reduces Fitness at Range Edges. *Trends Ecol. Evol.* **2021**, *36*, 533–544. [CrossRef]
58. Ouborg, N.; Vergeer, P.; Mix, C. The rough edges of the conservation genetics paradigm for plants. *J. Ecol.* **2006**, *94*, 1233–1248. [CrossRef]
59. Michalski, S.G.; Durka, W. High selfing and high inbreeding depression in peripheral populations of juncus atratus. *Mol. Ecol.* **2007**, *16*, 4715–4727. [CrossRef] [PubMed]
60. DeSilva, R.; Dodd, R.S. Patterns of Fine-Scale Spatial Genetic Structure and Pollen Dispersal in Giant Sequoia (*Sequoiadendron giganteum*). *Forests* **2021**, *12*, 61. [CrossRef]
61. Bacles, C.F.; Jump, A.S. Taking a tree's perspective on forest fragmentation genetics. *Trends Plant Sci.* **2011**, *16*, 13–18. [CrossRef] [PubMed]
62. Frankham, R. Challenges and opportunities of genetic approaches to biological conservation. *Biol. Conserv.* **2010**, *143*, 1919–1927. [CrossRef]
63. Pannell, J.R.; Dorken, M.E. Colonisation as a common denominator in plant metapopulations and range expansions: Effects on genetic diversity and sexual systems. *Landsc. Ecol.* **2006**, *21*, 837–848. [CrossRef]
64. Jangjoo, M.; Matter, S.F.; Roland, J.; Keyghobadi, N. Connectivity rescues genetic diversity after a demographic bottleneck in a butterfly population network. *Proc. Natl. Acad. Sci. USA* **2016**, *113*, 10914–10919. [CrossRef]

plants

MDPI

Article

Disentangling Species Delineation and Guiding Conservation of Endangered Magnolias in Veracruz, Mexico

Fabián Augusto Aldaba Núñez [1,*], Emily Veltjen [2,3], Esteban Manuel Martínez Salas [4] and Marie-Stéphanie Samain [1,2]

[1] Instituto de Ecología, A.C., Red de Diversidad Biológica del Occidente Mexicano, Pátzcuaro 61600, Mexico; mariestephanie.samain@gmail.com
[2] Systematic and Evolutionary Botany Lab, Department of Biology, Ghent University, 9000 Ghent, Belgium; emily.veltjen@ugent.be
[3] Ghent University Botanical Garden, Ghent University, 9000 Ghent, Belgium
[4] Herbario Nacional de México, Departamento de Botánica, Instituto de Biología, Universidad Nacional Autónoma de México, Mexico City 04510, Mexico; ems@ib.unam.mx
* Correspondence: fabian.aldaba@outlook.com

Abstract: The Mexican state of Veracruz has suffered very high deforestation rates in the last few decades, and despite the establishment of protected areas and conservation projects, primary forest is now mainly persisting in mostly small, scattered, fragmented remnants. New species of *Magnolia* section *Talauma* in this state have been described with little to no reference to the already existing ones, potentially resulting in over-splitting, obscuring their taxonomic delineation and conservation status, and consequently conservation programs. To study the conservation units and their genetic diversity, we here employ 15 microsatellite markers on a highly representative sampling of 254 individuals of what are presumed to be five *Magnolia* species. The results support at least three species and maximum five main conservation units. We propose downgrading the latter to four, given morphological, ecological, demographical, and geographical considerations. Two out of the three sympatrically occurring species in the rainforest in the Los Tuxtlas volcanic area have weak genetic evidence to be considered separate species. Similarly, the individuals in the Sierra de Zongolica in central Veracruz, who bear a very high morphological and genetic similarity to *Magnolia mexicana*, have weak genetic evidence to be recognised as a separate species. Nonetheless, the individuals could be identified as *Magnolia decastroi* based on morphology, and further research including the full range of this species is recommended.

Keywords: conservation units; genetic diversity; IUCN Red List conservation status; Magnoliaceae; microsatellite; neotropical trees; SSR; *Talauma*

Citation: Aldaba Núñez, F.A.; Veltjen, E.; Martínez Salas, E.M.; Samain, M.-S. Disentangling Species Delineation and Guiding Conservation of Endangered Magnolias in Veracruz, Mexico. *Plants* **2021**, *10*, 673. https://doi.org/10.3390/plants10040673

Academic Editor: Gregor Kozlowski

Received: 10 February 2021
Accepted: 25 March 2021
Published: 31 March 2021

1. Introduction

Biodiversity is being lost at an accelerated rate, often referred to as the sixth mass extinction [1,2]. This is particularly striking in plant diverse countries such as Mexico, which span a wide variety of ecosystems [3,4]. The latest assessment of plant richness in Mexico registers 297 families, 2854 genera, and 23,314 species, of which 11,600 are endemic [5]. Particularly in the state of Veracruz, there are 271 families, 1956 genera, and 8497 vascular plant species, of which 238 are endemics, representing around 27% of the national diversity, being the third most diverse state in terms of plants [5–7]. This plant biodiversity is threatened mainly by land conversion, which has resulted in the loss of 42% of the tropical ecosystems [3]. The state of Veracruz ranks first nationally in the loss of vegetation; besides, it is estimated that only 8.6% of this vegetation is conserved [8], which is mainly found in unprotected areas [7]. Within Veracruz, the areas of Sierra de Zongolica and Los Tuxtlas have been recognised for their great biological and ecological diversity [7,9–12], which in recent decades have been largely destroyed, mainly by primary economic activities (agriculture and livestock), despite the fact that a large part of the territory of Los Tuxtlas is formally protected [9,11,13,14]. Therefore, any study carried out in these areas is of vital importance in order to propose conservation strategies that mitigate the effects of anthropogenic development.

Representatives of the Magnoliaceae family are part of these threatened, small, declining tropical ecosystems in Veracruz, and have great potential to serve as flagship [15] and umbrella [16,17] species for conservation studies and management. The first is due to their striking flowers and interesting evolutionary history [18–20]. Their status as umbrella species owes to the fact that these trees are one of the main constituents of their forest ecosystem. Since [16] highlighted the urgent need for conservation work based on genetic research in Magnoliaceae, focus on Neotropical Magnolias is increasing, with very promising and hopeful results for the species and forest conservation [21–23]. Although certain progress is being made, less than 50% of the most threatened taxa are found in ex situ collections in botanical gardens and arboreta [24], and more research is needed on (Critically) Endangered species. To allow for effective conservation, there are three basic steps [16]: (1) delimitation and selection of priority taxa; (2) analysis of diversity of natural populations and ex situ collections, and (3) final selection of sampling sites.

Delimitation of taxa, the initial action of the abovementioned steps, is especially challenging for the Magnolias of Veracruz belonging to the genus *Magnolia* sect. *Talauma sensu* Figlar and Nooteboom [25]. This concerns the three recently described endemic and endangered species that occur in the region of Los Tuxtlas: *Magnolia lopezobradorii* A. Vázquez, *Magnolia sinacacolinii* A. Vázquez and *Magnolia zoquepopolucae* A. Vázquez; one in the Sierra de Zongolica area in Southern Veracruz: *Magnolia decastroi* A. Vázquez and M. Muñiz and one in the Uxpanapa area in Southern Veracruz: *Magnolia wendtii* A. Vázquez [26–28]. The first three species had been assessed as Data Deficient (DD) by the International Union for Conservation of Nature (IUCN) Red List, because they lacked population information and there were doubts about their taxonomic status. In contrast, *M. decastroi* was assessed as Endangered (EN) and *M. wendtii* as Critically Endangered (CR). These five species were segregated from *Magnolia mexicana* DC., previously considered to be widely distributed in Mexico: from Veracruz to Chiapas in the east and from Jalisco to Guerrero in Western Mexico [29]. There are even specimens from Guatemala identified as *M. mexicana* [30]. *Magnolia mexicana* s.s. (not including the five segregated species) was assessed as Vulnerable [31]. Most of these Magnolias have local uses that are threatening their survival, either through logging for construction of fence doors (*M. sinacacolinii*), construction of houses and furniture (*M. zoquepopolucae*), and collection of complete flowers for medicinal application against heart diseases (*M. mexicana*) (pers. obs.).

As a result of the very short protologue descriptions of four of the five recently described section *Talauma* species in Veracruz and the often contradicting carpel numbers in the available identification keys [32,33], their circumscription is no longer clear. Moreover, some of the species descriptions are based on few specimens [27,28], or in the case of *M. mexicana* that was described in 1893, it is based on a scientific illustration [34–37]. The concept of *M. mexicana* in particular becomes even more challenging, because the scientific illustration is said to be originating from Cuernavaca, in the state of Morelos in Central-Southern Mexico [34,37], yet it is known that *M. mexicana* is not native to this area. Hence, it is most likely a specimen cultivated by the Aztec culture due to its medicinal properties. Furthermore, the illustration nor the description mention some of the characteristics of the section *Talauma*, such as the stipule scar along the entire length of the petiole or the circumcised fruit dehiscence [7,29,38], together with leaf and petal morphological characteristics that have been observed in living specimens.

When the identity and relationships are complex, morphological data can be complemented with molecular characters [39,40]. Including both types of data offers a closer approximation to the relationships between them, whereby conflicts between the two can be resolved by analysis of total evidence [41]. Recently, SSR (Single Sequence Repeat) markers have been used to elucidate the genetic patterns of Neotropical Magnolias in fragmented environments and the data obtained have shown that these species still have ample gene flow within populations, yet little gene flow between populations [42].

These five species have not been studied from a molecular point of view (e.g., phylogenetically): only *M. decastroi* and *M. mexicana* have been poorly studied in Mexico, sometimes with a minimum sample size [43,44] and *M. mexicana* from one accession is sequenced in family-wide studies [18–20,45,46], but those identifications have a high probability of being incorrect given the former widespread concept of the species. In a first taxonomic review [47], morphological characteristics of around 300 accessions of these species were tested statistically for their morphological distinctness, and here no significant morphological differences were found between *M. lopezobradorii* and *M. zoquepopolucae*.

Here we aimed to assess the genetic diversity and structure of five *Magnolia* species (Figure 1) from Veracruz and surrounding regions, especially the state of Puebla (Figure 2), namely, *M. decastroi*, *M. lopezobradorii*, *M. mexicana*, *M. sinacacolinii*, and *M. zoquepopolucae*, employing 15 SSR markers, with the applied goal of developing effective conservation strategies tailored for the sampled localities under study. We specifically wanted to test: (1) do the SSR data support the five morphospecies?, (2) are the individuals at the different localities within the taxa exhibiting sufficient (past) gene flow and random mating to maintain (relatively to other studied Magnolias) healthy levels of genetic diversity?, (3) what should we consider the conservation units?, (4) which are priority taxa/localities for conservation management?, and (5) are the current ex situ collections a good genetic representation of the in situ diversity? The acquired insights of the demographic, distributional, genetic, morphological, and habitat data allowed us to re-assess their conservation status for the IUCN Red List, as well as to propose conservation strategies for each of the species.

Figure 1. Morphology of *Magnolia* sect. *Talauma* in Puebla and Veracruz, Mexico. (**A**) *Magnolia decastroi*: 1. Tree, 2. Voucher, 3. Fruit; 1–3 By F.A. Aldaba Núñez, 2019. (**B**) *Magnolia lopezobradorii*: 1. Tree, 2. Voucher, 3. Fruit; 1–2 By F.A. Aldaba Núñez, 2019, 3 By E.M. Martínez Salas, 2019. (**C**) *Magnolia mexicana*: 1. Tree, 2. Voucher, 3. Fruit, 4. Flower; 1–4 By F.A. Aldaba Núñez, 2019. (**D**) *Magnolia sinacacolinii*: 1. Tree, 2. Voucher, 3. Fruit, 4. Flower; 1–2 By F.A. Aldaba Núñez, 2019, 3–4 By E.M. Martínez Salas, 2019. (**E**) *Magnolia zoquepopolucae*: 1. Tree, 2. Voucher, 3. Fruit; 1–3 By F.A. Aldaba Núñez, 2019.

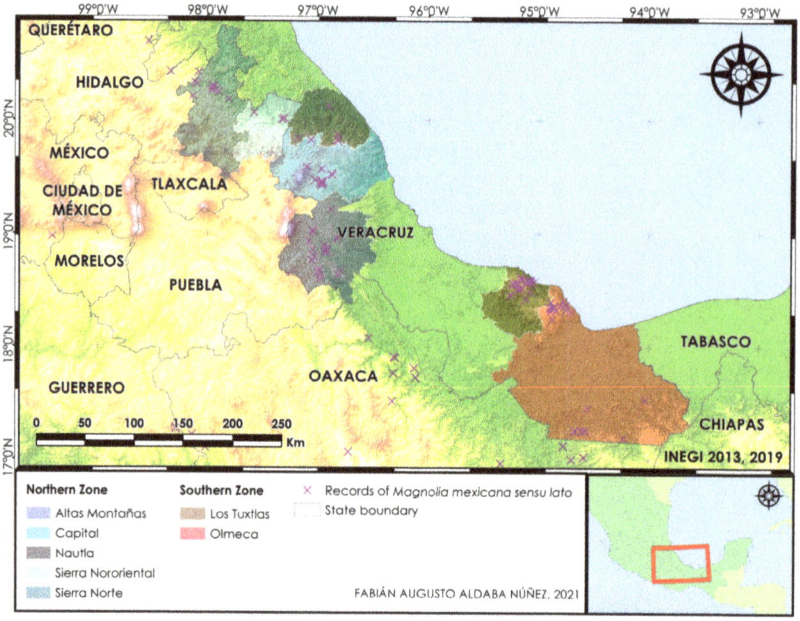

Figure 2. Map of study zones showing their natural regions; in Puebla and Veracruz states.

2. Results

2.1. Genetic Analysis and Characterisation

2.1.1. Null Alleles and Linkage Disequilibrium

In the analyses respecting the localities, several potential null-alleles were reported for five of the 15 SSR markers (Table 1). The locality FA-6 had potential null-alleles in four markers, and hence we suspect that the markers MA39_009, MA39_287, and MA42_421, which only had a potential null-allele detected for that locality, do not have true null-alleles. Instead, the strong signal of deviations of Hardy–Weinberg Proportions (HWP) used to determine null-allele presence comes forth from an underlying biological reason or sampling issues linked to this locality [48]. We do acknowledge that there is a chance that the null alleles are species-specific for FA-6, which is identified as *M. sinacacolinii*, as we only have one locality with a sample size N > 10 to study the presence of null alleles for that species. For locus MA42_471, two out of the 11 localities showed that the marker might express null alleles with null allele frequencies being 0.061 and 0.011 for the localities FA-12 (*M. decastroi*) and FA-10 (*M. mexicana*), respectively. However, because the two species have parallel, larger (N > 10) sampled localities, in which null alleles are not detected, and as the null allele frequencies are low, we decided to keep the markers in the subsequent analyses. Marker MA39_224 yielded null alleles for all study localities, with estimated frequencies between 0.122 and 0.390. As this pattern is consistent and the estimated null allele frequencies are high, we deleted marker MA39_224 from subsequent analyses and continued the downstream analyses with 14 SSR markers.

Table 1. Results of MICRO-CHECKER and ML-NullFreq analyses for null alleles on the Single Sequence Repeat (SSR) dataset of five *Magnolia* species in Puebla and Veracruz, Mexico. Only SSR loci for which potential null alleles were detected by MICROCHECKER are tabulated. N: sample size. NA: not available due to small sample size. Where a potential null allele was found the frequency as calculated by ML-NullFreq is given between brackets. Metadata for the locality abbreviations can be found in Table 5. Metadata of the SSR markers are specified in [42]. (C) marks cultivated sample localities as opposed to the wild sample localities.

Species	Locality	N	SSR Locus				
			MA39_009	MA39_224	MA39_287	MA42_421	MA42_471
M. decastroi	FA-12 (C)	30	No	YES (0.122)	No	No	YES (0.061)
	FA-13	33	No	YES (0.256)	No	No	No
M. lopezobradorii	FA-5	3	NA	NA	NA	NA	NA
	FA-7	27	No	YES (0.249)	No	No	No
	FA-8	3	NA	NA	NA	NA	NA
	LT-2	14	No	YES (0.219)	No	No	No
M. mexicana	FA-1 (C)	24	No	YES (0.202)	No	No	No
	FA-2	13	No	YES (0.187)	No	No	No
	FA-9	7	No	No (0.330)	No	No	No
	FA-10	31	No	YES (0.369)	No	No	YES (0.011)
	FA-11	4	No	No (0.202)	No	No	No
	FA-15 (C)	18	No	YES (0.396)	No	No	No
M. sinacacolinii	FA-6	28	YES (0.263)	YES (0.339)	YES (0.348)	YES (0.339)	No
	LT-3	4	NA	NA (0.278)	NA	NA	NA
M. zoquepopolucae	FA-3	5	No	YES (0.286)	No	No	No
	FA-4	3	NA	NA	NA	NA	NA
	FA-14	4	No	YES (0.387)	No	No	No
	LT-5	3	NA	NA	NA	NA	NA

In the analyses for Linkage Disequilibrium (LD), we found 161 out of 1 108 significant pairwise comparisons (all pairwise tests that could not be performed due to low allelic variation or small sample sizes were not included to determine the total number of pairwise comparisons). For 1 108 pairwise tests we expected there to be 55.4 [44,49] Type I errors on the nominal p-level of 5%, hence linkage was at hand in the dataset. Following sequential Bonferroni correction, six of the pairwise comparisons remained in LD. Five of the six pairs were between MA40_282 and MA39_236; and one was between MA40_282 and MA42_495. We removed marker MA40_282 from the dataset as it cannot be guaranteed that this marker is an independent sampling of the genome with respect to the other SSR markers. We thus executed further downstream analyses with 13 microsatellite markers.

2.1.2. Genetic Structure

According to the Structure analyses, the optimal ΔK value for the complete dataset was K = 2 (Figure 3A), which separates the 18 localities according to the two main sampled zones: the Northern Zone and the Southern Zone (Figure 2). The mean L(K) graph (Figure 3B) illustrates that the likelihood increases substantially when further subdivision is allowed at K = 3–5, and we expected 4-5 species based on [47]; hence, we explored the bar plots K = 3–5 in greater detail. The bar plot for K = 2 of the complete dataset is visualised in Figure 3C. When we studied the ten replicate bar plots from K=3, the 10 replicates were the following: 4 replicates clustered localities of *M. decastroi*, *M. mexicana* and the Southern Zone; 4 replicates clustered the localities of the Northern Zone, *M. lopezobradorii*–*M. zoquepopolucae* and *M. sinacacolinii*; 2 replicates clustered *M. decastroi*–*M. sinacacolinii*, *M. mexicana* and *M. lopezobradorii*–*M. zoquepopolucae*. When we studied the ten replicates from K = 4 and K = 5 (visualized in Figure 3D,E), 9/10 and 2/10 replicates indicate clusters according to species boundaries, respectively with a "conflicting" signal of individuals in the *M. mexicana* localities. When analyzing the Northern Zone and the Southern Zone separately in two new Structure analyses, the results are as follows. The optimal ΔK value for the Northern Zone was K = 2 (Figure A1), supported by the Mean L(K) plot result (Figure A2), splitting the *M. decastroi* and *M. mexicana* localities. The optimal ΔK value for the Southern Zone was K = 2 (Figure A3), supported by the Mean L(K) plot result (Figure A4), separating the *M. lopezobradorii* and *M. zoquepopolucae* localities from the *M. sinacacolinii* localities. For the dataset only comprising the wild localities, the optimal ΔK value was K=5 (Figure 3F), separating the 15 localities according to the previously defined species (Figure 3H).

Results from the DAPC analysis from the complete dataset (Figure 3A) revealed three main groups: the localities containing individuals identified as *M. decastroi–M. mexicana* and *M. lopezobradorii–M. zoquepopolucae* were clustered closely together according to the two most explanatory axes in the ordination space, while a third group is composed of the localities containing individuals identified as *M. sinacacolinii*. One hundred and fifty principal components (PC) were retained which explained 88.1% of the total variance of the data. When the Northern Zone was analysed separately, *M. decastroi* and *M. mexicana* conformed three groups: the first one comprising the wild localities of *M. mexicana* (i.e., FA-1 and FA-15), the second one including cultivated and wild localities of *M. decastroi* (i.e., FA-12 and FA-13), and the third group with only the wild localities of *M. mexicana* (i.e., FA-10, FA-11, FA-9, and FA-2; Figure 3B). In this analysis 50 PCs were retained which explained 97.1% of the total variance. Similarly, when analyzing the Southern Zone separately excluding the clearly differentiated localities identified as *M. sinacacolinii*, the localities of *M. zoquepopolucae* formed a different group from *M. lopezobradorii* (Figure 4C), whereby the 5 retained PCs explained 40.3% of the total variance. The three-grouping pattern recorded in the complete dataset was retrieved when only wild localities were analysed (Figure 4D). One hundred and fifty principal components (PC) were retained, which explained 88.5% of the total variance of the data. Finally, the separation between *M. decastroi* and *M. mexicana* observed in the complete dataset remained when only the wild localities were examined (Figure 4E), in which 60 PCs retained explained 88.2% of the total variance.

AMOVA showed that the proportion of the genetic variance explained among localities of all species was 65.95%, while the genetic variance within localities was 34.05% (results not shown). When localities were grouped according to the Northern and the Southern Zone (Figure 2) the percentage of variation explained by this grouping was 22.13%. When localities were grouped according to three species groups (*M. decastroi–M. mexicana*, *M. lopezobradorii–M. zoquepopolucae*, and *M sinacacolinii*), this explained 73% of the genetic variation in the dataset. When four species groups were considered (*M. decastroi*, *M. mexicana*, *M. lopezobradorii–M. zoquepopolucae*, and *M. sinacacolinii*), this declined to 67.85% and in five species groups, the explained variation was 67.72%.

Pairwise F_{ST} and D_{JOST} values between the 18 localities are tabulated in Table 2 and visualised in an accompanying heatmap in Figure 5. Pairwise F_{ST} values between localities varied between −0.04 and 046. Pairwise D_{JOST} values between localities varied between 0.00 and 0.79. The pairwise F_{ST} values and D_{JOST} for the *M. decastroi* localities, of which one is a cultivated locality (FA-12) and the other a wild one (FA-13), is 0.03 for both measures. The pairwise F_{ST} values and D_{JOST} between the wild *M. mexicana* localities (FA10, FA-11, FA-2, and FA-9) ranged between 0.06–0.22 and 0.06–0.17, respectively. The pairwise F_{ST} values and D_{JOST} between the wild *M. mexicana* localities (FA10, FA-11, FA-2, and FA-9) and the cultivated localities (FA-1 and FA-15) ranged between 0.06–0.26 and 0.02–0.20, respectively. The pairwise F_{ST} values and D_{JOST} between the *M. lopezobradorii* localities ranged between 0.08–0.22 and 0.06–0.27, respectively. The pairwise F_{ST} values and D_{JOST} between the *M. zoquepopolucae* localities ranged between −0.04 to 0.09 and −0.03 to 0.04, respectively. The pairwise F_{ST} values and D_{JOST} between the *M. sinacacolinii* localities were 0.15 and 0.23, respectively.

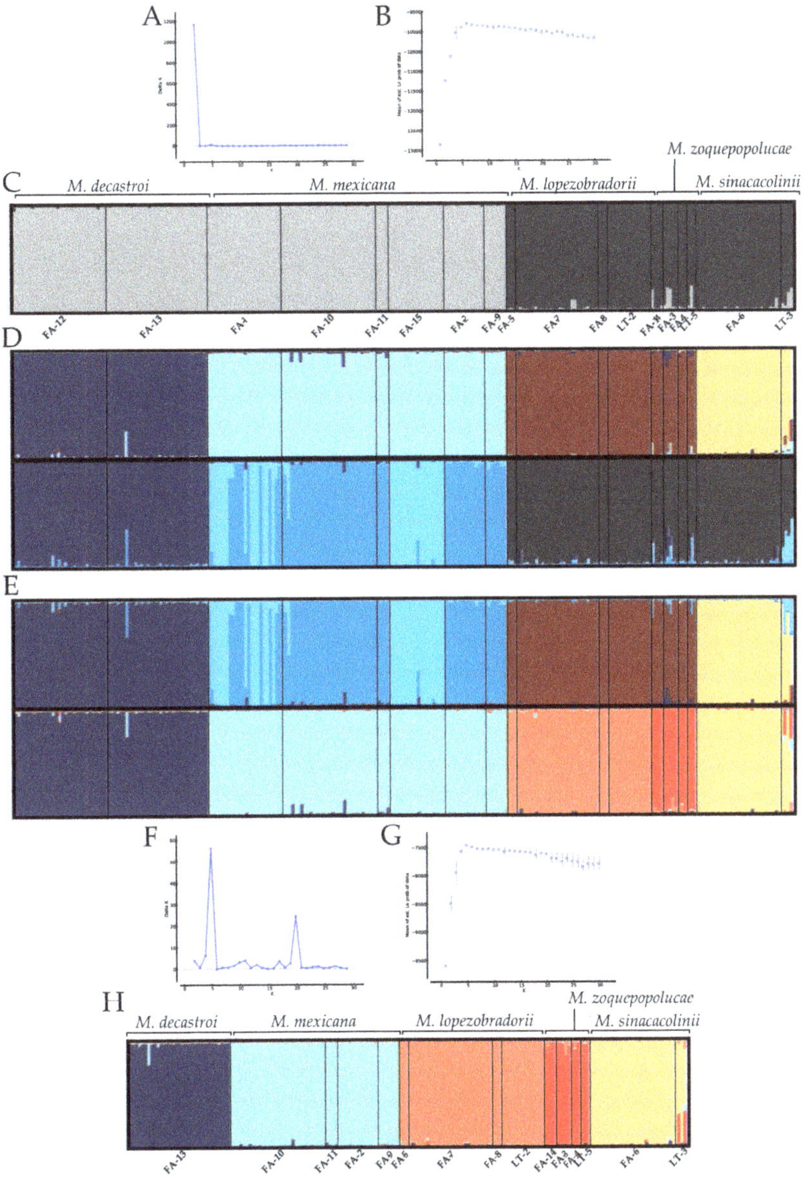

Figure 3. Structure bar plots of *Magnolia* sect. *Talauma* individuals in Puebla and Veracruz, Mexico. A–E: analyses on the complete dataset of 18 localities. F–H: analyses of the 15 wild localities only. (**A**) The delta K plot for the complete dataset. (**B**) The mean Ln(K) plot for the complete dataset. (**C**) Representative bar plot (out of ten replicates) for K = 2. (**D**) Representative bar plots for K = 4. The upper bar plot is found in 9/10 replicates, the lower bar plot in 1/10 replicates. (**E**) Representative bar plots for K = 5. The upper bar plot is found in 8/10 replicates, the lower bar plot in 2/10 replicates. (**F**) The delta K plot for the wild localities only. (**G**) The mean Ln(K) plot for the wild localities only. (**H**) Representative bar plot (out of ten replicates) for K = 5.

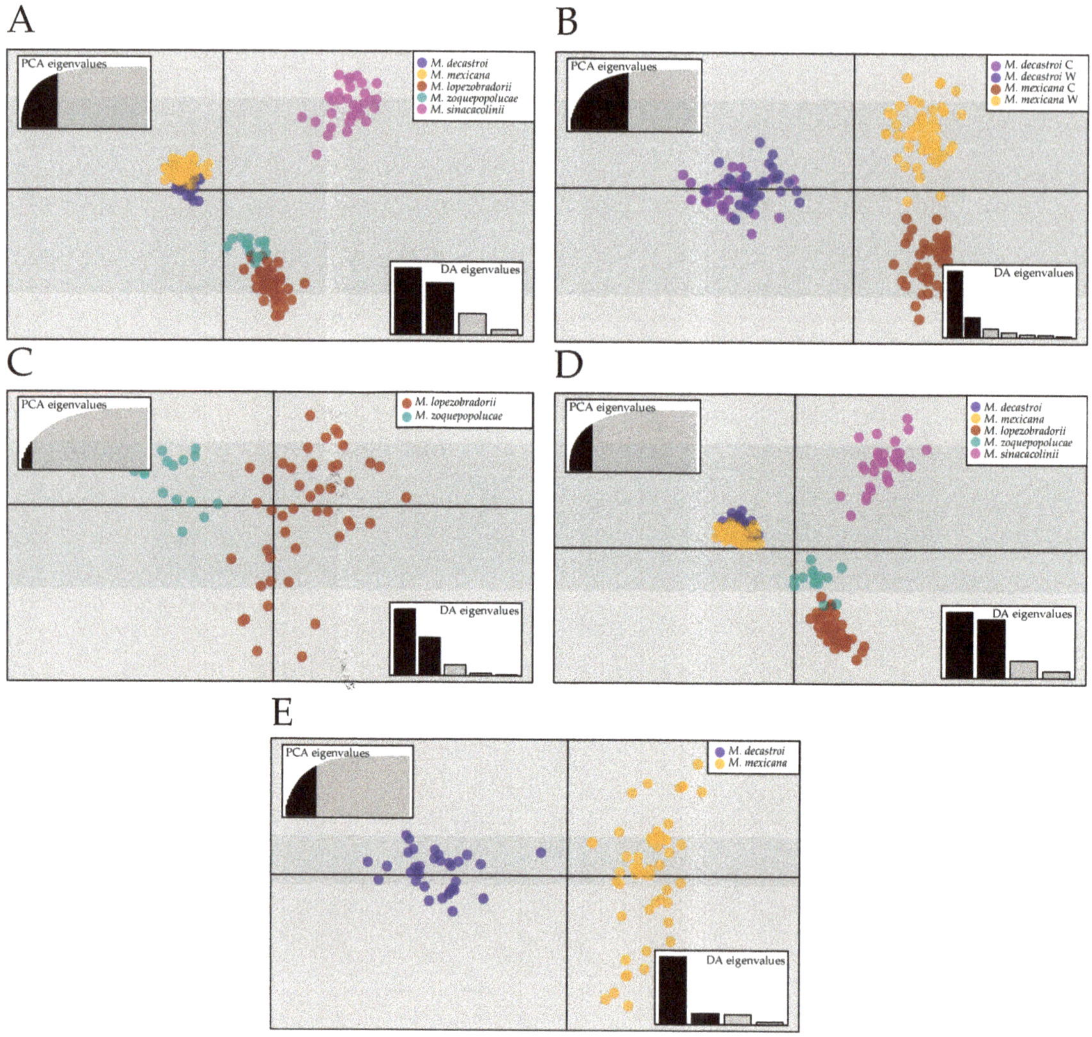

Figure 4. Discriminant Analysis of Principal Components (DAPC) of *Magnolia* sect. *Talauma* individuals in Puebla and Veracruz, Mexico. The axes represent the first two linear discriminants. The upper left graph (principal component analysis (PCA) eigenvalues) inset displays the variance explained by the principal component axes used for DAPC and the bottom-right inset (DA eigenvalues) displays in relative magnitude the variance explained by the two discriminant axes plotted. (**A**) DAPC graph of the complete dataset analysis, 150 principal components (PCs) retained. (**B**) DAPC graph of the *M. decastroi* and *M. mexicana* localities, 50 PCs retained. (**C**). DAPC graph of the *M. lopezobradorii* and *M. zoquepopolucae* localities, 5 PCs retained. (**D**) DAPC graph of the wild dataset, 150 PCs retained. (**E**) DAPC graph of the *M. decastroi* and *M. mexicana* wild localities, 60 PCs retained.

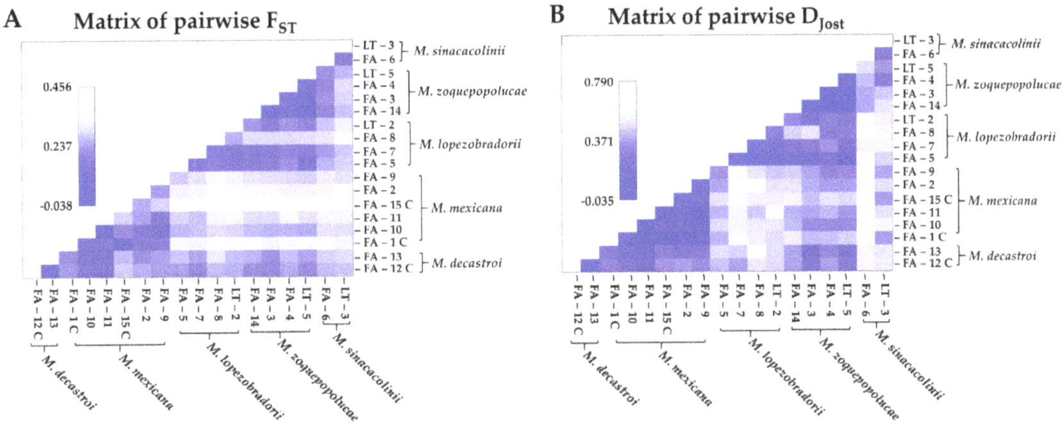

Figure 5. Pairwise F_{ST} and D_{JOST} values between the 18 localities of *Magnolia* species in Puebla and Veracruz, Mexico, visualized as a heatmap, C=cultivated. (**A**) Pairwise F_{ST} values. (**B**) Pairwise D_{JOST} values. Locality metadata can be found in Table 5.

When F_{ST} values were further compared with each other, only the localities with sample sizes higher than 10 were considered, as the two parameters are largely affected by unequal sample size [50,51]. The pairwise comparison of FA-13 (N = 33) vs. FA-10 (N = 31), which are wild localities of the species *M. decastroi* and *M. mexicana*, stood out as this is an interspecific comparison with allelic differentiation (D_{JOST}: 0.17) and fixation index (F_{ST}: 0.11) in the range of intraspecific comparisons. With the exception of the FA-13 vs. FA-10 pair, the allelic differentiation (D_{JOST}) showed smaller values for the intraspecific comparisons (D_{JOST}: 0.02–0.18), and larger values for the interspecific comparisons (D_{JOST}: 0.02–0.75). For the fixation index, there was no clear separation between intraspecific and interspecific values. The two localities of *M. decastroi* and FA-1 and FA-15 of *M. mexicana* showed little fixation (F_{ST}: 0.03 and 0.06, respectively). The locality pair with the highest intraspecific pairwise F_{ST} was that of FA-15 and FA-2 of *M. mexicana* (F_{ST}: 0.22). Due to small sample sizes of the *M. zoquepopolucae* localities, this species was omitted from a detailed study of the pairwise genetic differentiation at the level of localities, yet at the level of species (see further), it was included.

Pairwise F_{ST} and D_{JOST} values between the five presumed species, respecting cultivated and wild populations separately, are tabulated in Table 3 and visualised in Figure 6. Pairwise F_{ST} values (excluding the cultivated populations) varied between 0.11 and 0.26. Pairwise D_{JOST} values varied between 0.19 and 0.64. The lowest pairwise differences (excluding the cultivated populations) were between the wild localities of *M. decastroi* and *M. mexicana* (F_{ST}: 0.12; D_{JOST}: 0. 19); and between *M. lopezobradorii* and *M. zoquepopolucae* (F_{ST}: 0.11; D_{JOST}: 0.28). The highest pairwise differences (excluding the cultivated populations) were between the wild *M. mexicana* localities and *M. sinacacolinii* (F_{ST}: 0.26, D_{JOST}: 0.64).

Table 2. Pairwise F_{ST} and D_{JOST} values between the 18 localities of *Magnolia* species in Puebla and Veracruz, Mexico. Above the diagonal pairwise D_{JOST} values are tabulated. Below the diagonal pairwise F_{ST} values are tabulated. Locality metadata can be found in Table 5. In blue the intraspecific values. (C) marks cultivated sampling localities as opposed to wild sampling localities.

Species		*M. decastroi*		*M. mexicana*						*M. lopezobradorii*				*M. zoquepopolucae*				*M. sinacacolinii*	
	Localities	**FA-12**	**FA-13**	**FA-1**	**FA-10**	**FA-11**	**FA-15**	**FA-2**	**FA-9**	**FA-5**	**FA-7**	**FA-8**	**LT-2**	**FA-14**	**FA-3**	**FA-4**	**LT-5**	**FA-6**	**LT-3**
M. decastroi	FA-12 (30) (C)	–	0.03	0.30	0.23	0.19	0.36	0.38	0.35	0.45	0.39	0.52	0.49	0.39	0.23	0.39	0.25	0.57	0.53
	FA-13 (33)	0.03	–	0.23	0.17	0.15	0.31	0.25	0.29	0.51	0.46	0.63	0.49	0.38	0.19	0.34	0.21	0.56	0.52
M. mexicana	FA-1 (24) (C)	0.21	0.18	–	0.12	0.08	0.02	0.12	0.11	0.43	0.58	0.53	0.49	0.53	0.43	0.45	0.47	0.75	0.37
	FA-10 (31)	0.14	0.11	0.11	–	0.06	0.18	0.16	0.08	0.48	0.56	0.51	0.52	0.44	0.35	0.35	0.34	0.72	0.56
	FA-11 (4)	0.14	0.13	0.13	0.06	–	0.13	0.17	0.06	0.42	0.55	0.45	0.59	0.41	0.41	0.44	0.36	0.79	0.49
	FA-15 (18) (C)	0.26	0.24	0.06	0.19	0.26	–	0.17	0.20	0.50	0.59	0.53	0.54	0.54	0.45	0.52	0.48	0.73	0.36
	FA-2 (13)	0.23	0.20	0.15	0.13	0.22	0.22	–	0.14	0.39	0.66	0.72	0.50	0.51	0.42	0.39	0.39	0.65	0.48
	FA-9 (7)	0.22	0.22	0.16	0.11	0.14	0.26	0.21	–	0.35	0.58	0.50	0.46	0.46	0.40	0.33	0.33	0.70	0.39
M. lopezobradorii	FA-5 (3)	0.21	0.25	0.32	0.24	0.27	0.38	0.30	0.26	–	0.07	0.06	0.06	0.19	0.17	0.28	0.06	0.63	0.47
	FA-7 (27)	0.18	0.22	0.31	0.24	0.25	0.33	0.31	0.28	0.08	–	0.15	0.17	0.26	0.30	0.38	0.22	0.55	0.56
	FA-8 (3)	0.24	0.29	0.39	0.27	0.31	0.46	0.43	0.33	0.14	0.12	–	0.27	0.46	0.52	0.30	0.31	0.73	0.59
	LT-2 (14)	0.23	0.25	0.34	0.27	0.32	0.38	0.33	0.31	0.12	0.11	0.22	–	0.33	0.18	0.31	0.27	0.59	0.54
M. zoquepopolucae	FA-14 (4)	0.20	0.23	0.34	0.24	0.28	0.40	0.34	0.29	0.14	0.13	0.28	0.19	–	0.00	0.03	-0.03	0.39	0.49
	FA-3 (5)	0.18	0.20	0.32	0.23	0.28	0.37	0.32	0.28	0.16	0.15	0.28	0.14	0.06	–	0.00	0.01	0.39	0.44
	FA-4 (3)	0.23	0.25	0.34	0.23	0.32	0.39	0.31	0.27	0.20	0.16	0.28	0.20	0.09	0.08	–	0.04	0.43	0.25
	LT-5 (3)	0.18	0.21	0.33	0.22	0.28	0.39	0.31	0.26	0.10	0.14	0.28	0.17	-0.04	0.04	0.09	–	0.50	0.37
M. sinacacolinii	FA-6 (28)	0.25	0.26	0.35	0.28	0.32	0.38	0.32	0.31	0.24	0.21	0.29	0.25	0.17	0.20	0.17	0.19	–	0.23
	LT-3 (4)	0.25	0.28	0.31	0.27	0.31	0.38	0.35	0.28	0.26	0.25	0.36	0.30	0.25	0.26	0.27	0.24	0.15	–

Table 3. Pairwise F_{ST} and D_{JOST} values between the five *Magnolia* species studied in Puebla and Veracruz, Mexico. Above the diagonal pairwise D_{JOST} values are tabulated. Below the diagonal pairwise F_{ST} values are tabulated. Metadata can be found in Table 5; (C) marks cultivated sampling localities as opposed to wild (W) sampling localities.

Species	*M. decastroi* (C)	*M. decastroi* (W)	*M. mexicana* (C)	*M. mexicana* (W)	*M. lopezobradorii*	*M. zoquepopolucae*	*M. sinacacolinii*
M. decastroi (C)	–	0.04	0.32	0.25	0.38	0.33	0.55
M. decastroi (W)	0.03	–	0.26	0.19	0.45	0.32	0.54
M. mexicana (C)	0.23	0.21	–	0.1	0.52	0.49	0.68
M. mexicana (W)	0.15	0.12	0.1	–	0.51	0.38	0.64
M. lopezobradorii	0.18	0.20	0.29	0.2	–	0.28	0.54
M. zoquepopolucae	0.19	0.21	0.31	0.21	0.11	–	0.36
M. sinacacolinii	0.23	0.25	0.35	0.26	0.19	0.17	–

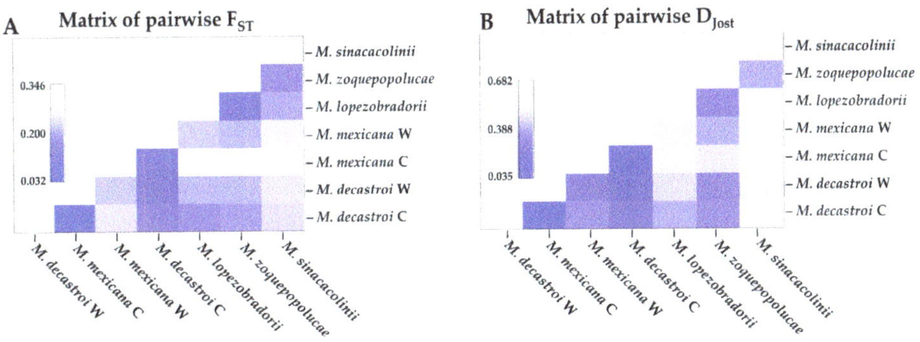

Figure 6. Pairwise F_{ST} and D_{JOST} values between the five *Magnolia* species studied in Puebla and Veracruz, Mexico, visualized as a heatmap. W = wild, C = cultivated. (**A**) Below the diagonal pairwise F_{ST} values are tabulated. (**B**) Pairwise D_{JOST} values are tabulated. Locality metadata can be found in Table 5.

2.1.3. Genetic Diversity

The genetic diversity data obtained are summarized per locality and per species in Table 4. The number of alleles (A) varied between 2.85 and 7.7. The allelic richness, rarified to 12 individuals (A_R (12)), varied between 2.69 and 6.55. The allelic richness, rarified to 24 individuals (A_R (24)), varied between 3.46 and 7.23. The number of private alleles (A_P) varied between 0 and 33. Observed heterozygosity (H_O) varied between 0.45 and 0.72. Expected heterozygosity (H_E) varied between 0.41 and 0.69.

When respecting the current species delimitations, we consequently saw the trend of lowest genetic diversity for localities identified as *M. mexicana* and the highest genetic diversity for localities identified as *M. sinacacolinii*, with the exception of the parameter P. *Magnolia sinacacolinii* had a very high number of private alleles (41) compared to the other species, for which A_P ranged from 0 to 12. Significant inbreeding was detected for FA-12 (*M. decastroi*), LT-2 (*M. lopezobradorii*), FA-10 (*M. mexicana*), and for both localities of *M. sinacacolinii*.

Comparing the genetic diversity between the localities, FA-15 had the lowest genetic diversity values. FA-1, FA-2, FA-10 (all *M. mexicana*), and LT-2 (*M. lopezobradorii*) had similar A_R(12) values (A_R(12) = 3–4) and localities FA-7 (*M. lopezobradorii*), FA-12 and FA-13 (both *M. decastroi*) had higher A_R (12) values (A_R(12) = 4.5–6) and the FA-6 (*M. sinacacolinii*) showed the highest number (A_R(12) = 6.5). Regarding private alleles, FA-6 (*M. sinacacolinii*) had the highest number (A_P = 33) and FA-7 the second highest (*M. lopezobradorii*, A_P = 4). Five out of the 18 localities showed significant signs of inbreeding (Table 4). Locality FA-6 was the locality with the highest, significant inbreeding coefficient (F_{IS} = 0.19).

Table 4. Genetic diversity measures estimated per sampling locality of different *Magnolia* species in Puebla and Veracruz, Mexico. N: sample size; C: number of cultivated individuals; N_G: (mean) number of genotyped individuals. A: (mean) number of alleles. A_R: allelic richness whereby the number of individuals to which the rarefaction is undertaken is between brackets. A_P: (mean) number of private alleles. H_O: (mean) observed heterozygosity. H_E: (mean) expected heterozygosity. F_{IS}: inbreeding coefficient. Significant deviations from Hardy–Weinberg Proportions (HWP): * ($p = 0.05$) are only calculated at the level of locality.

Species	Locality	N	Adults	C	N_G	P	A	A_R(12)	A_R(24)	A_R (14)	A_P	H_O	H_E	F_{IS}
M. decastroi	TOTAL	63	10	30	63	100%	5.7	NA	NA	4.85	8	0.67	0.67	0.01
	FA-12	30	0	30	30	100%	5.62	4.81	5.16	NA	2	0.66	0.67	0.03 *
	FA-13	33	10	0	33	100%	5.08	4.54	4.78	NA	1	0.68	0.64	−0.04
M. lopezobradorii	TOTAL	47	19	0	47	100%	7.54	NA	NA	5.95	12	0.65	0.71	0.09
	FA-5	3	2	0	3	100%	3.08	NA	NA	NA	0	0.62	0.5	0.08
	FA-7	27	3	0	27	100%	6.7	5.7	6.34	NA	4	0.70	0.69	0.00
	FA-8	3	3	0	3	92.31%	2.85	NA	NA	NA	2	0.67	0.49	−0.17
	LT-2	14	11	0	14	100%	4.15	4.09	NA	NA	0	0.56	0.59	0.09 *
M. mexicana	TOTAL	97	42	27	97	100%	5.62	NA	NA	4.57	5	0.52	0.60	0.14
	FA-2	13	5	0	13	100%	3.46	3.46	NA	NA	2	0.47	0.46	0.01
	FA-9	7	6	0	7	92.31%	2.92	NA	NA	NA	0	0.50	0.49	0.06
	FA-10	31	10	0	31	100%	4.46	4.13	4.32	NA	1	0.61	0.62	0.04 *
	FA-11	4	3	0	4	100%	3.00	NA	NA	NA	0	0.58	0.50	−0.01
	FA-1	24	10	8	24	92.31%	3.54	3.3	3.46	NA	0	0.47	0.49	0.06
	FA-15	18	8	18	18	84.62%	2.85	2.69	NA	NA	0	0.45	0.41	−0.09
M. sinacacolinii	TOTAL	32	16	0	30.77	92.31%	8.69	NA	NA	7.08	41	0.54	0.68	0.21
	FA-6	28	12	0	27	92.31%	7.7	6.5	7.23	NA	33	0.5	0.6	0.19 *
	LT-3	4	4	0	3.77	76.92%	3.15	NA	NA	NA	4	0.53	0.50	0.10 *
M. zoquepopolucae	TOTAL	15	9	0	14.69	100%	6.46	NA	8.00	6.39	8	0.68	0.6	−0.01
	FA-3	5	2	0	5	92.31%	3.85	NA	NA	NA	1	0.69	0.58	−0.08
	FA-4	3	2	0	3	84.62%	3.23	NA	NA	NA	2	0.72	0.54	−0.14
	FA-14	4	3	0	4	92.31%	3.7	NA	NA	NA	0	0.65	0.58	0.01
	LT-5	3	2	0	2.692	84.62%	2.92	NA	NA	NA	1	0.67	0.54	0.01

2.2. Assessment of Conservation Status

Based on our data, *M. sinacacolinii* and *M. zoquepopolucae* were assessed as Endangered (EN) and these assessments in the meantime have been published [52,53]. For both species, the data revealed that their current population trend is decreasing, and the main threats were habitat destruction, fragmentation of ecosystems, and extensive change in land use; especially shifting agriculture practices that are widespread among the local people, as well as selective logging along with conversion of forest for pasture (cattle ranching) and construction of transportation/service corridors. Area of occupancy (AOO) and extent of occurrence (EOO) both showed a continuing decline. In terms of diseases, symptoms resembling mosaic virus disease were observed on the leaves of some juvenile individuals in San Andrés Tuxtla and Soteapan (both municipalities are located at extremes of the distribution) of *M. zoquepopolucae* (pers. obs.).

3. Discussion

3.1. Disentangling the Species

Speciation is a continuous process whereby two separately metapopulation lineages acquire more differences or evidence, either morphological, (phylo)genetic, or other lines of evidence [54]. Using SSR data, we gathered molecular evidence to discuss where exactly in the continuum between populations and species our studied *Magnolia* individuals at the collection localities are found. If there is no gene flow occurring anymore between two localities for ample amount of evolutionary time, their populations will become more and more genetically differentiated over time [48,54]. However, a few migrants between such populations can reset such genetic differentiation, hence maintaining the concept of a metapopulation lineage or species [55,56].

Our Structure analysis on the complete dataset (Figure 3A–C) put forward the separation between the North and South, which is supported by the geography of the study area.

The Northern Zone corresponds with the Sierra Madre Oriental and the Huasteca climate zone, characterized by alkaline soils, whereas the Southern ones is located in the isolated volcanic mountain range of Los Tuxtlas with extremely humid climate, characterized by neutral soils, and surrounded by the Gulf coast plain with acid soils. The DAPC plot (Figure 4A,D) supported this pattern: along the primary, horizontal axis (ignoring the secondary, vertical axis), we observe indeed a gap between the Northern and Southern samples. Based on this first Structure result, one could argue that we thus have two main groups or metapopulations and hence two species. As we expected K to be 4 or 5 according to the described species, we observe in Figure 3D and 2E that this genetic structure is clouded by a strong genetic signal splitting of the *M. mexicana* localities into two genetic clusters, following a split between the cultivated localities FA-1 and FA-15 (Table 5) and the wild localities FA-10, FA-11, FA-9, and FA-2. Interestingly, when only analysing the wild localities (Figure 3F–H): excluding sampling localities FA-1, FA-15, and FA-12 (Table 5), the Structure analysis does find five genetic clusters with high confidence, corresponding to the five described species.

Next to the Structure result, the other analyses and parameters put forward the recognition of *M. sinacacolinii* as a separate species within the Southern Zone. Firstly, we saw in the DAPC plots (Figure 4A, D) that the secondary, vertical axis, which also had a significant contribution to the discrimination of the genetic clusters (evidenced by the DA values) clearly separates *M. sinacacolinii* from the cluster that exists of individuals belonging to *M. lopezobradorii* and *M. zoquepopolucae*. The potential null alleles (Table 1) and/or high inbreeding (Table 4) for *M. sinacacolinii* could explain the result of the Structure analysis on the complete dataset not detecting this species as a "significantly" different genetic cluster (Figure 3A), while in the DAPC analysis (Figure 4) this pattern was very clear, as Structure analyses are aimed to find random mating genetic clusters [57]. Secondly, the AMOVA test showed the highest percentage of variation explained by the grouping of the localities according to three groups (73%): *M. decastroi*–*M. mexicana*, *M. lopezobradorii*–*M. zoquepopolucae*, and *M sinacacolinii*, compared to grouping according to the Northern and the Southern Zone (22.13%). Lastly, the remarkably high number of private alleles (Table 4) highlights this species as being a very distinct entity. Our genetic results are strengthened by morphological and ecological data. Morphologically *M. sinacacolinii* is easily discriminated from the other four (alleged) species by tree architecture, leaf texture and pubescence and fruit morphology [47], which is very remarkable given the close geographic proximity of the species to *M. zoquepopolucae* and *M. lopezobradorii*. Ecologically *M. sinacacolinii* only grows within the Los Tuxtlas area at lower elevations in localities protected from the northern winds, in contrast with *M. lopezobradorii* and *M. zoquepopolucae* that occur at much higher elevations, both in localities protected from and exposed to these northern winds. Hence, based on genetic, morphological, and ecological evidence we state that there are (at least) three species in our dataset: one in the Northern Zone, and two in the Southern Zone.

Within the Northern Zone, the sample localities FA-12 and FA-13 identified as *M. decastroi* are a distinct genetic cluster (Figure 3D,E,H), but here the interspecific genetic differentiation (Table 3; Figure 6) between these localities and the other northern localities is less pronounced (F_{ST}: 0.12, D_{JOST}: 0.19) and in the range of the intraspecific genetic differentiation (Table 2; Figure 5). This result is even more striking when taking into consideration that the northernmost wild locality of *M. mexicana* (FA-10) and the most southernly located *M. decastroi* wild locality (FA-13) are the interspecific locality pair in the range of the other intraspecific values (Table 2, Figure 5). Hence, based on our SSR data FA-13 (and FA-12) would be considered a separate population from the other *M. mexicana* populations instead of a separate species. Morphological and physiogeographic data are not in accordance with this result. There is one morphological characteristic that distinguishes *M. decastroi* and *M. mexicana*: the pubescence of the flower bracts [47], which is difficult to observe because the flower bracts are deciduous. During the sampling it was first assumed that FA-12 and FA-13 were the southernmost localities of *M. mexicana*, and only after the SSR results showed that this particular population consistently was a separate genetic cluster (Figure 3) the pubescence was detected on three individuals in the field from which herbarium vouchers were collected. Physiogeographically, within the Northern Zone, the Sierra de Zongolica that holds the populations FA-12 and FA-13 corresponds to the Southern portion of the Sierra Madre Oriental that is isolated from the northernmost *Magnolia* localities by the Trans-Mexican Volcanic Belt and a much more humid climate. As the data are somewhat conflicting, we recommend that for a more final decision on the recognition of the species it would be necessary to genetically compare populations or individuals from the type locality of *M. decastroi* (slightly more to the South, in the Sierra Mazateca, Oaxaca, around 75 km southwards of the sampled localities), as well as other recently found localities in the same region with FA-12 and FA-13. Moreover, there might be more undocumented localities, hence, more explorations can provide more insight in gene flow between localities that are identified as *M. mexicana* and *M. decastroi*. Taken all together, based on the data gathered so far, it can be concluded that the studied wild FA-13 and cultivated FA-12 will most likely be synonymized with *M. mexicana*.

Similarly, the Southern Zone localities FA-3, FA-4, FA-14, and LT-5 identified as *M. zoquepopolucae* had very little genetic support for being treated as a separate species (Table 2, Figure 4) compared to the localities FA-5, FA-7, FA-8, and LT-2 identified as *M. lopezobradorii*. Firstly, in the Structure results of the complete dataset the two species are not retrieved in the K = 4 replicates (Figure 3D), and only retrieved in two of the ten replicates in the K = 5 cluster (Figure 3E). However, in the dataset comprising only the wild individuals the two species are found as two genetic clusters *cf.* the species (Figure 3H). Secondly, the pairwise genetic fixation between the species (F_{ST}: 0.11: Table 3; Figure 5) is in the lower range of that found in intraspecific comparisons (F_{ST}: 0.03–0.22: Table 2; Figure 4) and the pairwise allelic differentiation (D_{JOST}: 0.28) is in the range of the wild *M. decastroi–M. mexicana* pairwise comparison (D_{JOST}: 0.17&0.25) rather than the pairwise comparisons which we are positive to be interspecific (D_{JOST}: 0.46–0.75). These genetic results are confirmed by the absence of a significant morphological distinction found by [47]. The only argument left in favour to discriminate the two described species is the geography: the species are around 55 km apart and are located on different volcanoes. Taken all together, based on the data gathered so far, we conclude that the studied populations can be considered to be two genetically differentiating populations of the same species: *M. zoquepopolucae*.

Comparing the found measures of genetic differentiation of the two species complexes under consideration of being over-splitted (i.e., *M. decastroi–M. mexicana* F_{ST}: 0.12; D_{JOST}: 0.19 and *M. lopezobradorii–M. zoquepopolucae* F_{ST}: 0.11; D_{JOST}: 0.28) with other studies of Neotropical *Magnolia* populations, we found that the F_{ST} values were lower, or well within the lower half of what is considered intraspecific genetic differentiation. For example, in [21] *M. pedrazae* and *M. schiedeana* were reconsidered to be one species with F_{ST} values between the populations ranging between 0.053 and 0.283. In the study on Caribbean Magnolias of [42], intraspecific F_{ST} values ranged between 0.044 and 0.223. In [58], *M. nuevoleonensis* and *M. alejandrae* were proposed to be synonymised with *M. dealbata* with F_{ST} values that ranged between 0.21 and 0.43.

In the debate of species delineation, both in our study and in other SSR studies to date, we must acknowledge that, although the SSR markers are valuable in studying the stochastic processes shaping the populations' genetic diversity, it is only partial evidence. Genes and their adaptation to a specific environment can be what differentiates one species from another, while the structure patterns in neutral DNA still lags behind [21,22,42].

3.2. *Patterns of Gene Flow Between, and Inbreeding within the Wild Sample Localities*

Overall, we observed great variation in genetic differentiation among localities within the alleged five species (Figures 5 and 6, Tables 2 and 3), whereby the (wild) populations of *M. mexicana*, *M. lopezobradorii* and *M. sinacacolinii* showed levels of genetic differentiation of moderate and great genetic differentiation [59] for both the amount of genetic fixation and the amount of allelic differentiation. This means that the past and current gene flow among the sampled localities overall is low. As our current sampling comprised both adults and juveniles, we expect that in the fragmented landscape this result of overall genetic differentiation will become only increasingly pronounced if there are no conservation management actions to reverse this differentiation [22,60].

As gene flow between the populations appears to be limited, more inbreeding is expected. However, significant inbreeding in wild localities was detected only in the wild localities LT-2 (*M. lopezobradorii*), FA-10 (*M. mexicana*), and in the two *M. sinacacolinii* populations (FA-6 and LT-3) (Table 4). In all the other localities there was no signal for inbreeding at hand. This pattern of limited gene flow and little inbreeding is similar to the study of [42] and can be attributed to the evolutionary resilience of the tree habit of strongly overlapping generations [61] and potentially the reproductive biology of the species promoting outcrossing [62]. Only the localities of *M. sinacacolinii* probably have reached a threshold of a too small population size, for the populations to remain genetically resilient to inbreeding.

It is preferable for plants to maintain high levels of genetic variation within their populations, as their sessile nature can lead to the evolution of locally adapted ecotypes [63]. However, in many woody plant and tropical tree species, high levels of genetic variation are reported to be found within populations [49,64–66], while a small fraction of diversity is observed between populations. On the other hand, species with a wide distribution range have greater genetic variability within populations than species with a more restricted distribution [66,67].

3.3. Defining the Conservation Units of the Magnolias of Veracruz

Conservation units, also called management units [68], can either be populations within a species or can even be synonymised with the species as a whole [69,70]. Based on our data we recommend recognising maximally five conservation units *cf.* the genetic clusters retrieved by the Structure analysis on the wild localities (Figure 3H). Each of these genetic clusters represents a collection of localities currently identified as one described species (Figure 3). We recommend to enhance gene flow among the different sample localities within the five genetic clusters (Table 4) and treat each described species as one conservation unit, not divided further in separate managed localities or populations. This because of various reasons: (1) the intensive sampling executed for this study retrieved a low number (N < 10) of *Magnolia* trees at 9 of the 15 sampled wild localities (Table 4); (2) between the localities within the five genetic clusters there is up to great intraspecific genetic differentiation (Figure 5, Table 2); and (3) in 4 out of the 15 sampled wild localities there is significant inbreeding detected (Table 4).

Although geographically distinct and at one point described as two species [26,27], we recommend to recognise only four conservation units. This by managing the eight localities (Table 5) that are now identified to contain individuals of *M. lopezobradorii* and *M. zoquepopolucae* as one. Although the data do clearly support them to be two genetically differentiated populations that are not randomly mating (Figure 3F–H), the collected demographic data of the sampled localities is too precarious (Table 4 and Table 5). We recommend translocating between the two populations because the 15 individuals at the four localities identified as *M. zoquepopolucae* are a relict population of which most are adults (i.e., there is no recruitment) (Table 4). Chances of finding more individuals and/or localities of this genetic cluster are very low (as opposed to the *M. decastroi* genetic cluster, see next paragraph).

Lastly, we recommend further investigation to consider managing the wild population FA-12 identified as *M. decastroi* together with the populations of *M. mexicana*, i.e., the other sampled populations in the Northern Zone (Figure 2). We recommend an additional molecular (conservation genetic, or phylogenetic) study that expands the sampling and includes the type population of *M. decastroi*. In the meanwhile, the wild population FA-13 containing 33 individuals with no significant inbreeding (Table 4) could best be managed separately, as one conservation unit.

3.4. Priorities for Magnolia Conservation in Veracruz

We notice that, overall, *M. mexicana* had the lowest genetic diversity, while *M. sinacacolinii* was the most genetically diverse (Table 4). Interestingly, when taking into account the IUCN status of the species, *M. mexicana* is denoted as VU (Vulnerable), while the other three species (i.e., *M. decastroi*, *M. zoquepopolucae*, and *M. sinacacolinii*) are EN (Endangered) (*M. lopezobradorii* is DD). This illustrates that even though the number of individuals and species is (estimated to be) larger—which are the main parameters for calculating the IUCN Red List status [31,71], the genetic diversity of those species might actually be more alarming, likely due to a century-long collection of flowers for medicinal uses [72].

The interesting pattern observed in the *M. sinacacolinii* FA-6 locality, i.e., high genetic diversity, high number of private alleles, but inbreeding (Table 4), could be due to the population structure, where the adult individuals still harbour much genetic diversity, significantly different from the other studies species (i.e., private alleles); yet, recently a reproductive event of a few more related individuals, or perhaps even geitonogamy, delivered that this genetic variation that is found more in a homozygotic state.

It is hard to state which has priority for conservation as each of the four conservation units have a different set of challenges ahead, which threaten their existence. However, out of the four proposed conservation units, *M. sinacacolinii* is flagged the most, as this species has strong inbreeding detected in both populations and only two (known) localities of which one only has four (known) individuals (Table 4).

3.5. Ex Situ Collections Versus In Situ Populations

The Structure result (Figure 3C,E) was striking, as the division among the *M. mexicana* individuals in two genetic clusters was unexpected. FA-15 is a completely cultivated locality, introduced through the Francisco Javier Clavijero Botanical Garden of the Instituto de Ecología, A.C. (JBFJC). It appears that genetically, FA-1 is a mixture of the wild individuals and the FA-15 individuals. Villagers in the area commented that *M. mexicana* used to be abundant, but its population size has been reduced mainly due to northern winds. The JBFJC data (pers. comm.) indicate that the arboretum individuals were brought from Northern Veracruz, near locality FA-9, which could correspond to extinct populations. This could be confirmed during our field work, as many localities from where individuals of *M. mexicana* had been recorded according to herbarium vouchers, were no longer found, due to deforestation and coffee plantations.

For *M. mexicana*, the allelic diversity [A_R (12)] in the localities that consist of ex situ individuals (FA-1 and FA-15) is lower than compared to the wild populations (Table 4), although the difference is not that pronounced. This stresses the importance of sampling a good variety of mother trees to capture the genetic diversity of the population [61,70]. This is exemplified by the *M. decastroi ex situ* collection FA-12 compared to FA-13: here the allelic richness of the ex situ collection is higher than the in situ (sampled) population with two private alleles (A_P). However, for the FA-12 population, the inbreeding coefficient (F_{IS}) was significant, likely caused by more kinship among the ex situ population.

3.6. Implications for Conservation

Based on the genetic data, we now define three species with certainty and updated their IUCN Red List statuses of two of them accordingly. The importance of adhering to the Red List guidelines lies in the fact that it is the world's leading instrument of its kind. It provides alerts on the state of the world's biodiversity; its applications at the national level enable decision-makers to consider the best options for the conservation of species [31]. The current IUCN Red List assessments still respect the five species delimitation based on the species descriptions.

We propose a preliminary conservation strategy for the four proposed conservation units, based on three main guidelines: diffusion, protection, and propagation. We urge that for efficient conservation, local people are included to achieve an integrated strategy so that they become decision-makers and are involved in the preservation of endangered plant species [73,74]. The first guideline: diffusion is aimed to ensure that knowledge of the species reaches more inhabitants and local organizations in the areas where they are naturally distributed, for example, through information posters and talks to local people. The second guideline: protection aims at ensuring that out of the currently known individuals, no further trees are lost. The third guideline: propagation aims at increasing the genetic diversity and number of individuals at localities. For this purpose, a method of manual propagation by seed has already been developed for species of the *Talauma* section from Cuba that has worked for other *Magnolia* species in various Latin American countries, and agreements have been made with various organizations that also have experience cultivating Magnolias in Mexico [75,76]. It is important that these three guidelines are carried out together and are seen as a process, although depending on each species or conservation unit, it may be necessary to place more emphasis on one than the other. However, in general, it can be stated that the most important actions are diffusion and protection, protecting what is known to remain, while trying to inform the local communities.

Given that there is still adequate genetic diversity present (Table 4), it is proposed to propagate the studied species both in situ and ex situ, which are contemplated in different protection strategies, such as the botanical garden conservation strategy [77], as well as in the Mexican strategy for plant conservation [73]. For the inclusion of species in ex situ collections, arboreta in national botanical gardens can be considered. This is currently executed at the JBFJC which has already successfully propagated other plant groups [78–80]. Because the genetic diversity within the three conservation units appears to be limited by gene flow (Figures 5 and 6, Tables 3 and 4), we propose that translocations between localities can be executed and, preferably, that their outcome is monitored. Although we risk undoing local adaptation and outbreeding [81,82] as we only quantified the populations with neutral genetic data [42], the genetic consequences of fragmentation and subsequent loss of genetic diversity are far greater [70]. As a matter of urgency, two actions are proposed in the Southern Zone: collecting seeds from the small populations and add them to the large populations, while focusing on protecting these larger populations. In the Northern Zone, we suggest focusing on the small populations and reforest them from the other populations to make them larger again. Although we currently only propose to translocate seeds between localities or populations, future work should consider connecting the forest fragments in the landscape, so that gene flow within the conservation units occurs naturally.

Finally, it is proposed to apply all of the above strategies to *M. wendtii* (including genetic evaluations, taxonomy, and conservation), the only species from Veracruz in the section *Talauma* that could not be included in this study. More exploration work is needed, primarily in the border area of Oaxaca and Veracruz, as only a small population has so far been identified in the latter state, but the rainforest in between these two states is extremely fragmented.

4. Materials and Methods

4.1. Study Species, Study Zones, and Sampling

Five species of *Magnolia* sect. *Talauma* were studied (Figure 1), of which four were recently described [26,28] and segregated from *M. mexicana*, namely, *M. decastroi*, *M. lopezobradorii*, *M. sinacacolinii*, and *M. zoquepopolucae*. The only species belonging to this section in Veracruz that was not included was *M. wendtii* from southern Veracruz. *Magnolia wendtii* is only known from an area with an extremely high deforestation degree, and the sample number was too small to include.

Two zones were considered according to the natural distribution of the species: the Northern Zone which corresponds to a part of the Sierra Madre Oriental [83–85] in the states of Puebla and Veracruz, encompassing five natural regions; in contrast, the Southern Zone comprises the natural regions of Los Tuxtlas and Olmeca [11] in Veracruz (Figure 2). Nine field trips were conducted between February 2016 and January 2020, three of them in areas unexplored for *Magnolia* (especially around the municipalities of Xalapa and Coatepec in Veracruz). The entire distribution area of the five species was covered, visiting two states,

12 municipalities, and 31 localities (Table 5), covering areas without previous records for *Magnolia* (central Veracruz) and others that had not been visited since the 1980s (northern Puebla and southern Veracruz).

A total of 254 young leaf samples were collected for molecular analyses (approximately 5 cm^2 of the leaf avoiding veins) and dried in silica gel. These belonged to 18 localities, of which three localities contained cultivated individuals. A first cultivated locality, FA-15 was in situ (private home in Coatepec). The second cultivated locality, FA-1 was in situ (public parks in Xalapa) and the third cultivated locality was ex situ (greenhouse of the Instituto Tecnológico Superior de Zongolica, ITSZ). The first two are identified as *M. mexicana* and the third as *M. decastroi* (Table 5). To correctly identify the individuals at the localities, 145 herbarium vouchers were collected, representing at least one individual at each sample locality (55 collection numbers with their respective duplicates), which will be deposited in the herbaria of the Instituto de Ecología, A.C., Centro Regional del Bajío (IEB), National herbarium of Mexico (MEXU), and Instituto de Ecología, A.C. (XAL).

Table 5. Details of the localities collected for each species for *Magnolia* in the states of Puebla and Veracruz, Mexico. Locality coordinates have been omitted because of conservation concerns but can be obtained from the corresponding author. Voucher specimens will be deposited in the herbaria IEB, MEXU, and XAL (acronyms are according to [86]). [1] Locality was a seedling nursery. N: Sample size.

Species	Locality	State	Municipality	N	Voucher
M. decastroi	FA-12	Veracruz	Zongolica	30	NA [1]
	FA-13			33	Aldaba 224
M. lopezobradorii	FA-5	Veracruz	San Andrés Tuxtla	3	Aldaba 241
	FA-7			27	Aldaba 242
	FA-8			3	Aldaba 245
	LT-2			21	Samain & Martínez 2016–03
M. mexicana	FA-2	Puebla	Cuetzalan del Progreso	13	Aldaba 215
	FA-10		Xicotepec	31	Aldaba 219
	FA-11		Hueytamalco	4	Aldaba 202
	FA-1	Veracruz	Xalapa	24	Aldaba 210
	FA-9		Yecuatla	7	Aldaba 218
	FA-15		Coatepec	18	Aldaba 227
M. sinacacolinii	FA-6	Veracruz	Catemaco	29	Aldaba 235
	LT-3		San Andrés Tuxtla	4	Samain & Martínez 2016-07
M. zoquepopolucae	FA-3	Veracruz	Soteapan	5	Aldaba 239
	FA-4			3	Aldaba 240
	FA-14			4	Aldaba 247
	LT-5			3	Samain & Martínez 2016–12

The aim was to sample 30 individuals from each locality, and when this was not possible, all individuals were collected. In case the number of individuals exceeded 30, individuals were selected randomly covering the whole area. In each locality, tree height, GPS coordinates, habitat description, phenology (if the tree was flowering or fruiting), age class (adult or juvenile, based on whether it had reproductive organs or scars left by them), and DBH (diameter at breast height) were recorded for each individual. In total, 121 individuals were classified as adults and 157 as juveniles.

In order to classify the sampled localities according to the described *Magnolia* species (Table 5) and to obtain a complete overview of the morphological variation of the species involved, 136 herbarium vouchers have been studied. The following herbaria in the states of Puebla and Veracruz were visited: Centro de Investigaciones Tropicales, CHAPA, CIB, CORU, ENCB, FCME, FEZA, HUAP, IEB, IZTA, Estación de Biología Tropical Los Tuxtlas, MEXU, QCA, UAMIZ, XAL, XALU and ZON; complemented by a study of digitally available collections in F and MA (acronyms are according to [86]). Photographs were taken of all specimens and loans were requested from each of these herbaria. The detailed visual evaluation of the phenotypic traits of these specimens, as well as our own collections, have resulted in a list of 35 characters to distinguish the species [47]. Moreover, since the protologue of the recently described species was mainly based on differences in the number of carpels, this feature has been statistically analysed by [47].

4.2. DNA Extraction and PCR

DNA extraction was performed using the CTAB method modified by [87]. A total of 181 existing microsatellites created from *M. cubensis*, *M. dealbata*, *M. lacandonica*, and *M. mayae* were evaluated [42]. DNA quality was assessed using a spectrophotometer NanoDrop 1000 Spectrophotometer (Thermo Fisher Scientific, Waltham, MA, USA). Forward primers were linked to a universal strand to achieve multiplex pooling. The universal tags used (T3, M13, Hill, and Neo) were those recommended by [42].

PCR reactions were prepared under the following conditions: denaturation at 95 °C for 15 min followed by 35 cycles of denaturation at 94 °C for 30 s, annealing at 57 °C for 1.30 min, extension at 72 °C for 1.30 min and final extension at 72 °C for 10 min, extension at 72 °C for 1.30 min and final extension at 72 °C for 10 min. Each Master Mix used for the reaction contained: 0.2 µM forward primer, 0.2 µM reverse primer, DNA (diluted in 1 × TE buffer) and QIAGEN Multiplex PCR Kit. The total PCR volume was 5 µL, of which 1 µL was diluted DNA (1/10), 2 µL of Qia Multiplex PCR master Mix, and 2 µL of primer mix (forward and reverse primers). When testing SSR primers for amplification of a single PCR product, the PCR products were run on 1% agarose gel for 1 h at 115 V and 400 mA. Subsequently, the gel was stained in ethidium bromide for 25 min, placed under UV light and the digital image was captured. Of the SSR primers delivering a single product, fragment analyses were performed by ABI 3130XL fragment analyser (Thermo Fisher Scientific, Waltham, MA, USA) using the GeneScanTM 500 LIZ^{TM} (Thermo Fisher Scientific, Waltham, MA, USA) as a ladder in "singleplexes" and after verification, de novo designed multiplexes. The products were genotyped in Geneious v. 8.1.9 [88].

4.3. Genetic Analysis and Characterisation

The software Convert v. 1 [89], Create v. 1.38 [90] and PGDSpider v. 2.1.1.5 [91] were used to convert both data sets to the different formats used by the other programs mentioned in the following sections.

4.3.1. Null Alleles and Linkage Disequilibrium

Null allele detection was carried out using Microchecker v. 2.2.3 [92], setting the maximum expected size of the allele: 400, confidence interval: 95%, 1000×, not including the alleles with a zero value. To calculate the frequencies of the potential null-alleles we used ML-Null Freq v.1 [93] with 1 000 randomisations.

The linkage disequilibrium (LD) was tested by exact probability test using Genepop v. 4.3 [94,95] applying the following parameters: number of dememorization steps: 10,000, number of batches: 1000, iterations per batch: 50,000; sequential Bonferroni correction was applied to correct the nominal *p*-value of 0.05 for multiple testing [96].

4.3.2. Genetic Structure

Genetic structure analyses were carried out using Structure v. 2.3.4 [97,98]. We decided to use two datasets. In the first one, called the complete dataset, all sampled individuals (both wild and cultivated) were considered. In the second one, only the individuals of the 15 localities with exclusively wild individuals were maintained (Table 4). For both datasets, the number of genetic groups K was set to run from 1 to 30, with 10 replicates each. The upper bound of K = 30 was chosen to allow for substructure within the 15 or 18 sample localities. Each run was performed using 100,000 iterations as burn-in and 100,000 repetitions of the Markov chain Monte Carlo (MCMC) after the burn-in. The ancestry model was the admixture model. The allele frequency model was set to allele frequencies independent, as we expected there to be different species in the dataset, which have been separated for a substantial amount of evolutionary time. After the complete dataset was run, we repeated the Structure analysis for the two main obtained genetic clusters (GC) to further investigate substructure. For these analyses we used the same parameter settings, except that the upper limit of K was set to be twice the number of sampling locations corresponding to each GC obtained, and the allele frequencies set to be correlated. We determined the optimal K of each of the eight structure runs, using the online resource of Structure Harvester [99] whereby we examined the ΔK plots [57] and the mean likelihood plots. Bar plots were visualised using DISTRUCT v. 1.1 [100].

A discriminant principal component analysis (DAPC) in R [101] using the adegenet R package [102] was carried out to further investigate the number and relationship of the genetic clusters following the method proposed by [103] and the recommendation of [104]. For both datasets, 150 Principal Components (PCs) were retained. The number of PCs to retain for the eigenvalues of the principal component analysis (PCA) was determined using cross validation.

Analyses of Molecular Variance (AMOVA) were performed, defining different groups. Firstly, we performed an AMOVA on all the individuals, not defining any groups. Next, AMOVA was run dividing the populations into two, three, four or five groups, according to the Structure and DAPC results and discussion on the number of true species. Significance of AMOVA components was tested with 1 000 permutations in Arlequin v. 3.5.2.2 [105].

To quantify the genetic differentiation among the localities and among the genetic clusters, we ran two analyses using the diveRsity R package [106]. One analysis was run respecting the localities (i.e., 18 "populations") and one was run respecting the five species and separating cultivated and wild localities in *M. decastroi* and *M. mexicana*. Pairwise F_{ST} [50] and D_{JOST} [51] were calculated using 1000 bootstrap replicates.

4.3.3. Genetic Diversity

Allele richness (A_R), number of alleles (A) and inbreeding coefficient (F_{IS}) were calculated in FSTAT v. 2.9.3.2 [107]; expected (H_E) and observed heterozygosity (H_O), population polymorphism (P), and private alleles (A_P) were evaluated in GenAlEx v. 6.5 extension [108] for Microsoft Excel; and deviations from Hardy–Weinberg equilibrium (HWE) were tested in Genepop v. 4.3 [94] with the following parameters: number of dememorization steps: 10,000, number of batches: 200, iterations per batch: 50,000.

4.4. Assessment of Conservation Status

IUCN Red List categories and criteria [71] were applied to define the conservation status of the resulting species (taxonomic changes were not yet formalized [47]). Comments from local people regarding the increase or decrease of individuals were considered, as well as using herbarium records used to search for individuals. Area of Occupancy (AOO) and Extent of Occurrence (EOO) were calculated in GeoCAT [109]. Threats observed in the habitats of each species were also detailed following the IUCN classification scheme [110]. All data collected on distribution, population, use, threats, conservation, etc., were captured in the IUCN Species Information Service (SIS) database to generate the final assessments.

5. Conclusions

In conclusion, we find genetic support for at least three out of the five studied described species, and we propose four main conservation units. The genetic evidence indicates over-splitting is most likely at hand and we recommend a formal taxonomic revision of the species, with emphasis on the *M. decastroi–M. mexicana* complex and the *M. lopezobradorii–M. zoquepopolucae* complex. Localities are exhibiting variable, case-specific levels of genetic differentiation, yet most can be classified as moderate or great, which indicates low (past) gene flow. Five of the 18 studied localities showed genetic signatures of inbreeding. The 13 populations with no signs of inbreeding indicate that random mating was maintained within the majority of populations. *Magnolia sinacacolinii* was flagged as the highest priority conservation unit, given that the species had signs of inbreeding in both its populations and a low number of known localities and individuals. However, the other three conservation units are also in need of urgent conservation management: the *M. mexicana* and *M. decastroi* conservation units had the highest intraspecific genetic differentiation reported and lowest genetic variability and the *M. zoquepopolucae–M. lopezobradorii* conservation unit have 6/8 relict localities that are not exhibiting gene flow between the two sampled volcanoes in Los Tuxtlas. We recommend to genetically characterise more populations of *M. decastroi* to make further tailored decisions on their conservation management. The three evaluated ex situ collections maintain a moderate to good representation of the in situ genetic diversity. The (partly updated) IUCN Red List status for the five studied species are the following: *M. decastroi*, *M. sinacacolinii*, and *M. zoquepopolucae*: Endangered (EN); *M. mexicana*: Vulnerable (VU); *M. lopezobradorii*: Data Deficient (DD). *M. wendtii* is still assessed as Critically Endangered (CR) and we were only able to find a few individuals, hence it is necessary to implement immediate in situ and ex situ conservation actions.

The studied *Magnolia* sect. *Talauma* species of Veracruz and Puebla are hereby put forward as flagship and umbrella species for conservation in the region. In this research, valuable localities were genetically quantified, which can guide conservation management, such as choice of mother trees for collection of seeds for both in situ reforestation by translocations and establishing and genetically enriching ex situ collections. It is proposed to implement a conservation strategy based on three guidelines (diffusion, protection, and propagation) in conjunction with local people, and public and private institutions. The information generated about the genetic diversity of the localities will allow guided reforestation of these species so that the survival of new localities is not affected by low genetic diversity.

Author Contributions: Conceptualization, M.-S.S. and E.M.M.S.; field work, F.A.A.N., E.M.M.S., and M.-S.S., data generation and analyses, E.V. and F.A.A.N.; writing—original draft preparation, F.A.A.N.; writing—review and editing, all authors; visualization, F.A.A.N.; supervision, M.-S.S.; project administration, E.V.; funding acquisition, M.-S.S. and E.V. All authors have read and agreed to the published version of the manuscript.

Funding: This research was funded by the Fondation Franklinia, project number 2019-03; the Instituto de Ecología, A.C., project numbers 20005-30895 and 20006-11337; the Consejo Nacional de Ciencia y Tecnología (Conacyt) through a master grant, CVU 916022, and private funding from the last author.

Institutional Review Board Statement: Not applicable.

Informed Consent Statement: Not applicable.

Data Availability Statement: The data presented in this study are available on request from the corresponding author. The data is not publicly available do its usage in the ongoing study.

Acknowledgments: We are grateful to the Mexican authorities for issuing the collection permit SGPA/DGGFS/1063/18. We thank the following persons who helped us in the field: Adolfo Sebástian Servín Díaz, Álvaro Campos Villanueva, Alejandra Celeste Dolores Fuentes, Ángel Cervantes, Ángel Mena, Braulio Malagar, Brenda Graciela Clemente Genoveva, Carlos Gustavo Iglesias Delfín, Carmen Hernández, Claudia Mateo, Héctor David Jimeno Sevilla, Jesús Eduardo Quintero Melecio, José Hernández, José Luis Abrajam Velasco, Josefina Guerrero, Juan Cázares Hernández, Marisol Alicia Zurita Solís, Martín Mata Rosas, Mateo Reyes Hernández, Moisés Castro Payno, Nataly Cruz Yepez, Noemí Platas, Rafael Hernández, Ricardo Saldaña, Roberto Valdés, Romárico Valdés, Salvador Guzmán Díaz, Samuel Constantino Rayón Méndez, and Saul Herrera Toto. We are also very grateful to the colleagues who helped us in the lab: Andy Vierstraete and Pieter Asselman.

Conflicts of Interest: The authors declare no conflict of interest. The funders had no role in the design of the study; in the collection, analyses, or interpretation of data; in the writing of the manuscript, or in the decision to publish the results.

Appendix A

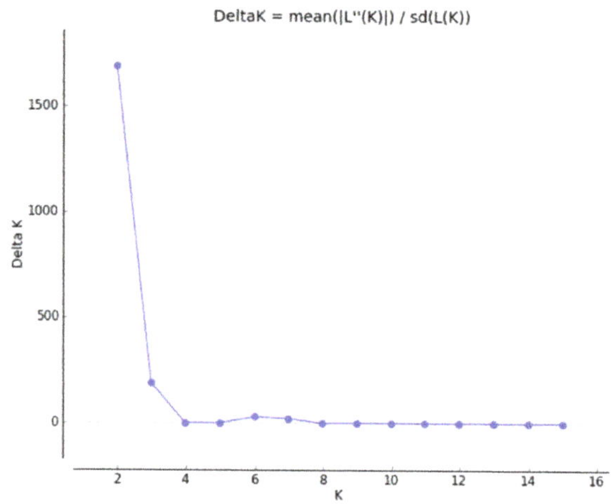

Figure A1. Delta K plot for the Northern Zone.

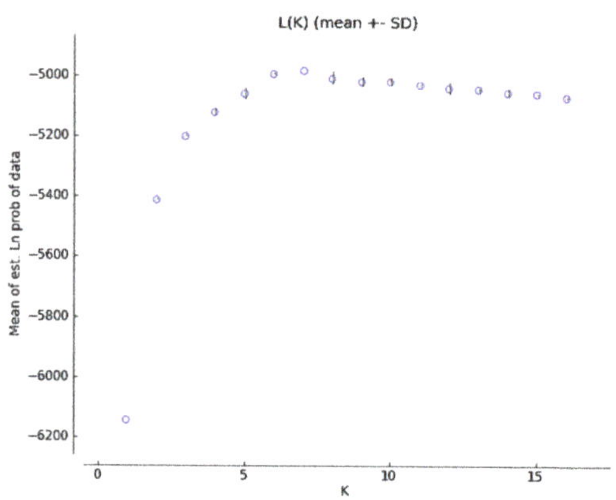

Figure A2. Mean Ln(K) plot for the Northern Zone.

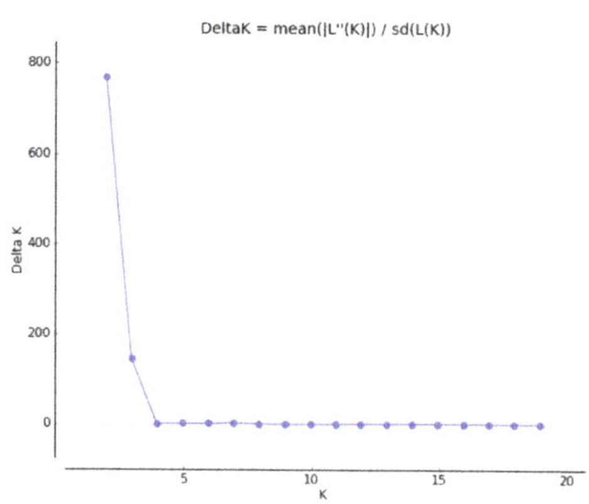

Figure A3. Delta K plot for the Southern Zone.

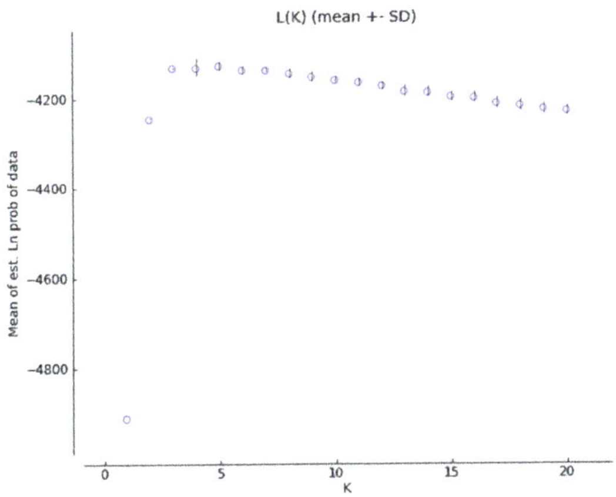

Figure A4. Mean Ln(K) plot for the Southern Zone.

References

1. Ceballos, G.; Ehrlich, P.R.; Barnosky, A.D.; García, A.; Pringle, R.M.; Palmer, T.M. Accelerated Modern Human-induced Species Losses: Entering the Sixth Mass Extinction. *Sci. Adv.* **2015**, *1*, 9–13. [CrossRef] [PubMed]
2. Barnosky, A.D.; Matzke, N.; Tomiya, S.; Wogan, G.O.U.; Swartz, B.; Quental, T.B.; Marshall, C.; McGuire, J.L.; Lindsey, E.L.; Maguire, K.C.; et al. Has the Earth's Sixth Mass Extinction Already Arrived? *Nature* **2011**, *471*, 51–57. [CrossRef] [PubMed]
3. Morrone, J.J. Regionalización Biogeográfica y Evolución Biótica de México: Encrucijada de la Biodiversidad del Nuevo Mundo. *Rev. Mex. Biodivers.* **2019**, *90*, 1–68. [CrossRef]
4. Llorente-Bousquets, J.; Ocegueda, S. Estado del Conocimiento de la Biota. In *Capital Natural de México: Conocimiento Actual de la Biodiversidad*; Comisión Nacional para el Conocimiento y Uso de la Biodiversidad (CONABIO): Tlalpan, Mexico, 2008; Volume I, pp. 283–322, ISBN 978-607-7607-03-8.
5. Villaseñor, J.L. Checklist of the Native Vascular Plants of Mexico. *Rev. Mex. Biodivers.* **2016**, *87*, 559–902. [CrossRef]
6. Lorea-Hernández, F.G.; Durán Espinosa, C.; Gallardo Hernández, C.; Peredo Nava, M. La Diversidad de las Plantas con Semillas de la flora Veracruzana. In *La Biodiversidad en Veracruz: Estudio de Estado*; Comisión Nacional para el Conocimiento y Uso de la Biodiversidad (CONABIO): Tlalpan, Mexico, 2012; Volume II, pp. 201–211.
7. Gómez-Pompa, A.; Krömer, T.; Castro-Cortés, R. *Atlas de la Flora de Veracruz: Un Patrimonio Natural en Peligro*; Comisión del Estado de Veracruz para la Conmemoración de la Independencia Nacional y de la Revolución Mexicana: Tlalnepantla, México, 2010; ISBN 9786079513177.
8. Ellis, E.A.; Martínez Bello, M.; Monroy Ibarra, R. Focos Rojos para la Conservación de la Biodiversidad. In *La Biodiversidad en Veracruz*; Cruz Aragón, A., Ed.; Comisión Nacional para el Conocimiento y Uso de la Biodiversidad (CONABIO): Tlalpan, Mexico, 2011; Volume I, pp. 351–367.
9. Arroyo-Rodríguez, V.; Mandujano, S.; Benítez-Malvido, J. Diversidad y Estructura de la Vegetación en Fragmentos de Selva de Los Tuxtlas. In *Biodiversidad de Veracruz. Estudio de Estado*; Cruz Angón, A., Ed.; Comisión Nacional para el Conocimiento y Uso de la Biodiversidad (CONABIO): Tlalpan, Mexico, 2011; pp. 235–242.
10. Castillo-Hernández, L.A.; Flores-Olvera, H. Floristic Composition of the Cloud Forest of the Bicentenario Reserve, Zongolica, Veracruz, México. *Bot. Sci.* **2017**, *95*, 1–25. [CrossRef]
11. González-Soriano, E.; Dirzo, R.; Vogt, R. *Historia Natural de Los Tuxtlas*; Instituto de Biología, UNAM: Mexico City, Mexico, 1997; ISBN 968-36-5646-3.
12. Vázquez-Torres, M.; Campos-Jimenez, J. *Árboles de la Región de Los Tuxtlas*; Gobierno del Estado de Veracruz, Secretaría de Educación del Estado de Veracruz, Comisión del Estado de Veracruz de Ignacio de la Llave para la Conmemoración del Bicentenario de la Independencia Nacional y del Centenario de la Revolución: Xalapa, Veracruz, 2010; ISBN 9786073300070.
13. Diario Oficial de la Federación. Secretaría de Medio Ambiente, Recursos Naturales y Pesca. 1998. Decreto por el que se Declara Área Natural Protegida con el Carácter de Reserva de la Biósfera de la Región Denominada Los Tuxtlas, Ubicada en los Municipios de Angel R. Cabada, Catemaco, Mecayapan, Pajapan, San Andrés Tuxtla, Santiago Tuxtla, Soteapan y Tatahuicapan de Juárez, en el Estado de Veracruz, con una Superficie Total de 155 122 hectáreas. Available online: https://www.dof.gob.mx/nota_detalle.php?codigo=4900167&fecha=23/11/1998 (accessed on 11 December 2020).
14. Salazar-Arteaga, H. Estructura y Conectividad del Paisaje en la Reserva de la Biósfera Los Tuxtlas. Master's Thesis, Instituto de Ecología, A.C., Xalapa, Mexico, 2015.
15. Rivers, M.; Beech, E.; Murphy, L.; Oldfield, S. *The Red List of Magnoliaceae: Revised and Extended*; Botanic Gardens Conservation International, 2016; ISBN 9781905164646.

16. Cires, E.; De Smet, Y.; Cuesta, C.; Goetghebeur, P.; Sharrock, S.; Gibbs, D.; Oldfield, S.; Kramer, A.; Samain, M.-S. Gap Analyses to Support ex situ Conservation of Genetic Diversity in *Magnolia*, a Flagship Group. *Biodivers. Conserv.* **2013**, *22*, 567–590. [CrossRef]
17. Simberloff, D. Flagships, Umbrellas, and Keystones: Is Single-species Management Passe in the Landscape Era? *Biol. Conserv.* **1998**, *83*, 247–257. [CrossRef]
18. Azuma, H.; García-Franco, J.G.; Rico-Gray, V.; Thien, L.B. Molecular Phylogeny of the Magnoliaceae: The Biogeography of Tropical and Temperate Disjunctions. *Am. J. Bot.* **2001**, *88*, 2275–2285. [CrossRef]
19. Wang, Y.-B.; Liu, B.-B.; Nie, Z.-L.; Chen, H.-F.; Chen, F.-J.; Figlar, R.B.; Wen, J. Major Clades and a Revised Classification of *Magnolia* and Magnoliaceae Based on Whole Plastid Genome Sequences via Genome Skimming. *J. Syst. Evol.* **2020**, *58*, 673–695. [CrossRef]
20. Nie, Z.L.; Wen, J.; Azuma, H.; Qiu, Y.L.; Sun, H.; Meng, Y.; Sun, W.B.; Zimmer, E.A. Phylogenetic and Biogeographic Complexity of Magnoliaceae in the Northern Hemisphere Inferred from Three Nuclear Data Sets. *Mol. Phylogenet. Evol.* **2008**, *48*, 1027–1040. [CrossRef]
21. Rico, Y.; Gutiérrez Becerril, B.A. Species Delimitation and Genetic Structure of Two Endemic *Magnolia* species (section *Magnolia*; Magnoliaceae) in Mexico. *Genetica* **2019**, *147*, 57–68. [CrossRef] [PubMed]
22. Hernández, M.; Palmarola, A.; Veltjen, E.; Asselman, P.; Testé, E.; Larridon, I.; Samain, M.S.; González-Torres, L.R. Population Structure and Genetic Diversity of *Magnolia cubensis* subsp. *acunae* (Magnoliaceae): Effects of Habitat Fragmentation and Implications for Conservation. *Oryx* **2020**. [CrossRef]
23. Castillo, R.E.; Encarnación, Y.; Peguero, B.; Clase, T.; Gratzfeld, J. *Plan de Acción de Conservación integrada de las Magnolias (Magnoliaceae) Amenazadas de República Dominicana—Magnolia domingensis, M. hamorii, y M. pallescens*; Fundación PROGRESSIO y Jardín Botánico Nacional Dr. Rafael M. Moscoso: Santo Domingo, Dominican Republic, 2018.
24. *BGCI Global Survey of ex situ Magnoliaceae Collections*; Botanic Gardens Conservation International: London, UK, 2008.
25. Figlar, R.B.; Nooteboom, H.P. Notes on Magnoliaceae IV. *Blumea J. Plant Taxon. Plant Geogr.* **2004**, *49*, 87–100. [CrossRef]
26. Vázquez-García, J.A.; De Castro-Arce, E.; Múñiz-Castro, M.Á.; Cházaro-Basáñez, M.d.J. *Magnolia zoquepopolucae* (subsection *Talauma*, Magnoliaceae), a New Species from Sierra de Santa Marta, Veracruz, Mexico. *Phytotaxa* **2012**, *57*, 51–55. [CrossRef]
27. Vázquez-García, J.A.; Múñiz-Castro, M.Á.; Arroyo, F.; Pérez-Castañeda, Á.J.; Serna, M.; Cuevas Guzmán, R.; Domínguez-Yescas, R.; De Castro-Arce, E.; Gurrola-Díaz, C.M. Novelties in Neotropical *Magnolia* and an Addendum Proposal to the IUCN Red List of Magnoliaceae. In *Recursos forestales en el Occidente de México: Diversidad, Manejo, Producción, Aprovechamiento y Conservación*; Salcedo-Pérez, E., Hernández-Álvarez, E., Vázquez-García, J.A., Escoto-García, T., Días-Echavarría, N., Eds.; Universidad de Guadalajara: Guadalajara, Mexico, 2013; pp. 458–496, ISBN 9786078072569.
28. Vázquez-García, J.A.; Múñiz-Castro, M.Á.; De Castro-Arce, E.; Murguía Araiza, R.; Nuño Rubio, A.T.; Cházaro-Basáñez, M.d.J. Twenty New Neotropical Tree Species of *Magnolia* (Magnoliaceae). In *Recursos Forestales en el Occidente de México: Diversidad, Manejo, Producción, Aprovechamiento y Conservación*; Salcedo-Pérez, E., Hernández-Álvarez, E., Vázquez-García, J.A., Escoto-García, T., Días-Echavarría, N., Eds.; Universidad de Guadalajara: Guadalajara, México, 2012; pp. 91–130.
29. Pennington, T.D.; Sarukhán, J. *Arboles Tropicales de México. Manual para la Identificación de las Principales Especies*; Fondo de Cultura Económica: Mexico City, Mexico, 2005; ISBN 9789703216437.
30. Vázquez-García, J.A.; Tribouillier-Navas, E.; Archila, F.; Véliz-Pérez, M.E. A Conspectus of *Magnolia* (Magnoliaceae) in Guatemala: Novelties and Conservation. *Phytotaxa* **2019**, *427*, 221–238. [CrossRef]
31. IUCN Red List of Threatened Species. Version 2020-3. Available online: https://www.iucnredlist.org/ (accessed on 8 February 2021).
32. Vázquez-García, J.A.; Gómez-Domínguez, H.; López-Cruz, A. *Magnolia perezfarrerae*, a New Species and a Key to Mexican Species of *Magnolia* (section *Talauma*, subsection *Talauma*, Magnoliaceae). *Bot. Sci.* **2013**, *91*, 417–425. [CrossRef]
33. Domínguez-Yescas, R.; Vázquez-García, J.A. Flower of the heart, *Magnolia yajlachhi* (subsect. *Talauma*, Magnoliaceae), a New Species of Ceremonial, Medicinal, Conservation and Nurse Tree Relevance in the Zapotec Culture, Sierra Norte de Oaxaca, Mexico. *Phytotaxa* **2019**, *393*, 21–34. [CrossRef]
34. McVaugh, R. *Botanical Results of the Sessé & Mociño Expedition. VII. A Guide to Relevant Scientific Names of Plants*; Hunt Institute for Botanical Documentation; Carnegie Mellon University: Pittsburgh, PA, USA, 2000.
35. Sutherland, C.N. Material tipo de la colección de Sessé y Mociño en el Real Jardín Botánico de Madrid. *An. del Jardín Botánico Madrid* **1997**, *55*, 375–418.
36. Germán-Ramírez, M.T. Iconografía Inédita de la Flora Mexicana, Obra de Sessé y Mociño, en el Acervo Histórico del Herbario Nacional de México, MEXU. *Bot. Sci.* **1994**, *97*, 67. [CrossRef]
37. McVaugh, R. El Itinerario y las Colectas de Sessé y Mociño en México. *Bot. Sci.* **1969**, *30*, 137–142. [CrossRef]
38. Hernández-Cerda, M.E. Magnoliaceae. *Flora de Veracruz* **1988**, *14*, 1–38.
39. Duminil, J.; Di Michele, M. Plant Species Delimitation: A Comparison of Morphological and Molecular Markers. *Plant Biosyst.* **2009**, *143*, 528–542. [CrossRef]
40. González, D. El Uso de Secuencias Génicas para Estudios Taxonómicos. *Bot. Sci.* **1997**, *157*, 137. [CrossRef]
41. Rentaría-Alcántara, M. Breve Revisión de los Marcadores Moleculares. In *Ecología molecular*; Eguiarte, L.E., Souza, V., Aguirre, X., Eds.; Secretaría de Medio Ambiente y Recursos Naturales de los Estados Unidos Mexicanos (SEMARNAT): Mexico City, Mexico, 2007; pp. 541–566.

42. Veltjen, E.; Asselman, P.; Hernández Rodríguez, M.; Palmarola Bejerano, A.; Testé Lozano, E.; González Torres, L.R.; Goetghebeur, P.; Larridon, I.; Samain, M.-S. Genetic Patterns in Neotropical Magnolias (Magnoliaceae) Using *de novo* Developed Microsatellite Markers. *Heredity (Edinb.)* **2019**, *122*, 485–500. [CrossRef]
43. Medrano-Hernández, J.M.; Rodríguez de la O, J.L.; Reyes-Trejo, B.; Peña-Ortega, M.G. Molecular Characterization Using ISSR Primers of *Magnolia mexicana* DC. from Two Regions in Zongolica, Veracruz, Mexico. *Rev. Chapingo Ser. Ciencias For. y del Ambient.* **2017**, *23*, 427–436. [CrossRef]
44. Arteaga-Rios, L.D.; Mejía-Carraza, J.; Piña-Escutia, J.L.; González-Díaz, J.G.; Rivera-Colín, A. Comparación Molecular y Morfológica entre Ejemplares de *Magnolia mexicana* "Yoloxóchitl" del Estado de México y Veracruz. *Polibotánica* **2020**, *48*, 107–124. [CrossRef]
45. Li, J.; Conran, J.G. Phylogenetic Relationships in Magnoliaceae subfam. Magnolioideae: A Morphological Cladistic Analysis. *Plant Syst. Evol.* **2003**, *242*, 33–47. [CrossRef]
46. Kim, S.; Suh, Y. Phylogeny of Magnoliaceae Based on Ten Chloroplast DNA Regions. *J. Plant Biol.* **2013**, *56*, 290–305. [CrossRef]
47. Aldaba Núñez, F.A. Sistemática, Diversidad Genética y Conservación de *Magnolia* en Veracruz, México. Master's Thesis, Instituto de Ecología, A.C., México, Xalapa, Mexico, 2020.
48. Waples, R.S. Separating the Wheat from the Chaff: Patterns of Genetic Differentiation in High Gene Flow Species. *J. Hered.* **1998**, *89*, 438–450. [CrossRef]
49. Hamrick, J.L.; Godt, M.J.W.; Sherman-Broyles, S.L. Factors Influencing Levels of Genetic Diversity in Woody Plant Species. *New For.* **1992**, *6*, 95–124. [CrossRef]
50. Weir, B.S.; Cockerham, C.C. Estimating F-statistics for the Analysis of Population Structure. *Evolution* **1984**, *38*, 1358–1370. [CrossRef] [PubMed]
51. Jost, L. GST and its Relatives do not Measure Differentiation. *Mol. Ecol.* **2008**, *17*, 4015–4026. [CrossRef]
52. Aldaba Núñez, F.A.; Fuentes, A.C.D.; Martínez Salas, E.; Samain, M.-S. *Magnolia sinacacolinii*; The IUCN Red List of Threatened Species 2020: Gland, Switzerland, 2020. e.T67513605A176997455. [CrossRef]
53. Aldaba Núñez, F.A.; Fuentes, A.C.D.; Martínez Salas, E.; Samain, M.-S. *Magnolia zoquepopolucae*; The IUCN Red List of Threatened Species 2020: Gland, Switzerland, 2020. e.T67513642A176997582. [CrossRef]
54. de Queiroz, K. The General Lineage Concept of species, Species Criteria, and the Process of Speciation and Terminological Recommendations. *Endless Forms Species Speciat.* **1998**, 57–75.
55. Lacy, R.C. Loss of Genetic Diversity from Managed Populations: Interacting Effects of Drift, Mutation, Immigration, Selection, and Population Subdivision. *Conserv. Biol.* **1987**, *1*, 143–158. [CrossRef]
56. Wang, J. On the Measurements of Genetic Differentiation among Populations. *Genet. Res. (Camb.)* **2012**, *94*, 275–289. [CrossRef]
57. Evanno, G.; Regnaut, S.; Goudet, J. Detecting the Number of Clusters of Individuals Using the Software STRUCTURE: A Simulation Study. *Mol. Ecol.* **2005**, *14*, 2611–2620. [CrossRef] [PubMed]
58. Chávez-Cortázar, A.; Oyama, K.; Zavala-Ochoa, M.; Mata-Rosas, M.; Veltjen, E.; Samain, M.-S.; Quesada, A.M. Conservation Genetics of Relict Tropical Species of *Magnolia* (section *Macrophylla*). *Conserv. Genet.* **2021**, *22*, 259–273. [CrossRef]
59. Hartl, D.; Clark, A. *Principles of Population Genetics*; Sinauer Associates, Inc.: Sunderland, MA, USA, 1997.
60. Aparicio, A.; Hampe, A.; Fernández-Carrillo, L.; Albaladejo, R.G. Fragmentation and Comparative Genetic Structure of Four Mediterranean Woody Species: Complex interactions between life history traits and the landscape context. *Divers. Distrib.* **2012**, *18*, 226–235. [CrossRef]
61. Petit, R.J.; Hampe, A. Some Evolutionary Consequences of Being a Tree. *Annu. Rev. Ecol. Evol. Syst.* **2006**, *37*, 187–214. [CrossRef]
62. Tamaki, I.; Setsuko, S.; Tomaru, N. Estimation of Outcrossing Rates at Hierarchical Levels of Fruits, Individuals, Populations and Species in *Magnolia stellata*. *Heredity (Edinb.)* **2009**, *102*, 381–388. [CrossRef]
63. Antonovics, J. The Effects of a Heterogeneous Environment on the Genetics of Natural Populations: The Realization That Environments Differ Has Had a Profound Effect on Our Views of the Origin and Role of Genetic Variability in Populations. *Am. Sci.* **1971**, *59*, 593–599. [PubMed]
64. Hamrick, J.L.; Linhart, Y.B.; Mitton, J.B. Relationships between Life History Characteristics and Electrophoretically Detectable Genetic Variation in Plants. *Annu. Rev. Ecol. Syst.* **1979**, *10*, 173–200. [CrossRef]
65. Hamrick, J.L.; Murawski, D.A. The Breeding Structure of Tropical Tree Populations. *Plant Species Biol.* **1990**, *5*, 157–165. [CrossRef]
66. Loveless, M.D. Isozyme Variation in Tropical Trees: Patterns of Genetic Organization. *New For.* **1992**, *6*, 67–94. [CrossRef]
67. López-Barrera, G. Diversidad y Estructura Genética de *Brosimum alicastrum*. Master's Thesis, Universidad Nacional Autónoma de México, Mexico City, Mexico, 2018.
68. Coates, D.J.; Byrne, M.; Moritz, C. Genetic Diversity and Conservation Units: Dealing with the Species-population Continuum in the Age of Genomics. *Front. Ecol. Evol.* **2018**, *6*, 1–13. [CrossRef]
69. Taylor, B.L.; Dizon, A.E. First Policy then Science: Why a Management Unit Based Solely on Genetic Criteria Cannot Work. *Mol. Ecol.* **1999**, *8*, 11–16. [CrossRef] [PubMed]
70. Veltjen, E. The Caribbean *Magnolia* Species (Magnoliaceae): Assessment of the Genetic Diversity and the Underlying Evolutionary History. Ph.D. Thesis, Ghent University, Ghent, Belgium, 2020.
71. IUCN. *Documentation Standards and Consistency Checks for IUCN Red List Assessments and Species Accounts. Version 2*; IUCN: Cambridge, UK, 2013.

72. Reyes-Chilpa, R.; Guzmán-Gutiérrez, S.L.; Campos-Lara, M.; Bejar, E.; Osuna-Fernández, H.R.; Hernández-Pasteur, G. On the First Book of Medicinal Plants Written in the American Continent: The *Libellus Medicinalibus Indorum Herbis* from Mexico, 1552. A review. *Bol. Latinoam. y del Caribe Plantas Med. y Aromat.* **2020**, *20*, 1–27. [CrossRef]
73. CONABIO. *Estrategia Mexicana para la Conservación Vegetal 2012-2030*; Comisión Nacional para el Conocimiento y Uso de la Biodiversidad (CONABIO): Mexico City, Mexico, 2012; ISBN 9786077607687.
74. CONABIO-CONANP-SEMARNAT. *Estrategia Mexicana para la Conservación Vegetal: Objetivos y Metas*; CONABIO-CONANP-SEMARNAT: Mexico City, Mexico, 2008; ISBN 9786077607687.
75. Hernández, M.; Testé, E.; Palmarola, A.; Albelo, N.; Moscoso, J.L.; Valle, O.; González-Torres, L.R. Conservación del «Mantequero» en Guamuhaya, a Diez Años de los Primeros Pasos. *Bissea* **2020**, *14*, 2.
76. Palmarola, A.; Hernández, M. Proyecto de Conservación de Magnolias Cubanas. *Bissea* **2016**, *10*, 148. [CrossRef]
77. Rodríguez-Acosta, M. *Estrategia para los Jardines Botánicos Mexicanos*; Asociación Mexicana de Jardines Botánicos, A. C.: Mexico City, Mexico, 2000; ISBN 968-7369-04-3.
78. Vovides, A.P. The Mexican Living Cycad Collection at the Jardín Botánico Francisco Javier Clavijero, Xalapa. In Proceedings of the 4th Global Botanic Gardens Congress, Dublin, Irish Republic, June 2010; pp. 1–9.
79. Palacios-Ríos, M. Cultivation of Tropical Pteridophytes in the Jardín Botánico "Francisco J. Clavijero" of Xalapa, Veracruz, México. In *Fern Horticulture: Past, Present and Future Perspectives*; Thomas, B.A., Ide, J., Eds.; Fern Gazette-Intercept: Andover, MA, USA, 1992; pp. 263–266.
80. Salazar, V.M.; Mata, M. Micropropagación y Conservación de Orquídeas Mexicanas en el Jardín Botánico Clavijero. *Lankesteriana* **2003**, *7*, 151–153. [CrossRef]
81. Edmands, S.; Timmerman, C.C. Modeling Factors Affecting the Severity of Outbreeding Depression. *Conserv. Biol.* **2003**, *17*, 883–892. [CrossRef]
82. Hufford, K.M.; Krauss, S.L.; Veneklaas, E.J. Inbreeding and Outbreeding Depression in *Stylidium hispidum*: Implications for Mixing Seed Sources for Ecological Restoration. *Ecol. Evol.* **2012**, *2*, 2262–2273. [CrossRef]
83. Peakbagger Sierra Madre Oriental. Available online: https://www.peakbagger.com/range.aspx?rid=172 (accessed on 1 August 2020).
84. Peakbagger Sierra Madre del Sur. Available online: https://www.peakbagger.com/range.aspx?rid=175 (accessed on 1 August 2020).
85. Stacy, L. *Mexico and The United States*; Marshall Cavendish: New York, NY, USA, 2002.
86. Thiers, B. Index Herbariorum: A Global Directory of Public Herbaria and Associated Staff. Available online: http://sweetgum.nybg.org/science/ih/ (accessed on 18 December 2018).
87. Larridon, I.; Walter, H.E.; Guerrero, P.C.; Duarte, M.; Cisternas, M.A.; Hernández, C.P.; Bauters, K.; Asselman, P.; Goetghebeur, P.; Samain, M.-S. An Integrative Approach to Understanding the Evolution and Diversity of *Copiapoa* (Cactaceae), a Threatened Endemic Chilean Genus from the Atacama Desert. *Am. J. Bot.* **2015**, *102*, 1506–1520. [CrossRef] [PubMed]
88. Kearse, M.; Moir, R.; Wilson, A.; Stones-Havas, S.; Cheung, M.; Sturrock, S.; Buxton, S.; Cooper, A.; Markowitz, S.; Duran, C.; et al. Geneious Basic: An Integrated and Extendable Desktop Software Platform for the Organization and Analysis of Sequence Data. *Bioinformatics* **2012**, *28*, 1647–1649. [CrossRef]
89. Glaubitz, J.C. CONVERT: A User-friendly Program to Reformat Diploid Genotypic Data for Commonly Used Population Genetic Software Packages. *Mol. Ecol. Notes* **2004**, *4*, 309–310. [CrossRef]
90. Coombs, J.A.; Letcher, B.H.; Nislow, K.H. Create: A Software to Create Input Files from Diploid Genotypic Data for 52 Genetic Software Programs. *Mol. Ecol. Resour.* **2008**, *8*, 578–580. [CrossRef]
91. Lischer, H.E.L.; Excoffier, L. PGDSpider: An Automated Data Conversion Tool for Connecting Population Genetics and Genomics Programs. *Bioinformatics* **2012**, *28*, 298–299. [CrossRef] [PubMed]
92. Van Oosterhout, C.; Hutchinson, W.F.; Wills, D.P.M.; Shipley, P. MICRO-CHECKER: Software for Identifying and Correcting Genotyping Errors in Microsatellite Data. *Mol. Ecol. Notes* **2004**, *4*, 535–538. [CrossRef]
93. Kalinowski, S.T.; Taper, M.L. Maximum Likelihood Estimation of the Frequency of Null Alleles at Microsatellite Loci. *Conserv. Genet.* **2006**, *7*, 991–995. [CrossRef]
94. Raymond, M.; Rousset, F. GENEPOP (Version 1.2): Population Genetics Software for Exact Tests and Ecumenicism. *J. Hered.* **1995**, *86*, 248–249. [CrossRef]
95. Rousset, F. GENEPOP'007: A Complete Re-implementation of the GENEPOP Software for Windows and Linux. *Mol. Ecol. Resour.* **2008**, *8*, 103–106. [CrossRef] [PubMed]
96. Rice, W.R. Analyzing Tables of Statistical Tests. *Evolution (N. Y.)* **1989**, *43*, 223–225. [CrossRef]
97. Pritchard, J.K.; Stephens, M.; Donnelly, P. Inference of Population Structure Using Multilocus Genotype Data. *Genetics* **2000**, *155*, 945–959. [CrossRef] [PubMed]
98. Hubisz, M.J.; Falush, D.; Stephens, M.; Pritchard, J.K. Inferring weak population structure with the assistance of Sample Group Information. *Mol. Ecol. Resour.* **2009**, *9*, 1322–1332. [CrossRef] [PubMed]
99. Earl, D.A.; von Holdt, B.M. STRUCTURE HARVESTER: A Website and Program for Visualizing STRUCTURE Output and Implementing the Evanno Method. *Conserv. Genet. Resour.* **2012**, *4*, 359–361. [CrossRef]
100. Rosenberg, N.A. DISTRUCT: A Program for the Graphical Display of Population Structure. *Mol. Ecol. Notes* **2004**, *4*, 137–138. [CrossRef]

101. R Core Development Team. *R: A Language and Environment for statistical Computing*; R Foundation for Statistical Computing: Vienna, Austria, 2008; Volume 2, ISBN 3-900051-07-0.
102. Jombart, T.; Devillard, S.; Dufour, A.B.; Pontier, D. Revealing Cryptic Spatial Patterns in Genetic Variability by a New Multivariate Method. *Heredity (Edinb.)* **2008**, *101*, 92–103. [CrossRef]
103. Jombart, T.; Devillard, S.; Balloux, F. Discriminant Analysis of Principal Components: A New Method for the Analysis of Genetically Structured Populations. *BMC Genet.* **2010**, *11*, 94. [CrossRef]
104. Miller, J.M.; Cullingham, C.I.; Peery, R.M. The influence of *a priori* grouping on inference of Genetic Clusters: Simulation Study and literature Review of the DAPC Method. *Heredity (Edinb.)* **2020**, *125*, 269–280. [CrossRef]
105. Excoffier, L.; Lischer, H.E.L. Arlequin Suite ver 3.5: A New Series of Programs to perform Population Genetics Analyses under Linux and Windows. *Mol. Ecol. Resour.* **2010**, *10*, 564–567. [CrossRef]
106. Keenan, K.; Mcginnity, P.; Cross, T.F.; Crozier, W.W.; Prodöhl, P.A. DiveRsity: An R Package for the Estimation and Exploration of Population Genetics Parameters and Their Associated Errors. *Methods Ecol. Evol.* **2013**, *4*, 782–788. [CrossRef]
107. Goudet, J. FSTAT (Version 1.2): A Computer Program to Calculate F-Statistics. *J. Hered.* **1995**, *86*, 485–486. [CrossRef]
108. Peakall, R.; Smouse, P.E. GenALEx 6.5: Genetic Analysis in Excel. Population Genetic Software for teaching and Research -An Update. *Bioinformatics* **2012**, *28*, 2537–2539. [CrossRef] [PubMed]
109. Bachman, S.; Moat, J.; Hill, A.W.; de laTorre, J.; Scott, B. Supporting Red List Threat Assessments with GeoCAT: Geospatial Conservation Assessment Tool. *Zookeys* **2011**, *150*, 117–126. [CrossRef] [PubMed]
110. IUCN Threats Classification Scheme (Version 3.2). Available online: https://www.iucnredlist.org/resources/threat-classification-scheme (accessed on 2 June 2020).

plants

MDPI

Article

Towards Conservation of the Remarkably High Number of Daisy Trees (Asteraceae) in Mexico

Rosario Redonda-Martínez [1,*], Patricio Pliscoff [2,3,4], Andrés Moreira-Muñoz [5], Esteban Manuel Martínez Salas [6] and Marie-Stéphanie Samain [1]

1 Instituto de Ecología, A.C., Red de Diversidad Biológica del Occidente Mexicano, Pátzcuaro 61600, Michoacán, Mexico; mariestephanie.samain@gmail.com
2 Departamento de Ecología, Facultad de Ciencias Biológicas, Pontificia Universidad Católica de Chile, Alameda 340, Santiago 8331150, Chile; pliscoff@uc.cl
3 Instituto de Geografía, Facultad de Historia, Pontificia Universidad Católica de Chile, Geografía y Ciencia Política, Avenida Vicuña Mackenna 4860, Macul, Santiago 7820436, Chile
4 Center of Applied Ecology and Sustainability (CAPES), Pontificia Universidad Católica de Chile, Santiago 8331150, Chile
5 Instituto de Geografía, Facultad de Ciencias del Mar y Geografía, Pontificia Universidad Católica de Valparaíso, Avenida Brasil 2241, Valparaíso 2340000, Chile; andres.moreira@pucv.cl
6 Departamento de Botánica, Instituto de Biología, Universidad Nacional Autónoma de México, Herbario Nacional de México, Mexico City 04510, Mexico; ems@ib.unam.mx
* Correspondence: r.redonda.martinez@gmail.com

Abstract: Mexico is floristically the fourth most species-rich country in the world, and Asteraceae is the most diverse vascular plant family in this country. The species exhibits a wide range of growth forms, but the tree-like habit, appropriately named daisy trees, is heavily underestimated, even though slightly different tree definitions are handled. Very little is known about their precise species number or conservation status in Mexico, so we update here the list of known Mexican daisy tree species, summarize their very diverse uses, present a general panorama of their present and future distribution, and discuss their conservation status. A bibliographic review and herbarium study were carried out, carefully curated taxonomical ocurrence maps were prepared for each species, and a climatic suitability modelling approach was used to characterise the spatial patterns of Mexican Asteraceae trees. With 149 daisy tree species, the country ranks second at a global level; within the country, their greatest diversity is found in central and western Mexico. A decrease in diversity is estimated in areas that currently host the highest species richness, whereas the hotspot regions are estimated to show an increase in species diversity, so climate change is not a threat to all Mexican daisy tree species.

Keywords: biogeographic provinces; Compositae; endemism; nectariferous plants; ornamental species; protected areas; species distribution modelling; traditional medicine

Citation: Redonda-Martínez, R.; Pliscoff, P.; Moreira-Muñoz, A.; Martínez Salas, E.M.; Samain, M.-S. Towards Conservation of the Remarkably High Number of Daisy Trees (Asteraceae) in Mexico. *Plants* **2021**, *10*, 534. https://doi.org/10.3390/plants10030534

Academic Editor: Gregor Kozlowski

Received: 9 February 2021
Accepted: 7 March 2021
Published: 12 March 2021

1. Introduction

With more than 23,000 vascular plant species, Mexico is floristically the fourth most species-rich country in the world, after Brazil, China, and Colombia [1,2]; 11,600 of the Mexican plant species are endemic [1]. Asteraceae is the most diverse family of vascular plants in Mexico, with 417 genera and 3050 native species, of which 1988 are endemic, representing about 65% of the family in Mexico [3]. In the flora of North America, 418 genera and 2413 species of this family are registered [4]; in Brazil, 310 genera (64 endemic, 17 exotic) and 2113 species (42 introduced) are registered [5]; in the flora of China, 248 genera (18 endemic, 49 introduced) and 2336 species (1145 endemic, 109 introduced) are registered [6]; in Colombia, 258 genera and 1302 species are registered [7]; while in Ecuador, 217 genera and 918 species (360 endemic) are registered [8]. Its representatives are found practically everywhere on the planet, except in Antarctica and polar regions with permanent ice [9,10].

In Mexico, they are distributed from sea level in coastal dunes to the alpine grasslands of mountainous regions at more than 4000 m elevation [3].

Asteraceae is characterized by its inflorescences called head or capitulum, that simulate a flower that contains numerous florets with unilocular, bicarpellate, inferior ovary, and syngeneic stamens. Their diversity and distribution are due, amongst others, to effective dispersal mechanisms of its fruits by the pappus, the modified calyx (in some cases, the apex of the cypsela lengthens, forming a hook-like structure, as, e.g., in dandelion, *Taraxacum officinale*, and several Mutisieae, which functions as an aerodynamic structure similar to a propeller to disperse the fruits with the help of the wind), and short life cycles in most of its members. The latter characteristic allows them to colonize disturbed environments or sites where the original vegetation has been removed, thus being essential elements of secondary vegetation, ruderal and weeds in various crops [10,11].

Members of the Asteraceae exhibit a wide range of growth forms, including short-lived annual or perennial herbs, subshrubs, shrubs, trees, and even climbing, epiphytic and (sub)aquatic plants [10]. In the particular case of trees, there are some studies dealing with Mexican species. Standley [12] was one of the first to document the diversity of woody species in Mexico; in the case of Asteraceae, he recorded mainly shrubs and only 14 tree or tree-like species. Other studies where tree species have been included correspond to the taxonomic reviews of some tribes [13,14], genera [15–21], or sections of these [22,23]. The most recent publications that include arborescent Asteraceae [24–26] consider only 36, 62, and 41 species, respectively. Even though at least some of these discrepancies might be due to different tree definitions applied, we consider that a considerable cause is what we might call "Asteraceae tree blindness"; most people, including botanists, picture representatives of this family as annual herbs or short-lived perennials, contrasting with the surprisingly high number of woody species it contains. These trees with their beautiful and striking inflorescences are appropriately called daisy trees (Figure 1).

Very little is known about the conservation status of Mexican Asteraceae species in general, and of tree species of this family in particular. Therefore, within the framework of the Global Tree Assessment, in cooperation with the IUCN/SSC Global Tree Specialist Group and Botanic Gardens Conservation International, all arborescent Asteraceae species that are endemic or near-endemic to Mexico (i.e., those shared with the south of the United States of America north of Mexico, and those shared with Central America south of the country) are being assessed for the IUCN Red List. Therefore, we use here the tree definition agreed on by the IUCN/SSC Global Tree Specialist Group, which has also been applied by [25]: a woody plant, usually with a single stem growing to a height of at least 2 m, or if multi-stemmed, then at least one vertical stem 5 cm in diameter at breast height.

A recent exploratory study, including species distribution and spatial analyses of a comprehensive list of native Mexican trees, carried out by [26], included 41 arborescent Asteraceae species. However, based on our knowledge of Asteraceae on the one hand, and our ongoing red listing work on the other hand, we realized that (1) this number is heavily underestimated, even though we handle a slightly different tree definition, and (2) the data analyzed were obtained from the National Biodiversity Information System database of Mexico [27] which, although it compiles and georeferences information, has insufficient taxonomic curation. Moreover, during the preparation of our Red List assessments, we noticed that information on arborescent Asteraceae, as is also the case for tree species in general, is very scattered and knowledge quite limited. As a consequence, a first step in the conservation of these species is the compilation of relevant information in order to obtain a general overview of their distribution, threats, and conservation status.

Figure 1. Selection of Mexican arborescent species of Asteraceae. (**A**) *Montanoa hexagona* (Heliantheae), (**B**) *Pittocaulon praecox* (Senecioneae), (**C**). *Telanthophora grandiflora* (Senecioneae), (**D**) *Nahuatlea smithii* (Gochnatieae), (**E**) *Critoniopsis uniflora* (Vernonieae), (**F**) *Sinclairia glabra* (Liabeae). Photo credits: (**A–D**) Rosario Redonda-Martínez; (**E–F**) Fernando Araujo-Mondragón.

Based on our ongoing Red List assessments of endemic and near-endemic Mexican Asteraceae and a meta-analysis of carefully curated distribution data, the objectives of this study are the following: (1) to document the precise number of Asteraceae trees that are distributed in Mexico and update the list of Mexican arborescent Asteraceae; (2) to summarize their very diverse uses; (3) to present a general panorama of their present and future distribution, including characterization of climatic suitability; and (4) to discuss the impact on their conservation in protected areas and biogeographical provinces.

2. Results

2.1. Species List

The list, generated from the bibliographic review and study of herbarium specimens, includes 149 tree species of Asteraceae, distributed in three subfamilies and 12 tribes (Appendix A), with Asteroideae having 129 species, being the most diverse. The latter subfamily consists of the tribes Heliantheae (54 species), Eupatorieae (42), and Senecioneae (20), which contain the highest number of species, whereas the remaining six tribes are each represented by only one to four species. Following in order of importance, there is subfamily Vernonioideae, in which the tribe Vernonieae groups 16 taxa and Liabeae only one. Finally, the subfamily Gochnatioideae represented by the Gochnatieae tribe includes only three species (Appendix A, Table 1). At the tribe level, Eupatorieae, Heliantheae, Senecioneae, and Vernonieae represent about 89%, while the other eight are equivalent to the remaining 11% of the total number of daisy trees.

Table 1. Number of Mexican Asteraceae tree species grouped by subfamilies and tribes based on the most recent classification by Susanna et al. [28]. The percentage represents the number of species of each tribe with respect to the 149 that represent the family in Mexico.

Subfamily	Tribe	Species	Percentage
Gochnatioideae	Gochnatieae	3	2.01%
Vernonioideae	Liabeae	1	0.67%
	Vernonieae	16	10.73%
Asteroideae	Senecioneae	20	13.42%
	Astereae	4	2.68%
	Inuleae	1	0.67%
	Neurolaeneae	1	0.67%
	Millerieae	3	2.01%
	Coreopsideae	2	1.34%
	Bahieae	1	0.67%
	Heliantheae	55	36.9%
	Eupatorieae	42	28.18%

Some species reported as arborescent both in the literature and on the labels of herbarium specimens were excluded from the list because they have been synonimized, e.g., *Roldana cordovensis*.

2.2. *Uses of Daisy Trees*

Of the 149 daisy tree species, just under 50% have a registered use. Of the 65 potentially used species, 37 have medicinal purposes, the leaves or young branches being the most used parts. Regarding the diseases they cure or the healing properties attributed to them, 12 species stand out as anti-inflammatory, 11 are used to treat stomach diseases, and 10 for skin conditions, followed by five used as antiseptics and five as febrifugals; they are also used to treat oral, heart, kidney, rheumatism, and vertigo conditions. Moreover, eight species with various medicinal uses were recorded.

Their usefulness as nectariferous species also stands out, distinguishing two main groups of insects and a group of birds, for which they serve as food for honeybees (*Apis mellifera*), butterflies, and hummingbirds, with 17, 3, and 1 species, respectively.

Six species are applied as forage, and of these, four are used only for that purpose, mainly when they are found in arid or semi-arid zones. Other documented uses for Asteraceae trees are as a living fence, cut flower, artisanal, ceremonial, fuel, construction, insecticide, ritual, and shade for coffee [29] (Table 2).

2.3. *Diversity per Vegetation Type*

The Mexican daisy trees occur in practically all vegetation types and most grow in several vegetation types, although the majority show an affinity for temperate and humid environments. Hence, the highest number of species are recorded in pine forests (107 spp.), followed by oak forests (104 spp.) and cloud forests (90 spp.). However, dry areas also host an important diversity, as 85 species are found in low deciduous forests and 47 in crassicaule shrubland. It should be noted that the genus *Nahuatlea* is exclusively distributed in arid and semi-arid areas of Mexico, being an important part of the vegetation structure in the crassicaule and thorny shrublands of the south and central part of the country. Finally, disturbed sites also host a significant number of arborescent Asteraceae, with 86 species, thus demonstrating the importance of this family as dominant elements of secondary vegetation [29].

Table 2. Use of Mexican arborescent Asteraceae species.

Use	Category	Species
Medicinal	Oral diseases	3
	Heart diseases	1
	Stomach diseases	11
	Skin diseases	10
	Gynecological diseases	2
	Anticonceptive	2
	Anti-inflammatory	12
	Antiseptic	5
	Diuretic	2
	Fever reducer	5
	Reuma	4
	Vertigo	1
	Various	8
Nectariferous	Honeybees	17
	Butterflies	3
	Hummingbirds	1
Ornamental	Live fence	2
	Cut flower	2
	Decoration	8
Others	Artesanal	2
	Ceremonial	3
	Fuel	8
	Construction	3
	Forage	6
	Insecticide	1
	Ritual	8
	Shade for coffee	3

2.4. Distribution in Mexico

Asteraceae trees are found in almost the entire territory; however, the highest number of species is found in the center and south of the country, mainly in the Trans-Mexican Volcanic Belt (Hidalgo, Jalisco, Mexico City, Michoacán, Morelos, State of Mexico, Puebla, Veracruz), the Sierra Madre del Sur (Chiapas, Guerrero, Oaxaca) and the southern portion of the Sierra Madre Oriental (Puebla, Querétaro, Veracruz). (Figure 2).

2.5. Climatic Suitability Patterns

The climatic suitability patterns of Asteraceae tree species in Mexico were characterized using models of 86 species, 17 of which show an expansion of over 10% of their current range, whereas 33 species exhibit a contraction of over 50% of their current range; both cases occurs under future scenario (Appendix B). Figure 3 depicts the current and future climatic suitability patterns in Mexico and the difference between scenarios. In the current scenario (1970 to 2000), it is clearly observed that the greatest diversity is found in the west, center, and south of the country, with the states of Jalisco, Michoacán, Mexico, Guerrero, Oaxaca, and Chiapas being those that host the greatest diversity of daisy tree species. The future model (2080 to 2100) estimates a drastic decrease in the number of species in the aforementioned states, although it is more noticeable in Oaxaca. As can be seen on the map that summarizes current and future differences, this state, together with Guerrero, Chiapas, and Jalisco, are those that are estimated to lose the greatest diversity. However, the results also show that the mountain regions of Guerrero and Oaxaca belonging to the Sierra Madre del Sur (SMS) and the Sierra Norte de Oaxaca (SNO), the Tacaná Volcano

(TV) in Chiapas on the border with Guatemala, the south of the State of Mexico, northern Michoacán, and the western portion of Jalisco, corresponding to the Trans-Mexican Volcanic Belt (TMVB), will maintain a considerable diversity, indicating that these areas could function as Anthropocene refugia for daisy trees.

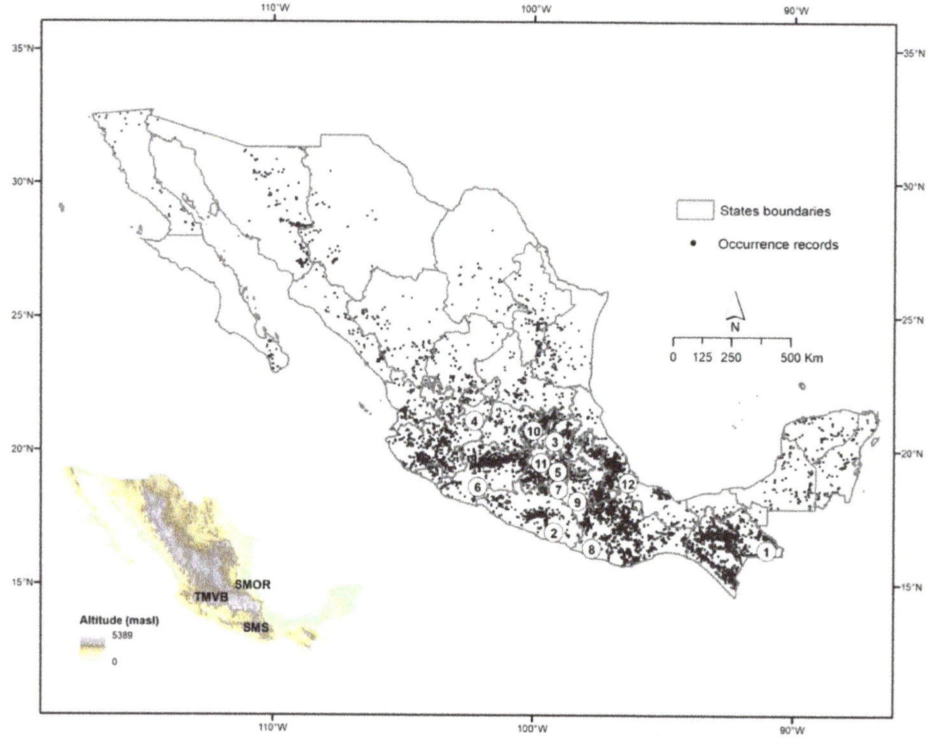

Figure 2. Occurrence records of Asteraceae tree species in Mexico. The states with the highest number of species are the following: 1 Chiapas, 2. Guerrero, 3. Hidalgo, 4. Jalisco, 5. Mexico City, 6. Michoacán, 7. Morelos, 8. Oaxaca, 9. Puebla, 10. Querétaro, 11. State of Mexico, 12. Veracruz. The map at the left shows the principal mountain regions mentioned in the text: Trans-Mexican Volcanic Belt (TMVB), Sierra Madre del Sur (SMS) and Sierra Madre Oriental (SMOr).

2.6. Protected Area Network and Biogeographic Provinces

The protected area network of Mexico (Figure 4) and biogeographic provinces (Figure 5) show an uneven distribution of climatic suitability. Observing spatial changes demonstrates that protected areas and provinces have a decrease in low suitability zones and an increase in high suitability zones in the future scenario, respectively. In the case of protected natural areas, the possible decrease that will occur in the future in protected areas such as the Sierra Gorda *s.l.* (Sierra Gorda and Sierra Gorda de Guanajuato), Los Tuxtlas, and Tehuacán-Cuicatlán Valley is notable. Some exceptions to this trend are the Flora and Fauna Protection Area Cuenca Alimentadora del Distrito Nacional de Riego 043, Estado de Nayarit, as well as the biosphere reserves of the Sierra de Manantlán (2), Monarch Butterfly (4) and El Triunfo (7). The models estimate in these areas that diversity could be maintained or increased in the long term, although this is uncertain. When modeling the climate change scenario on the map of biogeographic provinces, the results are similar. The current scenario shows that the greatest diversity is found along the Pacific coast, and in the Balsas Depression, Sierra Madre Occidental, Sierra Madre del Sur, Oaxaca, Altos de Chiapas, Soconusco, and Trans-Mexican Volcanic Belt. It is estimated that in the future, there will be a decrease in the Sierra Madre Occidental, Sierra Madre Oriental, and Altos de Chiapas. On the other hand, the difference between the two models indicates that there will be a small increase in the Trans-Mexican Volcanic Belt, Northern Altiplano (Chihuahuan desert) and Cape provinces, thereby maintaining the trend observed in the other models: a decrease in the sites that currently host the highest species richness, as well as an increase or no change in areas were the actual daisy tree diversity is considerable, such as the Trans-Mexican Volcanic Belt, Balsas Depression, Sierra Madre del Sur, Pacific coast, and Soconusco.

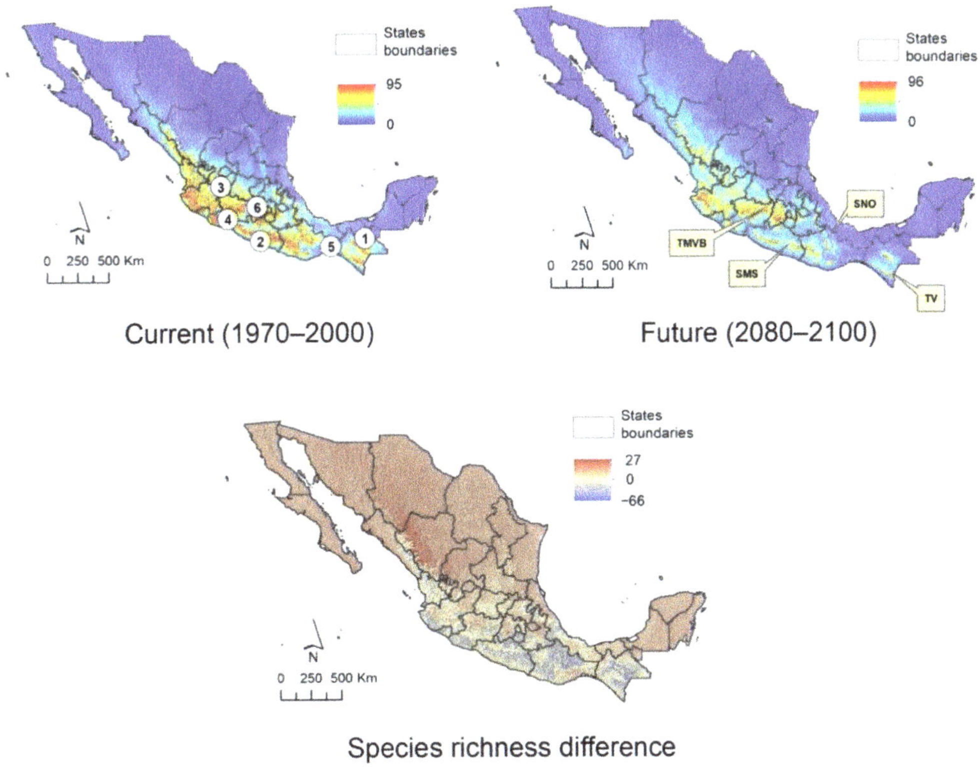

Figure 3. Effect of climate change on Mexican Asteraceae tree species in two different scenarios: current (1970 to 2000) and future (2080 to 2100) and the difference between both. States with the highest species richness: 1. Chiapas, 2. Guerrero, 3. Jalisco. 4. Michoacán, 5. Oaxaca, 6. State of Mexico. Regions identified as Anthropocene refugia: Trans-Mexican Volcanic Belt (TMVB), Sierra Madre del Sur (SMS), Sierra Norte de Oaxaca (SNO), and Tacaná Volcano (TV).

2.7. Mexican Daisy Tree Conservation

In the recently updated version of the Mexican decree of endangered species NOM-059-SEMARNAT-2010 [30], only 11 species of Asteraceae are included, of which none correspond to trees, despite the fact that some of them are only known from a few collections, or from the type collection only, and are distributed in areas with strong anthropogenic pressures derived from the change in land use, such as in the Uxpanapa-Chimalapas area in the states of Veracruz and Oaxaca. No Mexican species of Asteraceae are included in the updated appendices of the Convention of International Trade in Endangered Species of Wild Fauna and Flora [31], where only one species of Asteraceae is found—*Aucklandia costus* Falc. (cited as *Saussurea costus* (Falc.) Lipsch.)—due to its use in traditional Chinese medicine [32].

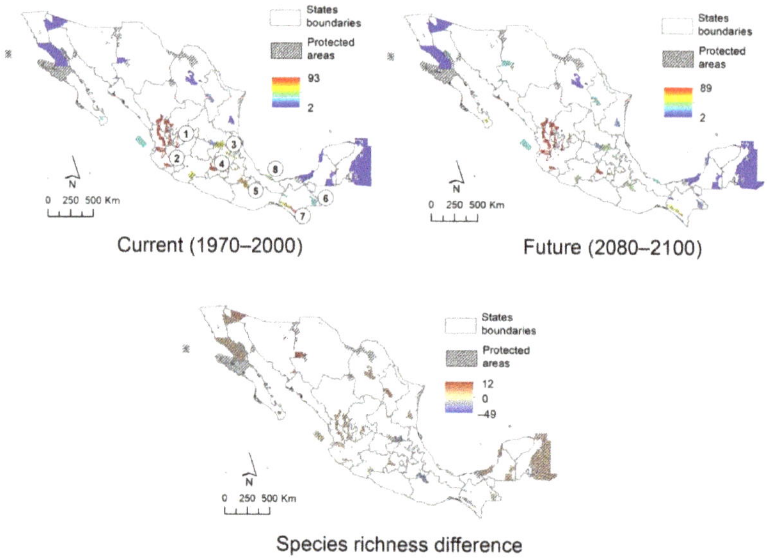

Figure 4. Actual and future scenarios for Mexican Asteraceae tree richness in the system of Natural Protected Areas at federal level. 1. Flora and Fauna Protection Area "Cuenca Alimentadora del Distrito Nacional de Riego 043, Estado de Nayarit", 2–8. Biosphere Reserves. 2 Sierra de Manantlán, 3. Sierra Gorda, 4. Monarch Butterfly, 5. Tehuacán-Cuicatlán Valley, 6. Montes Azules, 7. El Triunfo, 8. Los Tuxtlas.

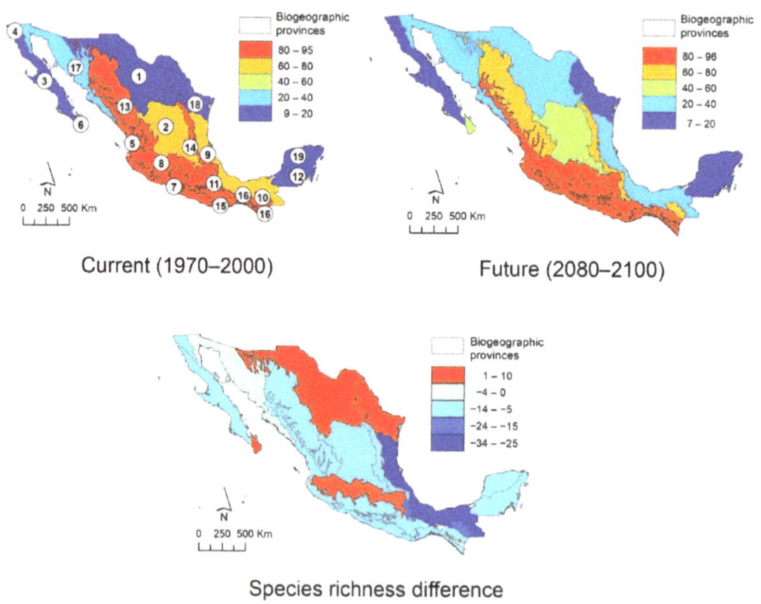

Figure 5. Estimation of the changes in the distribution of Mexican Asteraceae trees in the different biogeographical provinces of Mexico. 1. Northern Altiplano (Chihuahuense), 2. Southern Altiplano (Zacatecano–Potosina), 3. Baja California, 4. California, 5. Pacific Coast, 6. Cape, 7. Balsas Depression, 8. Trans-Mexican Volcanic Belt, 9. Gulf of Mexico, 10. Altos de Chiapas, 11. Oaxaca, 12. Petén, 13. Sierra Madre Occidental, 14. Sierra Madre Oriental, 15. Sierra Madre del Sur, 16. Soconusco, 17. Sonorense, 18. Tamaulipeca, 19. Yucatán.

With respect to our ongoing assessments of the conservation status of the Mexican Asteraceae trees for the IUCN Red List, less than 10 will be categorized as critically endangered (CR), an estimated 15 to 20 as endangered (EN), about 20 to 25 as vulnerable, and the remainder as species of least concern (LC). The assessments with final conservation statuses will be published on the IUCN Red List later this year.

3. Discussion

3.1. Daisy Tree Diversity in Mexico and Comparison with Other Diverse Areas

The study by Beech et al. [25] reported 3364 tree species for Mexico, positioning this country in the top 10 of the most tree species-rich countries, and due to our efforts in listing additional tree species since then, this has been increased to 3522 species [33]. Similarly, for native Mexican Asteraceae trees, in contrast to the previous studies and reports, we report a much higher number of arborescent Asteraceae taxa. Asteraceae is a main component of vegetation and bioregions along the Americas, with Mexico standing out as the most species-rich country for this family at a global level [1,2].

With respect to the tree species richness of Asteraceae in megadiverse countries and areas, Mexico and Central America rank second in the number of genera and species with 45 and 149, respectively. The first corresponds to Colombia as it is home to 169 species [7], followed by Brazil with 38 genera [34], and Ecuador, with 14 genera and 55 species, all of them in some risk category according to the IUCN Red List criteria [8]. Considering regional scales, the Amazonian (an area that includes the territory of nine South American countries: Bolivia, Brazil, Colombia, Ecuador, French Guyana, Guyana, Peru, Suriname, and Venezuela) is an area of high diversity with 37 genera and 107 species [35]. With respect to the number of Asteraceae trees registered in the flora of North America [4], eight genera and 10 species were found, of which five are mainly shared with the northern part of Mexico; hence, the high diversity of daisy trees in the country stands out. A similar situation occurs with the flora of China [6]; in this case, the numbers are even more contrasting, since in that area only five genera and nine species of native trees are registered. In addition, in China there are also two genera and three cultivated species that are native to Mexico or Central America and that have become naturalized in Chinese territory. These are the only daisy tree species shared between Mexico, Central America, and that region in Asia.

Tree-like Asteraceae are generally not prominently visible in the forests where they occur, as they grow in the understory or in open places, whereas they can reach up to 20 m in cloud forest, but they do not form a prominent part of the forest structure. In contrast, in scrubland vegetation and semi-evergreen low forest, they may be dominant, and an important part of the structure of the forest. It has been documented that Mexican coniferous forests show a relationship between forest structure and tree diversity [36]. In the particular case of Asteraceae, the highest number of arborescent species is found in pine forests (71.8%), while *Abies* forests concentrate just over 11% of the 149 species present in the country. Even considering that one and the same species can be found in various vegetation types, the percentages for pine and *Abies* forests are considerable.

3.2. Uses of Mexican Daisy Trees

A considerable amount of ruderal or malezoid Asteraceae are nectariferous and therefore are particularly important for honey-producing bees [37–40], or other pollinators, which are also attracted in addition to nectar, by the yellow colour of the flowers of many species [41]; hence, they do not depend on a single vector that carries out cross-pollination. The nectar produced by the Asteraceae is rich in glucose, fructose and sucrose [42], which encourages various groups of insects, including Hymenoptera, Diptera, Lepidoptera, and Coleoptera, to obtain food and assist in the pollination of these species [42–44]. Even several groups within Mutisieae are hummingbird-pollinated [45,46]. The family is of high economic importance as a honey supplier in several regions in Mexico. The worldwide known migratory phenomenon of the monarch butterfly occurs every year during the fall when millions of butterflies travel from the south of Canada and the north of the United States of America to Mexico to spend the winter season in the Monarch Butterfly Biosphere Reserve, located on the limits of Michoacán and the State of Mexico [47]. The presence of 103 Asteraceae species has been documented in its core zone [48]. From the illustrated flora of the Reserve [49], it can be observed that the butterflies feed on practically all Asteraceae that grow in the area.

Asteraceae are also an important source of food for honey-producing bees in Mexico, both European and native [37,38,40]. Among the species most used by these insects, there is a significant percentage of those that have been classified as "weeds" [40]. Indirectly, these plants are a source of income for beekeepers around the world. Mexico is among the top 10 honey producers worldwide, ranking fourth in exports of this product. In 2019 alone, 61.9 million tons of honey were produced, thus achieving an increase of just over six percent compared to the previous five-year period [50]. Eight states account for 70% of the national production, with Yucatán, Campeche, Jalisco, and Chiapas as the main producers [51].

Asteraceae are used in traditional medicine to treat various conditions such as the treatment of stomach and respiratory diseases, since around 6000 species contain sesquiterpene lactones, chemical compounds with antimicrobial, antiprotozoal, anti-inflammatory and cytotoxic properties [44,52,53]. Others are used as food, whether they are cultivated to obtain leaves or meristems, roots, tubers, heads, seeds to produce oil, natural dyes, bio-insecticides, or as ornamental or florist plants [52]. Although only eight Mexican daisy tree species are used for ornamental purposes, it is important to highlight the potential that other species of the family could have, since, considering their importance as species that produce nectar and pollen, they would be helpful in reducing the loss of bees and other pollinator groups.

The dahlia deserves a special mention as it has been the national flower of Mexico since 1963 [54], since the country has the largest number of wild and endemic species, with 38 and 35, respectively. These are a source of germplasm for the more than 50,000 varieties grown around the world [55]. In addition to the dahlia, there are other wild species with ornamental potential due to their visible inflorescences, among which the following stand out: *Montanoa bipinnatifida* and *Bartlettina sordida*. The first is highly appreciated in Mexico [56,57], Spain [58], Australia [59], and New Zealand [60], and the second in Spain [61]. Although there are no published data, some shrub or tree species of Asteraceae are used as living fences, in seasonal crops such as corn or beans, or in vegetables or gardens, among which the following stand out: *Barkleyanthus salicifolius*, *Baccharis heterophylla*, *Baccharis salicifolia*, *Montanoa tomentosa*, *M. leucantha*, and *M. grandiflora*. Asteraceae species associated with corn (milpa) are mainly *Tithonia tubiformis*, *Cosmos bipinnatus*, *C. sulphureus*, *Bidens odorata*, *B. pilosa*, *Melampodium perfoliatum*, *Simsia amplexicaulis*, and *Viguiera dentata*, weedy plants that farmers allow to grow alongside the milpa to serve as crop protection, as the loss of the harvest is lessened in the case of a grasshopper or locust plague [29] (pers. obs.).

3.3. Distribution, Including Characterization of Climatic Suitability

The largest quantity of daisy trees is found in the west, center, and south of the country, particularly where the important mountain ranges of the Sierra Madre Occidental, Sierra Madre del Sur, and Trans-Mexican Volcanic Belt converge. Based on the results obtained from the climate suitability model in the future (2080 to 2100), populations will tend to decrease in the sites that are currently particularly rich in tree-like Asteraceae species, e.g., Chiapas, Sierra Norte de Oaxaca, as well as the northern portion of the Sierra Madre Occidental, in the territory occupied by the states of Durango and Sinaloa. There will also be a considerable decrease in the number of species in the Sierra Madre del Sur, particularly in Guerrero, a state that currently ranks fourth in species richness at a national level [3]. If these predictions materialize, several populations of species that are currently found in sites considered Pleistocene refugia [62–64] would be lost. This may be due to the fact that their ecosystems would not withstand a scenario of abrupt climate change, such as the one that is estimated to occur in the next 80 years [65]. In this way, the states with the greatest diversity of Asteraceae—Oaxaca, Jalisco, Durango, Guerrero, and Michoacán [3]—would lose a significant number of species and endemism (Figures 2–4).

3.4. Conservation

The occurrence points of the known records, as well as the potential distribution models of Asteraceae trees compared with the areas occupied by the main natural protected areas present in the country, show a tendency to reduce their presence in some areas that are currently particularly rich in species; however, in other zones they will remain or increase; some of these correspond to protected natural areas.

Mexico has 182 protected natural areas distributed in maritime and continental territory. Of these, 67 correspond to National Parks, 44 are Biosphere Reserves, 40 Flora and Fauna Protection Areas, 18 Sanctuaries, eight Natural Resources Protection Areas, and five Natural Monuments [66]. Those that are located in continental territory are equivalent to 10.88% of the country's land surface [66]. One of the terrestrial protected areas with the largest territorial extension is the Flora and Fauna Protection Area Cuenca Alimentadora del Distrito Nacional de Riego 043, Estado de Nayarit, located in the west of the country, comprising part of the territory of the states of Aguascalientes, Jalisco, Durango, Nayarit, and Zacatecas [66]. It is home to around 11 types of vegetation, more than 2000 species of vascular plants and at least two endemic daisy tree species [67,68]. The territorial extension and biological diversity of this protected area is considerable. In the particular case of Asteraceae trees, the climatic suitability models estimate that the diversity of species will remain in the western part of the country, a situation similar to what could occur in the Sierra de Manantlán, one of the most important biosphere reserves in the western region with an area of 139,577.12 ha [66]. The latter was recognized as a biosphere reserve for the biological diversity that it houses in its territory, which includes vegetation of dry, temperate, and humid environments, in addition to being the main water source for more than 430,000 inhabitants of southern Jalisco and northern Colima [69].

The biosphere reserves are distributed throughout the country; 70% have territorial extensions of more than 100 ha or a high species diversity [66]. In the center of the country, the Sierra Gorda stands out with an extension of 383,567.45 ha, located on the limits of Querétaro, Guanajuato, San Luis Potosí and Hidalgo, and Sierra Gorda de Guanajuato (236,882.76 ha) [66], which together occupy the seventh place in size of all the protected natural areas of Mexico [70]. In its territory, there are dry shrublands and temperate forests, which host a great diversity of plants, many of them endemic to the Sierra Madre Oriental [70]. This region has three hydrological sub-basins and a dam declared a Ramsar site, as it is a wetland of global importance [70]. The climate suitability models estimate a slight decrease in the number of tree species of Asteraceae (Figure 4). Among those that would be affected are *Baccharis heterophylla*, *Barkleyanthus salicifolius*, *Critonia morifolia*, *Koanophyllon albicaulis*, and *Nahuatlea hypoleuca*, species that fortunately are not restricted to Mexico, since the first four also occur in Central America and the fifth in the south of the United States of America.

One of the most important protected areas in central Mexico is the Monarch Butterfly Biosphere Reserve with an area of 56,259.05 ha [66]. In its territory, there are pine forests, *Abies* forests, and oak forests that contribute in an important way to the carbon capture from industrial areas and favour the recharge of aquifers that provide water to the metropolitan area of Mexico City, as well as various areas of the states of Michoacán and Mexico [47]. Our models estimate that the daisy tree diversity in this region will be maintained in the next century, if this also happens with the ecosystems where they are found, as well as with the environmental services that they provide to this area of the country.

The Tehuacán-Cuicatlán Valley Biosphere Reserve, with a surface area of 490,186.87 ha, is the largest of all those found in arid and semi-arid zones [66]. In this case, the climatic suitability models estimate a possible decline in species (Figures 3–5). Although the number of trees that are currently in its territory is minimal, several species dominate and give structure to the vegetation of some areas, among them: *Baccharis heterophylla*, *Montanoa leucantha*, *M. tomentosa*, *Nahuatlea hypoleuca*, *N. smithii*, *Parthenium tomentosum*, *Pittocaulon praecox*, *P. velatum*, *Roldana eriophylla*, and *R. oaxacana*. Of these, *Nahuatlea smithii* and *Roldana eriophylla* are practically endemic to this region and could be at risk if the scenario predicted by the models takes place.

In the case of tropical ecosystems and particularly rainforests, the main biosphere reserves are: Montes Azules (331,200 ha) and El Triunfo (119,177.29 ha) in Chiapas, and Los Tuxtlas (155,122.47 ha), in Veracruz [66]. Interestingly, the three are in territory belonging to Pleistocene refugia; Montes Azules corresponds to the refugium called Lacandonia and El Triunfo to Soconusco [62]. Both are considered primary Pleistocene refugia; i.e., these zones maintained constant temperature and precipitation conditions during the dry and cold periods that occurred during this period, which allowed them to safeguard common species in humid tropical forests [62]. The climate suitability models estimate different situations in the event of abrupt climatic events, since a probable reduction would occur in Montes Azules, while in El Triunfo, the Asteraceae tree populations would remain. This may be due to the climatic conditions that currently prevail in each of these regions, since in Montes Azules the current oscillation of temperature and precipitation is lower (average annual temperature between 22 and 24 °C; average annual precipitation 2000 to 3000 mm [71]), compared to the values recorded in El Triunfo (mean annual temperature of 18 to 22 °C, with a mean annual rainfall of 1000 to 3000 mm [72]). However, both sites have invaluable biological potential and value. Montes Azules protects 20% of the plant species present in Mexico; recent calculations indicate that one hectare of this reserve protects about 160 tree species and about 700 vascular plants [73]. In El Triunfo, there are important extensions of mountain cloud forest, considered one of the ecosystems that hosts the greatest diversity of trees in North and Central America [74]. The Los Tuxtlas reserve is located in an area recognized as a secondary Pleistocene refugium, because it corresponds to an area that only managed to preserve itself from the drastic drop in temperature or precipitation, during the alternation of cold and dry periods of the Pleistocene [62].

Los Tuxtlas homes coniferous forest, oak forest, mountain cloud forest, high evergreen forest, and mangroves, where around 3000 plant species have been documented, and it is also one of the five areas with the highest amount of tree endemism in Mexico [75]. For this site, the climatic suitability models estimate a decrease in the number of daisy trees. Considering the importance of Los Tuxtlas as an area of endemism for tree species, it would be interesting to explore what happens with other angiosperm groups. In this area, the average annual temperature ranges between 22 and 26 °C and rainfall varies from 1500 to 4500 mm [76]. Although the temperature is relatively constant, the precipitation has a considerable range of variation, which would support the proposal that it was a secondary Pleistocene refugium [62].

Apparently, the aforementioned biosphere reserves will allow that, given a scenario such as that estimated by our models, the probable reduction of the populations does not become catastrophic as in other areas of the country, which are particularly rich in diversity and Asteraceae endemisms, as occurs with the Tehuacán-Cuicatlán Valley.

Anthropocene refugia correspond to territories that meet the following qualities: being ecologically suitable areas to house the diversity units analyzed and having relatively low levels of observed and predicted anthropogenic pressure to allow their long-term persistence in this area, i.e., through several generations [77]. The main difference between Pleistocene and Anthropocene refugia is that the former are sites where organisms resisted and responded to glacial and interglacial oscillations of the late Quaternary, having the possibility of expanding their distribution once environmental stress conditions decreased [78]. In contrast, in order to characterize probable refugia from the Anthropocene, climate change derived from anthropogenic pressures is taken into account [77]. Therefore, the identification of Anthropocene refugia is useful to categorize, plan, and decide where to establish conservation areas for the group of interest [77].

In the case of daisy trees, the models allow the identification of some areas that meet sufficient characteristics to be considered Anthropocene refugia and, therefore, to be maintained or proposed as conservation areas, although some of them are already cataloged like this. This is the case of the western region of Jalisco where the Flora and Fauna Protection Zone Cuenca Alimentadora del Distrito Nacional de Riego 043, Estado de Nayarit and the Biosphere Reserve Sierra de Manantlán are located. Moreover, the Biosphere Reserves of the Monarch Butterfly, on the limits of Michoacán and the State of Mexico, and of the Tacaná Volcano and El Triunfo in Chiapas are also already existing conservation areas; the latter has previously been proposed as a Pleistocene refugium [62]. Based on the results obtained, other areas that could function as Anthropocene refugia correspond to the northern portion of Michoacán and southern part of the State of Mexico, which together with western Jalisco form part of the Trans-Mexican Volcanic Belt, an area that currently contains a high richness of Asteraceae trees, which, according to our models, is estimated to remain or increase (Figures 3–5). The mountainous regions of Guerrero and Oaxaca, corresponding to the Sierra Madre del Sur (Figure 3), also seem to meet the characteristics of Anthropocene refugia. Although there are currently no natural protected areas decreed in these areas, despite being sites with high biological diversity and a large number of endemisms, it is a fact that their geographical location and the difficulty of accessing them has kept them safe from human damage. Special mention should be made of the Sierra Norte de Oaxaca, an area particularly rich in diversity and endemism of Asteraceae [79], in which the models indicate that there are also adequate conditions to serve as an Anthropocene refugium, which is confirmed by the fact that it has also been considered as a secondary Pleistocene refugium [62]. However, unlike the mountainous region of central Oaxaca, the Sierra Norte has sufficient infrastructure to access its territory and forest management of the coniferous forests, although the other ecosystems remain almost intact. This shows that, despite the fact that this area is not recognized as a protected area at the federal level, the community forest management that the inhabitants of the region have carried out has been adequate and successful, since in addition to generating jobs and resources for the inhabitants of the region, the forest area has increased in the last four decades [80–83].

Based on the aforementioned, the sites identified as priority areas to conserve the diversity and endemism of daisy trees are the following: Trans-Mexican Volcanic Belt (including Protected Areas of western of Jalisco), Sierra Madre del Sur, Sierra Norte de Oaxaca, El Triunfo and Tacaná Volcano. All these regions have previously been identified as diversity hotspots of other groups of plants [84,85] and animals [86].

As a consequence, whether the efforts and proposals to conserve nature are federal or local, everything seems to indicate that the establishment, maintenance and conservation of protected natural areas that currently exist in Mexico have been adequate. However, the ideal would be to keep them intact in the long term or, as far as possible, to extend their territory in order to safeguard a greater number of species, both Asteraceae and other families of angiosperms.

In conclusion, relatively few Mexican daisy tree species are currently seriously threatened by climate change or other factors, as most species are widely distributed. Direct exploitation for human use is also not generally a threatening factor. Mexico ranks first at the global level with respect to daisy diversity and second with respect to daisy tree diversity. As mentioned above, Asteraceae are ecologically successful, and the same goes for tree-like representatives of this family. However, it will be important to include those endemic species whose IUCN Red List assessment indicates that they are endangered or critically endangered in the NOM-059-SEMARNAT-2010, as several of these, e.g., *Ageratina chimalapana*, *Lepidonia wendtiana*, *Mixtecalia teitaensis*, *Montanoa revealii*, and *Verbesina sousae*, occur in areas subject to anthropogenic pressures that puts their survival at risk, either due to changes in land use, excessive or unplanned tourism, and the extraction of stone material.

4. Materials and Methods

4.1. Study Area

Mexico is located in North America, between the extreme coordinates 32°43′06″ and 19°32′25″ latitude N and 114°43′22″and 84°38′30″ longitude W. It limits to the north with the United States of America and to the south with Belize and Guatemala. It has a territorial extension of 1,960,189 km^2; which positions it in sixth place among the American countries and as 14th in the world. Nearly half of the country is located below the Tropic of Cancer, which favours the presence of temperate and cold climates in the north, as well as temperate and warm climates in the south. In addition, the geographical position of the country, the presence of large mountain ranges such as the Sierra Madre Oriental to the east, the Sierra Madre Occidental to the west and the Trans-Mexican Volcanic Belt that crosses the territory from east to west in the central-southern area of the country, together with geological, edaphic, and microclimatic variations, favour the existence of a great diversity of vegetation types and therefore a high biodiversity [87]. Mexico shares floristic diversity with neighbouring countries; in the northern part it has affinities with some regions of the United States of America, in the southern part with Central America, and to a lesser extent with South America. The species analyzed in this work are those that are endemic or near-endemic to Mexico, i.e., those shared with the south of the United States of America north of Mexico, and those shared with Central America south of the country.

For the purpose of this study, we only focus on the occurrence, distribution pattern uses, and conservation of the daisy tree species within the Mexico territory.

4.2. Compilation of Taxonomic List and Species Information (Use and Habitat)

From a bibliographic review and the consultation of herbarium specimens available online, a list of tree species of Asteraceae was made (Appendix A), considering in this category those that have been described as trees, arborescent or small trees and even some that have been registered as shrubs, but that sometimes also develop a tree habit according to the definition we used. This list was generated based on the revision of regional floras, such as: Flora Novogaliciana [56], Flora of Chiapas [14], Flora Mesoamericana [88], treatments of the Flora del Valle de Tehuacán-Cuicatlán [89–91], Flora del Bajío y de Regiones Adyacentes [57]; taxonomic reviews at tribe level (Eupatorieae [13]), genus level [15–21,92,93], and section level [22,23]. Moreover, online available collections were also consulted: the National Herbarium of Mexico (MEXU) [29], United States National Herbarium (US) [94], herbarium of the Missouri Botanical Garden (MO) [95], and several herbaria of northern Mexico and Arizona whose collections are available via the portal SEinet Arizona-New Mexico Chapter [96].

In addition to the compilation of Mexican species, we looked up the number of daisy tree species in at least five other megadiverse countries: Brazil [5], China [6], Colombia [7], Ecuador [8], and the United States of America [4].

The most recent classification of Asteraceae for subfamilies and tribes is used in Appendix A [28]. The names of some genera and species are based on specialist reviews and criteria, for example, *Ageratina* [13,97], *Critoniopsis* [98], and *Pachythamnus* [99].

4.3. Compilation of Species Ocurrence Geographical Data

The geographical coordinates were obtained from GBIF [100] via GeoCAT [101], carrying out an exhaustive curation of the information, which consisted mainly of eliminating records of human observations without support by vouchers or photos, as well as those that lacked information regarding the collector or herbarium where the voucher is located, or that only presented decimal coordinates. Moreover, duplicate records in the same locality were also deleted. Records that did not correspond to the known distribution of the species were also eliminated, based on expert knowledge, either due to possible misidentifications or because they correspond to specimens grown outside the natural distribution area of a certain species, or invasive groups in areas other than their natural distribution.

When the records were minimal or there was no information available in GBIF [100], this was complemented with data obtained from the labels of herbarium specimens available online through the digital platforms of the National Herbarium of Mexico (MEXU) [29], the National Herbarium of the United States of America (US) [94] and several herbaria of Northern Mexico and Arizona, USA [96]. The localities that lacked geographical coordinates were georeferenced using Google Earth Pro [102]. To locate some little-known localities, the Historical Archive of geostatistical localities [103], Mapcarta [104], and Pueblos de México [105] were used. The data for each species were generated in Excel tables in comma delimited text format (.csv).

4.4. Spatial Analyses

A species distribution modelling (SDM) approach was used to characterise the spatial patterns of tree Asteraceae species in Mexico. SDM allows identifying the geographic areas with the highest climatic suitability in the current period and projects this suitability in future scenarios. The modelling approach was maximum entropy using Maxent software [106]. To model climate adequacy, the compiled occurrence database is used together with a set of explanatory climate variables. Only species with more than 50 unique occurrences were selected for modelling, to ensure good performance [107]. In this study, the climate variables were obtained from the Worldclim 2 database [108] for the current period (1970 to 2000). The variables were selected from the set of 19 bioclimatic variables available in Worldclim at a spatial resolution of 1 km, which were analysed for their degree of correlation in the Americas' total extent of occurrence. Correlation values between variables higher than 0.7 were excluded, obtaining a set of six variables with low correlation. The variables were temperature seasonality (BIO4), minimum temperature of coldest month (BIO6), temperature annual range (BIO7), annual precipitation (BIO12), precipitation seasonality (BIO15), and precipitation of coldest quarter (BIO19). Each species' projection was carried out by maintaining the default Maxent regularisation parameters (auto features) and avoiding extrapolation and clamping options. Records less than 1 km apart per species were excluded to avoid spatial autocorrelation. Occurrences were divided into a training set (70% of the total) and a test set (30%). Maxent probability models were projected over the entire distribution of the species occurrences and then were restricted to the continental area of Mexico by cropping the total raster extent area. This step was done to avoid the loss of potential climatic space combinations where species are present outside Mexico and improving final model accuracy [109]. Models were transformed into a binary format, using a threshold of maximum training sensitivity plus specificity [110]. The binary models per species were summed to obtain the current climate suitability pattern. Future projections were obtained using the global circulation model MIROC6 [111], which has been assessed to represent the average conditions of different climatic factors on a global scale [112]. From this global circulation model, the most extreme scenario SSP585 was selected for the future period 2081 to 2100. The same transformation procedure to binary and summation by species was repeated to obtain the future pattern of climate suitability. Three spatial analyses were performed with the current and future climate suitability models: the calculation of the difference between the future and current patterns, the extraction of the climate suitability in Mexico's protected areas, and finally the extraction for the biogeographic provinces of Mexico [113]. All spatial analyses were performed with ESRI Arcgis software (version 10.8).

4.5. IUCN Red List Assessments

Red List categories were applied according to the IUCN red list criteria [114] and all relevant information was completed in the IUCN SIS database for pending publication on the publicly available IUCN Red List. Data on species occurrence, uses and habitat are those that were obtained for the abovementioned analyses.

Author Contributions: Conceptualization, R.R.-M., E.M.M.S., M.-S.S., data generation and analyses, R.R.-M., P.P., M.-S.S.; writing—original draft preparation, R.R.-M., M.-S.S.; writing—review and editing, all authors; visualization, R.R.-M., P.P.; funding acquisition, R.R.-M., P.P., M.-S.S. All authors have read and agreed to the published version of the manuscript.

Funding: This research was funded by the Fondation Franklinia, project number 2019-04, the Instituto de Ecología, A.C., project numbers 20006-11337 and 20006-11617, and FONDECYT grant N°1181677.

Institutional Review Board Statement: Not applicable.

Informed Consent Statement: Not applicable.

Data Availability Statement: All data used in this study were obtained from databases and publicly available literature sources, all of them were cited in the references.

Acknowledgments: R.R.-M. thanks Fernando Araujo Mondragón for his help in capturing information and georeferencing localities in Vernonieae, as well as for providing photographs of some tree species; Karina Machuca Machuca and Aarón Hernández Miranda (Critoniinae) for searching and compiling the information required to carry out the IUCN Red List evaluations; Francisco Galán and Adrián García for the support provided to capture specimen information from online collections. We are very grateful to Sara Oldfield for the critical revision of the manuscript, as well as to the three anonymous reviewers, whose comments have greatly improved our work.

Conflicts of Interest: The authors declare no conflict of interest. The funders had no role in the design of the study; in the collection, analyses, or interpretation of data; in the writing of the manuscript, or in the decision to publish the results.

Appendix A

Table A1. List of Mexican endemic and near-endemic arborescent Asteraceae species.

Subfamily	Tribe	Species
Gochnatioideae	Gochnatieae	*Nahuatlea arborescens* (Brandegee) V.A. Funk
	Gochnatieae	*Nahuatlea hypoleuca* (DC.) V.A. Funk
	Gochnatieae	*Nahuatlea smithii* (B.L. Rob. & Greenm.) V.A. Funk
Vernonioideae	Liabeae	*Sinclairia glabra* (Hemsl.) Rydb.
	Vernonieae	*Critoniopsis baadii* (McVaugh) H. Rob.
	Vernonieae	*Critoniopsis heydeana* (J.M. Coult.) H. Rob.
	Vernonieae	*Critoniopsis leiocarpa* (DC.) H. Rob.
	Vernonieae	*Critoniopsis macvaughii* (S.B. Jones) H. Rob.
	Vernonieae	*Critoniopsis obtusa* (Gleason) H. Rob.
	Vernonieae	*Critoniopsis salicifolia* (DC.) H. Rob.
	Vernonieae	*Critoniopsis shannonii* (J.M. Coult.) H. Rob.
	Vernonieae	*Critoniopsis tomentosa* (Lex.) H. Rob.
	Vernonieae	*Critoniopsis triflosculosa* (Kunth) H. Rob.
	Vernonieae	*Critoniopsis uniflora* (Sch. Bip.) H. Rob.
	Vernonieae	*Critoniopsis villaregalis* (Carvajal) H. Rob.
	Vernonieae	*Lepidaploa polypleura* (S.F. Blake) H. Rob.
	Vernonieae	*Lepidonia salvinae* (Hemsl.) H. Rob. & V.A. Funk
	Vernonieae	*Lepidonia wendtiana* (B.L. Turner) Redonda-Mart. & Villaseñor
	Vernonieae	*Vernonanthura cordata* (Kunth) H. Rob.
	Vernonieae	*Vernonanthura patens* (Kunth) H. Rob.
Asteroideae	Astereae	*Baccharis glandulifera* G.L. Nesom
	Astereae	*Baccharis heterophylla* Kunth
	Astereae	*Baccharis lancifolia* Less.
	Astereae	*Baccharis salicifolia* (Ruiz & Pav.) Pers. subsp. *monoica* (G.L. Nesom) Joch. Müll.
	Bahieae	*Peucephyllum schottii* A.Gray
	Coreopsideae	*Dahlia imperialis* Roezl ex Ortgies
	Coreopsideae	*Electranthera mutica* (DC.) Mesfin, D.J.Crawford & Pruski
	Eupatorieae	*Ageratina areolaris* (DC.) Gage ex B.L.Turner
	Eupatorieae	*Ageratina cerifera* (McVaugh) R.M. King & H. Rob.

Table A1. *Cont.*

Subfamily	Tribe	Species
	Eupatorieae	*Ageratina chiapensis* (B.L. Rob.) R.M. King & H. Rob.
	Eupatorieae	*Ageratina chimalapana* B.L. Turner
	Eupatorieae	*Ageratina cylindrica* (McVaugh) R.M. King & H. Rob.
	Eupatorieae	*Ageratina glabrata* (Kunth) R.M. King & H. Rob.
	Eupatorieae	*Ageratina grandifolia* (Regel) R.M. King & H. Rob.
	Eupatorieae	*Ageratina havanensis* (Kunth) R.M. King & H. Rob.
	Eupatorieae	*Ageratina ligustrina* (DC.) R.M. King & H. Rob.
	Eupatorieae	*Ageratina mairetiana* (DC.) R.M.King & H.Rob.
	Eupatorieae	*Ageratina vernalis* (Vatke & Kurtz) R.M. King & H. Rob.
Asteroideae	Eupatorieae	*Amolinia heydeana* (B.L.Rob.) R.M.King & H.Rob.
	Eupatorieae	*Bartlettina luxii* (B.L. Rob.) R.M. King & H. Rob.
	Eupatorieae	*Bartlettina pansamalensis* (B.L. Rob.) R.M. King & H. Rob.
	Eupatorieae	*Bartlettina pinabetensis* (B.L. Rob.) R.M. King & H. Rob.
	Eupatorieae	*Bartlettina platyphylla* (B.L. Rob.) R.M. King & H. Rob.
	Eupatorieae	*Bartlettina prionophylla* (B.L. Rob.) R.M. King & H. Rob.
	Eupatorieae	*Bartlettina sordida* (Less.) R.M. King & H. Rob.
	Eupatorieae	*Bartlettina tuerckheimii* (Klatt) R.M. King & H. Rob.
	Eupatorieae	*Bartlettina williamsii* R.M. King & H. Rob.
	Eupatorieae	*Chromolaena collina* (DC.) R.M. King & H. Rob.
	Eupatorieae	*Chromolaena glaberrima* (DC.) R.M.King & H.Rob.
	Eupatorieae	*Critonia breedlovei* R.M. King & H. Rob.
	Eupatorieae	*Critonia conzatti* (Greenm.) R.M. King & H. Rob.
	Eupatorieae	*Critonia daleoides* DC.
	Eupatorieae	*Critonia hebebotrya* DC.
	Eupatorieae	*Critonia hospitalis* (B.L. Rob.) R.M. King & H. Rob.
	Eupatorieae	*Critonia iltisii* R.M. King & H. Rob.
	Eupatorieae	*Critonia morifolia* (Mill.) R.M. King & H. Rob.
	Eupatorieae	*Critonia paneroi* B.L.Turner
	Eupatorieae	*Critonia quadrangularis* (DC.) R.M. King & H. Rob.
	Eupatorieae	*Critonia sexangularis* (Klatt) R.M. King & H. Rob.
	Eupatorieae	*Critonia tuxtlae* R.M. King & H. Rob.
	Eupatorieae	*Critoniadelphus microdon* (B.L. Rob.) R.M. King & H. Rob.
	Eupatorieae	*Critoniadelphus nubigenus* (Benth.) R.M. King & H. Rob.
	Eupatorieae	*Koanophyllon albicaule* (Sch. Bip. ex Klatt) R.M. King & H. Rob.
	Eupatorieae	*Koanophyllon galeottii* (B.L. Rob.) R.M. King & H. Rob.
	Eupatorieae	*Koanophyllon palmeri* (A. Gray) R.M. King & H. Rob.
	Eupatorieae	*Koanophyllon pittieri* (Klatt) R.M. King & H. Rob.
	Eupatorieae	*Koanophyllon revealii* B.L. Turner
	Eupatorieae	*Kyrsteniopsis nelsonii* (B.L. Rob.) R.M. King & H. Rob.
	Eupatorieae	*Pachythamnus crassirameus* (B.L. Rob.) R.M. King & H.Rob.
	Heliantheae	*Clibadium arboreum* Donn. Sm.
	Heliantheae	*Clibadium leiocarpum* Steetz in Seemann
	Heliantheae	*Clibadium surinamense* L.
	Heliantheae	*Dendroviguiera puruana* (Paray) E.E. Schill. & Panero
	Heliantheae	*Dendroviguiera quinqueradiata* (Cav.) E.E. Schill. & Panero
	Heliantheae	*Dendroviguiera sphaerocephala* (DC.) E.E. Schill. & Panero
	Heliantheae	*Dendroviguiera splendens* (Panero & E.E. Schill.) E.E. Schill. & Panero
	Heliantheae	*Lagascea palmeri* (B.L. Rob.) B.L. Rob.
	Heliantheae	*Lasianthaea fruticosa* (L.) K.M. Becker
	Heliantheae	*Montanoa andersonii* McVaugh
Asteroideae	Heliantheae	*Montanoa bipinnatifida* (Kunth) K. Koch
	Heliantheae	*Montanoa frutescens* (Mairet ex DC.) Hemsl.
	Heliantheae	*Montanoa grandiflora* Alamán ex DC.
	Heliantheae	*Montanoa hexagona* B.L. Rob. & Greenm.
	Heliantheae	*Montanoa imbricata* V.A. Funk
	Heliantheae	*Montanoa karwinskii* DC.
	Heliantheae	*Montanoa leucantha* (Lag.) S.F. Blake

Table A1. *Cont.*

Subfamily	Tribe	Species
	Heliantheae	*Montanoa revealii* H. Rob.
	Heliantheae	*Montanoa speciosa* DC.
	Heliantheae	*Montanoa tomentosa* Cerv.
	Heliantheae	*Parthenium fruticosum* Less. ex Schltdl. & Cham.
	Heliantheae	*Parthenium schottii* Greenm. ex Millsp. & Chase
	Heliantheae	*Parthenium tomentosum* DC.
	Heliantheae	*Perymenium grande* Hemsl.
	Heliantheae	*Perymenium hintonii* McVaugh
	Heliantheae	*Podachaenium chiapanum* B.L. Turner & Panero
	Heliantheae	*Podachaenium eminens* (Lag.) Sch. Bip. ex Sch. Bip.
	Heliantheae	*Podachaenium standleyi* (Steyerm.) B.L.Turner & Panero
	Heliantheae	*Rensonia salvadorica* S.F. Blake
	Heliantheae	*Rojasianthe superba* Standl. & Steyerm.
	Heliantheae	*Squamopappus skutchii* (S.F. Blake) R.K. Jansen, N.A. Harriman & Urbatsch
	Heliantheae	*Tetrachyron orizabaensis* (Klatt) Wussow & Urbatsch
	Heliantheae	*Tithonia koelzii* McVaugh
	Heliantheae	*Tithonia longiradiata* (Bertol.) S.F. Blake
	Heliantheae	*Verbesina apleura* S.F. Blake
	Heliantheae	*Verbesina breedlovei* B.L. Turner
	Heliantheae	*Verbesina culminicola* McVaugh
	Heliantheae	*Verbesina fastigiata* B.L. Rob. & Greenm.
	Heliantheae	*Verbesina furfuracea* McVaugh
	Heliantheae	*Verbesina guatemalensis* B.L. Rob. & Greenm.
	Heliantheae	*Verbesina hypargyrea* B.L. Rob. & Greenm.
	Heliantheae	*Verbesina hypoglauca* Sch. Bip. ex Klatt
	Heliantheae	*Verbesina klattii* B.L. Rob. & Greenm.
	Heliantheae	*Verbesina lanata* B.L. Rob. & Greenm.
	Heliantheae	*Verbesina montanoifolia* B.L. Rob. & Greenm.
	Heliantheae	*Verbesina oncophora* B.L. Rob. & Seaton
	Heliantheae	*Verbesina oligantha* B.L. Rob.
	Heliantheae	*Verbesina ovatifolia* A. Gray
	Heliantheae	*Verbesina perymenioides* Sch. Bip. ex Klatt
	Heliantheae	*Verbesina platyptera* Sch. Bip. ex Klatt
	Heliantheae	*Verbesina sousae* J.J. Fay
Asteroideae	Heliantheae	*Verbesina sphaerocephala* A. Gray
	Heliantheae	*Verbesina turbacensis* Kunth
	Heliantheae	*Verbesina villaregalis* McVuagh
	Heliantheae	*Wamalchitamia aurantiaca* (Klatt) Strother
	Inuleae	*Pluchea sericea* (Nutt.) Coville
	Millerieae	*Desmanthodium perfoliatum* Benth.
	Millerieae	*Rumfordia floribunda* DC.
	Millerieae	*Schistocarpha longiligula* Rydb.
	Neurolaeneae	*Neurolaena macrophylla* Greenm.
	Senecioneae	*Barkleyanthus salicifolius* (Kunth) H. Rob. & Brettell
	Senecioneae	*Lepidospartum squamatum* (A. Gray) A. Gray
	Senecioneae	*Mixtecalia teitaensis* Redonda-Mart., García-Mend., & D. Sandoval
	Senecioneae	*Pittocaulon filare* (McVaugh) H. Rob. & Brettell
	Senecioneae	*Pittocaulon praecox* (Cav.) H. Rob. & Brettell
	Senecioneae	*Pittocaulon velatum* (Greenm.) H. Rob. & Brettell
	Senecioneae	*Roldana albonervia* (Greenm.) H. Rob. & Brettell
	Senecioneae	*Roldana angulifolia* (DC.) H. Rob. & Brettell
	Senecioneae	*Roldana barba-johannis* (DC.) H. Rob. & Brettell
	Senecioneae	*Roldana eriophylla* (Greenm.) H. Rob. & Brettell
	Senecioneae	*Roldana gentryi* H. Rob. & Brettell

Table A1. *Cont.*

Subfamily	Tribe	Species
	Senecioneae	*Roldana greenmanii* H. Rob. & Brettell
	Senecioneae	*Roldana neogibsonii* (B.L. Turner) B.L. Turner
	Senecioneae	*Roldana oaxacana* (Hemsl.) H.Rob. & Brettell
	Senecioneae	*Roldana schaffneri* (Sch. Bip. ex Klatt) H. Rob. & Brettell
	Senecioneae	*Telanthophora cobanensis* (J.M. Coult.) H. Rob. & Brettell
	Senecioneae	*Telanthophora grandifolia* (Less.) H.Rob. & Brettell
	Senecioneae	*Telanthophora jaliscana* H. Rob. & Brettell
	Senecioneae	*Telanthophora standleyi* (Greenm.) H. Rob. & Brettell
	Senecioneae	*Telanthophora uspantanensis* (J.M. Coult.) H. Rob. & Brettell

Appendix B

Table A2. Species list and number of records used for the climatic suitability models. I. Records in whole distribution area, II. Records in Mexico, III. Current suitable area (km^2), IV. Future suitable area (km^2), V. Future range expansion in Mexico (km^2), VI. Future range stability in Mexico (km^2), VII. Future range contraction in Mexico (km^2), VIII. Future range expansion in Mexico (%), IX. Future range stability in Mexico (%), X. Future range contraction in Mexico (%).

ID	Species	I	II	III	IV	V	VI	VII	VIII	IX	X
1	*Ageratina areolaris* (DC.) Gage ex B.L.Turner	552	538	439,361	258,128	37,770	167,455	184,977	9.0	39.8	43.9
2	*Ageratina chiapensis* (B.L. Rob.) R.M. King & H. Rob.	97	94	364,784	246,875	3687	10,875	237,846	1.0	3.0	65.2
3	*Ageratina grandifolia* (Regel) R.M. King & H. Rob.	68	68	412,400	139,148	1401	110,475	221,618	0.3	26.8	53.7
4	*Ageratina ligustrina* (DC.) R.M. King & H. Rob.	824	716	625,473	241,772	2746	191,062	309,808	0.4	30.5	49.5
5	*Ageratina mairetiana* (DC.) R.M.King & H.Rob.	577	567	387,878	198,528	12,833	146,160	165,799	3.3	37.7	42.7
6	*Ageratina vernalis* (Vatke & Kurtz) R.M. King & H. Rob.	106	102	557,923	183,660	3409	142,534	301,187	0.6	25.5	54.0
7	*Baccharis heterophylla* Kunth	502	499	650,100	444,872	45,738	307,814	211,270	7.0	47.3	32.5
8	*Baccharis lancifolia* Less.	71	58	372,699	119,928	4076	92,403	206,874	1.1	24.8	55.5
9	*Baccharis monoica* G.L. Nesom	146	55	336,151	128,384	3697	101,109	172,325	1.1	30.1	51.3
10	*Barkleyanthus salicifolius* (Kunth) H. Rob. & Brettell	416	411	871,464	563,600	41,211	402,173	283,498	4.7	46.1	32.5
11	*Bartlettina platyphylla* (B.L. Rob.) R.M. King & H. Rob.	77	37	279,330	118,849	3467	92,476	132,519	1.2	33.1	47.4
12	*Bartlettina sordida* (Less.) R.M. King & H. Rob.	179	176	311,664	64,849	258	51,960	199,004	0.1	16.7	63.9
13	*Bartlettina tuerckheimii* (Klatt) R.M. King & H. Rob.	146	132	420,782	100,855	95	81,591	258,139	0.0	19.4	61.3
14	*Chromolaena collina* (DC.) R.M. King & H. Rob.	742	608	862,983	918,322	189,840	531,576	156,402	22.0	61.6	18.1
15	*Chromolaena glaberrima* (DC.) R.M.King & H.Rob.	227	97	386,145	174,246	5085	136,090	176,815	1.3	35.2	45.8
16	*Clibadium arboreum* Donn. Sm.	509	443	312,076	118,778	9241	87,285	164,964	3.0	28.0	52.9
17	*Clibadium surinamense* L.	359	9	17,074	2561	1116	978	12,947	6.5	5.7	75.8
18	*Critonia daleoides* DC.	417	228	534,699	215,084	11,819	160,473	270,312	2.2	30.0	50.6
19	*Critonia hebebotrya* DC.	121	95	725,959	587,158	37,854	430,086	153,176	5.2	59.2	21.1
20	*Critonia hospitalis* (B.L. Rob.) R.M. King & H. Rob.	78	75	300,071	64,605	241	52,128	190,791	0.1	17.4	63.6
21	*Critonia morifolia* (Mill.) R.M. King & H. Rob.	802	301	532,910	281,050	36,494	188,731	241,060	6.8	35.4	45.2
22	*Critonia quadrangularis* (DC.) R.M. King & H. Rob.	147	112	823,247	1,190,328	288,440	648,289	5097	35.0	78.7	0.6
23	*Critonia sexangularis* (Klatt) R.M. King & H. Rob.	178	31	77,991	11,474	4005	5362	58,177	5.1	6.9	74.6
24	*Critoniadelphus nubigenus* (Benth.) R.M. King & H. Rob.	57	37	313,638	89,891	0	73,133	182,011	0.0	23.3	58.0
25	*Critoniopsis leiocarpa* (DC.) H. Rob.	315	221	311,807	103,670	1117	82,904	170,099	0.4	26.6	54.6
26	*Critoniopsis obtusa* (Gleason) H. Rob.	118	118	461,220	360,251	54,851	230,966	133,515	11.9	50.1	28.9
27	*Critoniopsis salicifolia* (DC.) H. Rob.	115	115	587,539	563,699	46,612	404,981	68,241	7.9	68.9	11.6
28	*Critoniopsis tomentosa* (Lex.) H. Rob.	225	225	426,071	293,206	27,519	206,722	136,001	6.5	48.5	31.9
29	*Critoniopsis uniflora* (Sch. Bip.) H. Rob.	170	267	686,446	717,520	121,094	443,992	103,775	17.6	64.7	15.1
30	*Dahlia imperialis* Roezl ex Ortgies	143	197	536,126	473,635	33,319	345,751	84,478	6.2	64.5	15.8
31	*Dendroviguiera quinqueradiata* (Cav.) E.E. Schill. & Panero	114	42	218,168	33,681	0	27,472	150,172	0.0	12.6	68.8
32	*Dendroviguiera sphaerocephala* (DC.) E.E. Schill. & Panero	110	114	182,080	173,000	36,608	101,219	44,420	20.1	55.6	24.4
33	*Desmanthodium perfoliatum* Benth.	94	110	328,888	198,100	5352	154,402	111,546	1.6	46.9	33.9
34	*Electranthera mutica* (DC.) Mesfin, D.J.Crawford & Pruski	730	94	264,289	105,968	4255	81,480	132,739	1.6	30.8	50.2
35	*Critoniopsis triflosculosa* (Kunth) H. Rob.	364	679	540,285	336,732	18,823	252,324	183,402	3.5	46.7	33.9
36	*Koanophyllon albicaule* (Sch. Bip. ex Klatt) R.M. King & H. Rob.	501	407	694,526	919,970	189,819	548,328	9384	27.3	78.9	1.4
37	*Koanophyllon galeottii* (B.L. Rob.) R.M. King & H. Rob.	78	67	481,665	331,545	7205	262,540	128,231	1.5	54.5	26.6
38	*Koanophyllon palmeri* (A. Gray) R.M. King & H. Rob.	93	93	1,131,471	1,447,606	263,411	867,251	17,763	23.3	76.6	1.6
39	*Koanophyllon pittieri* (Klatt) R.M. King & H. Rob.	406	127	333,021	152,754	10,319	113,018	157,329	3.1	33.9	47.2
40	*Lasianthaea fruticosa* (L.) K.M. Becker	878	467	972,478	984,137	227,609	550,550	228,008	23.4	56.6	23.4
41	*Lepidaploa polypleura* (S.F. Blake) H. Rob.	122	121	247,682	36,979	0	30,114	171,540	0.0	12.2	69.3
42	*Lepidospartum squamatum* (A. Gray) A. Gray	334	27	132,657	74,277	0	54,872	43,738	0.0	41.4	33.0
43	*Montanoa frutescens* (Mairet ex DC.) Hemsl.	254	254	565,398	375,408	39,691	259,895	195,086	7.0	46.0	34.5
44	*Montanoa hexagona* B.L. Rob. & Greenm.	50	45	294,889	68,068	0	55,131	184,037	0.0	18.7	62.4
45	*Montanoa karwinskii* DC.	67	67	579,350	565,442	99,152	349,899	115,859	17.1	60.4	20.0
46	*Montanoa leucantha* (Lag.) S.F. Blake	826	826	1,004,151	833,182	63,699	588,544	202,978	6.3	58.6	20.2
47	*Montanoa revealii* H. Rob.	51	51	90,690	55,06	0	4477	69,263	0.0	4.9	76.4
48	*Montanoa speciosa* DC.	93	93	1,142,963	1,139,462	70,460	839,795	74,223	6.2	73.5	6.5
49	*Montanoa tomentosa* Cerv.	714	611	741,882	692,138	59,744	493,406	102,033	8.1	66.5	13.8
50	*Nahuatlea arborescens* (Brandegee) V.A. Funk	64	59	14,120	36,404	17,276	11,072	0	122.4	78.4	0.0
51	*Nahuatlea hypoleuca* (DC.) V.A. Funk	185	179	695,788	568,631	100,275	341,675	196,511	14.4	49.1	28.2
52	*Parthenium fruticosum* Less. ex Schltdl. & Cham.	58	58	268,619	625,789	309,213	171,803	36,718	115.1	65.1	13.7
53	*Parthenium tomentosum* DC.	302	302	619,554	1,277,564	529,608	468,614	19,320	85.5	75.6	3.1
54	*Perymenium grande* Hemsl.	329	176	454,542	224,368	743	181,786	187,402	0.2	40.0	41.2
55	*Peucephyllum schottii* A. Gray	237	15	164,105	182,993	17,727	118,339	3981	10.8	72.1	2.4
56	*Pittocaulon praecox* (Cav.) H. Rob. & Brettell	496	496	592,825	470,700	22,228	353,993	120,575	3.7	59.7	20.3
57	*Pittocaulon velatum* (Greenm.) H. Rob. & Brettell	163	162	509,460	637,622	138,987	369,228	40,493	27.3	72.5	7.9
58	*Pluchea sericea* (Nutt.) Coville	138	46	231,437	225,132	9077	158,609	14,129	3.9	68.5	6.1

Table A2. *Cont.*

ID	Species	I	II	III	IV	V	VI	VII	VIII	IX	X
59	*Podachaenium eminens* (Lag.) Sch. Bip. ex Sch. Bip.	683	564	619,757	294,788	7450	229,798	270,875	1.2	37.1	43.
60	*Rensonia salvadorica* S.F. Blake	96	52	126,496	22,543	0	18,456	84,920	0.0	14.6	67.
61	*Roldana albonervia* (Greenm.) H. Rob. & Brettell	441	441	448,148	251,806	10,237	191,290	168,160	2.3	42.7	37.
62	*Roldana angulifolia* (DC.) H. Rob. & Brettell	767	767	587,170	258,071	1712	204,876	264,237	0.3	34.9	45.
63	*Roldana barba-johannis* (DC.) H. Rob. & Brettell	821	817	550,253	228,428	1797	181,755	260,052	0.3	33.0	47.
64	*Roldana eriophylla* (Greenm.) H. Rob. & Brettell	138	138	500,630	608,966	94,727	396,654	7695	18.9	79.2	1.5
65	*Roldana gentryi* H. Rob. & Brettell	56	56	690,226	475,248	27,835	344,349	200,325	4.0	49.9	29.0
66	*Roldana schaffneri* (Sch. Bip. ex Klatt) H. Rob. & Brettell	264	209	436,186	118,254	1102	94,159	257,213	0.3	21.6	59.0
67	*Rumfordia floribunda* DC.	434	434	376,503	147,338	5695	112,434	190,713	1.5	29.9	50.7
68	*Schistocarpha longiligula* Rydb.	101	74	81,642	7568	494	5677	60,918	0.6	7.0	74.6
69	*Sinclairia glabra* (Hemsl.) Rydb.	495	390	522,077	402,878	53,658	267,867	153,473	10.3	51.3	29.4
70	*Telanthophora cobanensis* (J.M. Coult.) H. Rob. & Brettell	179	156	224,845	41,074	0	33,393	149,708	0.0	14.9	66.6
71	*Telanthophora grandifolia* (Less.) H.Rob. & Brettell	801	602	521,994	217,796	4899	169,326	249,901	0.9	32.4	47.9
72	*Telanthophora uspantanensis* (J.M. Coult.) H. Rob. & Brettell	160	157	375,618	68,252	0	55,137	248,274	0.0	14.7	66.1
73	*Tithonia longiradiata* (Bertol.) S.F. Blake	341	256	187,591	35,500	78	28,755	123,464	0.0	15.3	65.8
74	*Verbesina fastigiata* B.L. Rob. & Greenm.	343	343	526,821	363,759	47,946	242,226	182,200	9.1	46.0	34.6
75	*Verbesina guatemalensis* B.L. Rob. & Greenm.	88	9	144,376	43,261	0	35,374	82,631	0.0	24.5	57.2
76	*Verbesina hypargyrea* B.L. Rob. & Greenm.	58	50	298,605	131,918	22,497	84,905	157,747	7.5	28.4	52.8
77	*Verbesina hypoglauca* Sch. Bip. ex Klatt	149	141	488,252	171,872	12,271	125,163	265,497	2.5	25.6	54.4
78	*Verbesina klattii* B.L. Rob. & Greenm.	155	155	266,302	108,800	9019	77,940	136,101	3.4	29.3	51.1
79	*Verbesina lanata* B.L. Rob. & Greenm.	63	32	155,973	24,034	4323	15,346	111,813	2.8	9.8	71.7
80	*Verbesina montanoifolia* B.L. Rob. & Greenm.	80	80	504,621	339,200	24,377	244,898	157,276	4.8	48.5	31.2
81	*Verbesina oligantha* B.L. Rob.	91	90	437,165	277,890	5494	218,217	135,095	1.3	49.9	30.9
82	*Verbesina oncophora* B.L. Rob. & Seaton	292	292	431,477	196,816	6150	151,754	195,615	1.4	35.2	45.3
83	*Verbesina perymenioides* Sch. Bip. ex Klatt	215	203	463,276	279,876	2385	224,930	151,418	0.5	48.6	32.7
84	*Verbesina turbacensis* Kunth	658	396	482,089	157,618	6000	121,700	269,119	1.2	25.2	55.8
85	*Vernonanthura cordata* (Kunth) H. Rob.	221	220	472,583	382,946	25,378	281,877	99,033	5.4	59.6	21.0
86	*Vernonanthura patens* (Kunth) H. Rob.	748	206	360,749	56,005	786	44,745	248,219	0.2	12.4	68.8

References

1. Villaseñor, J.L. Checklist of the native vascular plants of Mexico. *Rev. Mex. Biodivers.* **2016**, *87*, 559–902. [CrossRef]
2. Ulloa Ulloa, C.; Acevedo-Rodríguez, P.; Beck, S.; Belgrano, M.J.; Bernal, R.; Berry, P.E.; Brako, L.; Celis, M.; Davidse, G Forzza, R.C.; et al. An integrated assessment of the vascular plant species of the Americas. *Science* **2017**, *358*, 1614–1617. [CrossRe [PubMed]
3. Villaseñor, J.L. Diversidad y distribución de la familia Asteraceae en México. *Bot. Sci.* **2018**, *96*, 332–358. [CrossRef]
4. FNA, Flora of North America. Asteraceae. Available online: http://www.efloras.org/florataxon.aspx?flora_id=1&taxon_id=100 4#:~{}:text=With%20418%20genera%20and%202413,alpine%20habitats%20to%20salt%20marshes (accessed on 1 February 2021
5. Roque, N.; Magalhães Teles, A.; Naoki Nakajima, J. (Eds.) *A Família Asteraceae no Brasil Clasifição e Diversidade*; Editora d Universidade Federal da Bahia: Salvador, Brazil, 2017; p. 260. [CrossRef]
6. Flora of China. Asteraceae. Available online: http://www.efloras.org/florataxon.aspx?flora_id=2&taxon_id=10074 (accessed o 28 January 2021).
7. Avila, F.; Funk, V.A.; Diazgranados, M.; Díaz-Piedrahíta, S.; Vargas, O. Asteraceae. In *Catálogo de Plantas y Líquenes de Colombi* Bernal, R., Gradstein, S.R., Celis, M., Eds.; Instituto de Ciencias Naturales, Universidad Nacional de Colombia: Bogotá, Colombi Available online: http://catalogoplantasdecolombia.unal.edu.co/es/resultados/familia/Asteraceae/ (accessed on 2 February 2021
8. Barriga, P.; Toasa, G.; Montúfar, R.; Tye, A. Asteraceae. In *Libro Rojo de Plantas Endémicas del Ecuador*; León-Yáñez, S., Valencia, R Pitmam, N., Endara, L., Ulloa Ulloa, C., Navarrete, H., Eds.; Publicaciones del Herbario QCA, Pontificia Universidad Católica de Ecuador: Quito, Ecuador; Available online: https://bioweb.bio/floraweb/librorojo/ListaEspeciesPorFamilia/500047 (accesse on 29 January 2021).
9. Funk, V.A.; Randall, J.B.; Keeley, S.C.; Chan, R.; Watson, L.; Gemeinholzer, B.; Schilling, E.; Panero, J.L.; Baldwin, B.G García-Jacas, N.; et al. Everywhere but Antarctica: Using a supertree to understand the diversity and distribution of the Com positae. *Biol. Skrif.* **2005**, *55*, 343–373.
10. Funk, V.A.; Susanna, A.; Stuessy, T.F.; Robinson, H. Classification of Compositae. In *Systematics, Evolution and Biogeography of th Compositae*; Funk, V.A., Susanna, A., Stuessy, T.F., Bayer, R.J., Eds.; IAPT: Vienna, Austria, 2009; pp. 171–189.
11. Crawford, D.J.; Lowrey, T.K.; Anderson, G.J.; Bernardello, G.; Santos-Guerra, A.; Stuessy, T.F. Genetic diversity in Asteracea endemic to oceanic islands: Baker's Law and polyploidy. In *Systematics, Evolution and Biogeography of the Compositae*; Funk, V.A Susanna, A., Stuessy, T.F., Bayer, R.J., Eds.; IAPT: Vienna, Austria, 2009; pp. 139–151.
12. Standley, P.C. Trees and shrubs of Mexico (Bignoniaceae-Asteraceae). *Constr. U. S. Natl. Herb.* **1926**, *23*, 1401–1641.
13. Turner, B.L. The comps of Mexico. A systematic account of the Family Asteraceae, Vol. 1 Eupatorieae. *Phytol. Mem.* **1997**, *11* 1–272.
14. Strother, J.L. Compositae-Heliantheae s.l. In *Flora of Chiapas*; Daniel, T.F., Ed.; California Academy of Sciences: San Francisco, CA USA, 1999; Volume 5, pp. 1–232.
15. Cabrera, A.L. Revisión del género *Gochnatia* (Compositae). *Rev. Mus. La Plata Nueva Ser.* **1971**, *12*, 1–160.
16. Robinson, H.; Brettell, R.D. Studies in the Senecioneae (Asteraceae). I. A new genus *Pittocaulon*. *Phytologia* **1973**, *26*, 451–453.
17. Robinson, H.; Brettell, R.D. Studies in the Senecioneae (Asteraceae). V. The genera *Psacaliopsis*, *Barkleyanthus*, *Telantophora* anc *Roldana*. *Phytologia* **1974**, *27*, 402–439.
18. Funk, V.A. The Systematics of *Montanoa* (Asteraceae, Heliantheae). *Mem. N. Y. Bot. Gard.* **1982**, *36*, 1–133.

19. Turner, B.L. A recension of Mexican species of *Roldana* (Asteraceae: Senecioneae). *Phytologia* **2005**, *87*, 204–249.
20. Funston, A.M. Taxonomic revision of *Roldana* (Asteraceae: Senecioneae), a genus of Southwestern U.S.A., Mexico, and Central America. *Ann. Missouri Bot. Gard.* **2008**, *95*, 282–337. [CrossRef]
21. Funk, V.A.; Sancho, G.; Roque, N. *Nahuatlea*: A new genus of Compositae (Gochnatieae) from North America. *Phytokeys* **2017**, *91*, 105–124. [CrossRef] [PubMed]
22. Jones, S.B. Revision of *Vernonia* section *Eremosis* (Compositae) in North America. *Brittonia* **1973**, *25*, 86–115. [CrossRef]
23. Turner, B.L. Revision of *Verbesina* sect. *Pseudomontanoa* (Asteraceae). *Plant Syst. Evol.* **1985**, *150*, 237–262. [CrossRef]
24. Ricker, M.; Hernández, H.M.; Sousa, M.; Ochoterena, H. Tree and tree-like species of Mexico: Asteraceae, Leguminosae, and Rubiaceae. *Rev. Mex. Biodivers.* **2013**, *84*, 439–470. [CrossRef]
25. Beech, E.; Rivers, M.; Oldfield, S.; Smith, P. GlobalTreeSearch: The first complete global database of tree species and country distribution. *J. Sustain. For.* **2017**, *36*, 454–489. [CrossRef]
26. Téllez, O.; Mattana, E.; Diazgranados, M.; Kühn, N.; Castillo-Lorenzo, E.; Lira, R.; Montes-Leyva, L.; Flores Ortiz, C.M.; Way, M.; Dávila, P.; et al. Native trees of Mexico: Diversity, distribution, uses and conservation. *PeerJ* **2020**, *8*, e9898. [CrossRef]
27. Sistema Nacional de Información Sobre la Biodiversidad de México (SNIB), Comisión Nacional Para el Conocimiento y Uso de la Biodiversidad (CONABIO). Available online: https://www.snib.mx/ (accessed on 19 January 2021).
28. Susanna, A.; Baldwin, B.G.; Bayer, R.J.; Bonifacio, J.M.; García-Jacas, N.; Keeley, S.C.; Mandel, J.R.; Ortiz, S.; Robinson, H.; Stuessy, T.F. The classification of the Compositae: A tribute to Vicki Ann Funk (1947–2019). *Taxon* **2020**, *69*, 807–814. [CrossRef]
29. DGRU, Dirección General de Repositorios Universitartios, Universidad Nacional Autónoma de México. Portal de Datos Abiertos UNAM, Colecciones Universitarias, Herbario Nacional de México (MEXU). Available online: https://datosabiertos.unam.mx/biodiversidad/ (accessed on 27 November 2020).
30. Diario Oficial de la Federación, Norma Oficial Mexicana NOM-059-SEMARNAT-2010, Protección Ambiental-Especies Nativas de México de Flora y Faina Silvestres-Categorías de Riesgo y Especificaciones para su Inclusion, Exclusion o Cambio-Lista de Especies en Riesgo. Available online: https://dof.gob.mx/nota_detalle_popup.php?codigo=5173091 (accessed on 26 January 2021).
31. CITES, Convención Sobre el Comercio Internacional de Especies Amenazadas de Fauna y Flora Silvestres. Apéndices I, II y III en vigor a Partir del 26 de Noviembre de 2019. Available online: https://cites.org/sites/default/files/esp/app/2019/S-Appendices-2019-11-26.pdf (accessed on 26 January 2021).
32. Hempen, C.H.; Fischer, T. *A Materia Medica for Chinese Medicine*; Churchill Livingstone: London, UK, 2009; pp. 466–513. [CrossRef]
33. BGCI. *GlobalTreeSearch Online Database*; Botanic Gardens Conservation International: Richmond, UK, 2017.
34. CASTUERA-OLIVEIRA, L.; OLIVEIRA-FILHO, A.T.; EISENLOHR, P.V. Emerging hotspots of tree in Brazil. *Acta Bot. Bras.* **2020**, *34*, 117–134. [CrossRef]
35. Ter Steege, H.; Vaessen, R.W.; Cárdenas-López, D.; Sabtier, D.; Antonelli, A.; Mota de Oliveira, S.; Pitman, N.C.A.; Møller Jørgensen, P. The discovery of the Amazonian tree flora with an updated checklist of all know tree taxa. *Sci. Rep.* **2016**, *6*, 29549. [CrossRef]
36. Návar, J. Modeling tree diversity, stand structure and productivity of northern temperate coniferous forests of Mexico. *PeerJ* **2019**, *7*, e7051. [CrossRef] [PubMed]
37. Alaniz-Gutiérrez, L.; Ail-Catzim, C.E.; Villanueva-Gutiérrez, R.; Delgadillo-Rodríguez, J.; Ortiz-Acosta, M.E.; García-Moya, E.; Medina-Cervantes, T.S. Caracterización palinológica de mieles del Valle de Mexicali, Baja California, México. *Polibotánica* **2017**, *43*, 255–283. [CrossRef]
38. Araujo-Mondragón, F.; Redonda-Martínez, R. Flora melífera de la región centro-este del municipio de Pátzcuaro, Michoacán, México. *Acta Bot. Mex.* **2019**, *126*, e1444. [CrossRef]
39. Andrada, A.C. Flora utilizada por *Apis mellifera* L. en el sur del Caldenal (Provincia Fitogeográfica del Espinal), Argentina. *Rev. Mus. Argent. Cienci. Nat.* **2003**, *5*, 329–336. [CrossRef]
40. Santana-Michel, F.J.; Cervantes-Aceves, N.; Jiménez-Reyes, N. Flora melífera del estado de Colima, México. *Ibugana* **1998**, *6*, 251–277.
41. Calabria, L.M.; Emerenciano, V.P.; Scotti, M.T.; Mabry, T.J. Secondary chemistry of Compositae. In *Systematics, Evolution and Biogeography of the Compositae*; Funk, V.A., Susanna, A., Stuessy, T.F., Bayer, R.J., Eds.; IAPT: Vienna, Austria, 2009; pp. 73–88.
42. Torres, C.; Galetto, L. Are nectar sugar composition and corolla tube length related to the diversity of insects than visit Asteraceae flowers? *Plant Biol.* **2002**, *4*, 360–366. [CrossRef]
43. Torres, C.; Galetto, L. Importancia de los polinizadores en la reproducción de Asteraceae de Argentina Central. *Acta Bot. Venez.* **2008**, *31*, 473–494.
44. Camina, J.L.; Tourn, E.; Andrada, A.C.; Pellegrini, C.; Ashworth, L. Spatial and temporal distribution of floral rewards within the capitula: The case of *Hyalis argentea* (Asteraceae). *Bol. Soc. Argent. Bot.* **2019**, *54*, 17–27. [CrossRef]
45. Arroyo, M.T.; Robles, V.; Tamburrino, I.; Martínez-Harms, J.; Garreaud, R.D.; Jara-Arancio, P.; Pliscoff, P.; Copier, A.; Arenas, J.; Keymer, J.; et al. Extreme Drought Affects Visitation and Seed Set in a Plant Species in the Central Chilean Andes Heavily Dependent on Hummingbird Pollination. *Plants* **2020**, *9*, 1553. [CrossRef]
46. Moreira-Muñoz, A.; Scherson, R.A.; Luebert, F.; Román, M.J.; Monge, M.; Diazgranados, M.; Silva, H. Biogeography, phylogenetic relationships and morphological analyses of the South American genus *Mutisia* L.f. (Asteraceae) shows early connections of two disjunct biodiversity hotspots. *Org. Divers. Evol.* **2020**, *20*, 639–656. [CrossRef]

47. Secretaría del Medio Ambiente y Recursos Naturales. Reserva de la Biosfera Mariposa Monarca. Available online: https://www.gob.mx/semarnat/articulos/reserva-de-la-biosfera-mariposa-monarca-79228 (accessed on 15 January 2021).
48. Cornejo-Tenorio, G.; Casas, A.; Farfán, B.; Villaseñor, J.L.; Ibarra-Manríquez, G. Flora y vegetación de las zonas núcleo de la Reserva de la Biosfera Mariposa Monarca, México. *Bol. Soc. Bot. Mex.* **2003**, *73*, 43–62. [CrossRef]
49. Cornejo-Tenorio, G.; Ibarra-Manríquez, G. *Flora Ilustrada de la Reserva de la Biosfera Mariposa Monarca*; Comisión Nacional para el Conocimiento y Uso de la Biodiversidad (CONABIO), Universidad Nacional Autónoma de México(UNAM): Mexico City, Mexico, 2008; p. 441.
50. Secretaría de Agricultura y Desarrollo Rural. La miel Mexicana va Endulzando al Mundo. Available online: https://www.gob.mx/agricultura/articulos/la-miel-mexicana-va-endulzando-el-mundo?idiom=es (accessed on 15 January 2021).
51. Secretaría de Agricultura y Desarrollo Rural. Yucatán se Encuentra entre los Principales Productores de miel del país. Available online: https://www.gob.mx/agricultura/yucatan/articulos/yucatan-se-encuentra-entre-los-principales-productores-de-miel-del-pais?idiom=es (accessed on 15 January 2021).
52. Simpson, B.B. Economic importance of Compositae. In *Systematics, Evolution and Biogeography of the Compositae*; Funk, V.A. Susanna, A., Stuessy, T.F., Bayer, R.J., Eds.; IAPT: Vienna, Austria, 2009; pp. 45–58.
53. Ruiz-Reyes, E.; Suárez, M. Lactonas sesquiterpénicas. Diversidad estructural y sus actividades biológicas. *Rev. CENIC Cienc. Biol.* **2015**, *46*, 9–24.
54. Diario Oficial de la Federación. Secretaría de Agricultura y Ganadería. Decreto por el que se Declara Símbolo de la Floricultura Nacional la Flor de Dalia en todas sus Especies y Variedades. Available online: http://www.dof.gob.mx/nota_to_imagen_fs.php?codnota=4720563&fecha=13/05/1963&cod_diario=203468 (accessed on 16 January 2021).
55. Secretaría de Medio Ambiente y Recursos Naturales. Dalia, flor Representativa de México. Available online: https://www.gob.mx/semarnat/articulos/dalia-flor-representativa-de-mexico?idiom=es (accessed on 16 January 2021).
56. McVaugh, R. Compositae. In *Flora Novo-Galiciana*; Anderson, W.R., Ed.; The University of Muchigan Press: Ann Arbor, MI, USA, 1984; Volume 12, pp. 1–1157.
57. Rzedowski, J.; Calderón de Rzedowski, G.; Carrillo-Reyes, P. Compositae Tribu Helientheae II (géneros *Lagascea-Zinnia*). In *Flora del Bajío y de Regiones Adyacentes*; Rzedowski, J., Calderón de Rzedowski, G., Eds.; Instituto de Ecología, A.C., Centro Regional del Bajío: Pátzcuaro, Michoacán, Mexico, 2011; Volume 172, pp. 1–100.
58. Flora Ornamental de Barcelona. Available online: https://floraornamentaldebarcelona.com/2020/01/18/flora-del-parc-de-la-guineueta-3-arbustos-2-margaritero-montanoa-bipinnatifida/ (accessed on 8 October 2020).
59. Plant This. Available online: http://plantthis.co.nz/plant-information.asp?gardener=18906&tabview=features&plantSpot=1 (accessed on 8 October 2020).
60. Auckland Museum. Available online: https://www.aucklandmuseum.com/collections-research/collections/record/am_naturalsciences-object-780123 (accessed on 8 October 2020).
61. VivoPlant. Available online: https://vivoplant.com/comprar-plantas-arbustivas-exterior/69-257-especies-arbustivas-bartlettina-sordida.html (accessed on 16 January 2021).
62. Toledo, V.M. Los Cambios Climáticos del Pleistoceno y sus Efectos Sobre la Vegetación Tropical Cálida y Húmeda de México. Master's Thesis, Facultad de Ciencias, Universidad Nacional Autónoma de México, Mexico City, Mexico, 29 April 1976.
63. Torrescano-Valle, N.; Islebe, G.A. Holocene paleoecology, climate history and human influence in the southwestern Yucatan Peninsula. *Rev. Palaeobot. Palynol.* **2015**, *217*, 1–8. [CrossRef]
64. Escobar García, P.; Winkler, M.; Flatscher, R.; Sonnleitner, M.; Krejcikova, J.; Suda, J.; Hülber, K.; Schneeweiss, G.; Schönswetter, P. Extensive range persistence in peripheral and interior refugia characterizes Pleistocene range dynamics in a widespread Alpine plant species (*Senecio carniolicus*, Asteraceae). *Mol. Ecol.* **2012**, *21*, 1255–1270. [CrossRef] [PubMed]
65. Cuervo-Robayo, A.P.; Ureta, C.; Gómez-Albores, M.A.; Meneses-Mosquera, A.K.; Téllez-Valdés, O.; Martínez-Meyer, E. One hundred years of climate change in Mexico. *PLoS ONE* **2020**, *15*, e0209808. [CrossRef] [PubMed]
66. Comisión Nacional de Áreas Naturales Protegidas. Áreas Naturales Protegidas Decretadas. Available online: http://sig.conanp.gob.mx/website/pagsig/datos_anp.htm (accessed on 19 February 2021).
67. Secretaría de Medio Ambiente y Desarrollo Territorial de Jalisco. 15 Cuenca Alimentadora del Distrito Nacional de Riego 043 Nayarit. Available online: https://semadet.jalisco.gob.mx/medio-ambiente/biodiversidad/areas-naturales-protegidas/142 (accessed on 19 February 2021).
68. Sistema de Información y Monitoreo para la Conservación, Comisión Nacional de Áreas Naturales Protegidas. CADNR043 Estado de Nayarit. Available online: https://simec.conanp.gob.mx/ficha_pdf.php?anp=4&reg=11 (accessed on 19 February 2021).
69. Secretaría de Medio Ambiente y Recursos Naturales. Reserva de la Biosfera Sierra de Manantlán. Available online: https://www.gob.mx/semarnat/articulos/reserva-de-la-biosfera-sierra-de-manantlan (accessed on 19 February 2021).
70. Secretaría de Medio Ambiente y Recursos Naturales. Reserva de la Biosfera Sierra Gorda. Available online: https://www.gob.mx/semarnat/articulos/reserva-de-la-biosfera-sierra-gorda-celebra-su-19-aniversario (accessed on 19 February 2021).
71. Centro de Investigación en Ciencias de Información Geoespacial. Selva Lacandona, Mosaico del Paisaje, Paisaje Ecológico. Available online: http://mapas.centrogeo.org.mx/ciberatlas/lacandona/mosaico/clima.htm (accessed on 20 February 2021).
72. Instituto Nacional de Ecología, Plan de manejo de la Reserva de la Biosfera El Triunfo. Available online: http://www.paot.org.mx/centro/ine-semarnat/anp/AN14.PDF (accessed on 20 February 2021).

3. Secretaría de Medio Ambiente y Recursos Naturales. Reserva de la Biosfera Montes Azules. Available online: https://www.gob.mx/semarnat/articulos/reserva-de-la-biosfera-montes-azules-selva-lancandona-chiapas?idiom=es (accessed on 20 February 2021).
4. Secretaría de Medio Ambiente y Recursos Naturales. Reserva de la Biosfera El Triunfo. Available online: https://www.gob.mx/semarnat/articulos/reserva-de-la-biosfera-el-triunfo?idiom=es (accessed on 20 February 2021).
5. Secretaría de Medio Ambiente y Recursos Naturales. Reserva de la Biosfera Los Tuxtlas. Available online: https://www.gob.mx/semarnat/articulos/reserva-de-la-biosfera-los-tuxtlas?idiom=es (accessed on 21 February 2021).
6. Comisión Nacional de Áreas Naturales Protegidas. Programa de Conservación y Manejo Reserva de la Biosfera Los Tuxtlas, México. Available online: https://www.conanp.gob.mx/que_hacemos/pdf/programas_manejo/tuxtla_final.pdf (accessed on 21 February 2021).
7. Montsarrat, S.; Jarvie, S.; Svenning, J.C. Anthropocene refugia: Integrating history and predictive modelling to assess the space available for biodiversity in a human-dominated world. *Philos. Trans. R. Soc. B* **2019**, *374*, 20190219. [CrossRef]
8. Bennett, K.D.; Provan, J. What do we mean by 'refugia'? *Quat. Scient. Rev.* **2008**, *27*, 2449–2455. [CrossRef]
9. Suárez-Mota, E.; Villaseñor, J.L.; Ramírez-Aguirre, M.B. Sitios prioritarios para la conservación de la riqueza y el endemismo de la Sierra Norte de Oaxaca, México. *Acta Bot. Mex.* **2018**, *124*, 49–74. [CrossRef]
0. Aschentrupp Toledo, R. *Las Comunidades Indígenas de la Sierra Norte de Oaxaca*; Centro de Estudios Sociales y Opinión Pública (CESOP): Mexico City, Mexico, 2015; p. 15.
1. Ceballos Pérez, S.G. *Manejo Forestal Comunitario Sustentable en el Sierra Norte de Oaxaca*; Published by the Autor: Mexico City, Mexico, 2017; p. 310.
2. Sastre Merino, S. Análisis de la gestión forestal comunitaria y sus implicaciones sociales en Ixtlán de Juárez, Oaxaca (México). Bachelor's Thesis, Universidad Politécnica de Madrid, Madrid, Spain, 2008.
3. Gasca Zamora, J. Gobernanza y gestión comunitaria de recursos naturales en la Sierra Norte de Oaxaca. *Reg. Soc.* **2014**, *60*, 89–120. [CrossRef]
4. Villaseñor, J.L.; Delgadillo, C.; Ortiz, E. Hostspots from a multigroup perspective: Mosses and Senecios in the Transmexican Volcaniv Belt. *Biodivers. Conserv.* **2006**, *15*, 4045–4058. [CrossRef]
5. Suárez-Mota, M.E.; Téllez-Valdés, O. Red de áreas prioritarias para la conservación de la biodiversidad del Eje Volvánico Transmexicano analizando su riqueza florística y variabilidad climática. *Polibotánica* **2014**, *38*, 67–93.
6. Critical Ecosystem Partnership Fund. Región Norte del Hotspot de Biodiversidad de Mesoamérica, Belice, Guatemala, México. Available online: https://www.cepf.net/sites/default/files/final.spanish.mesoamerica.northernmesoamerica.ep_.pdf (accessed on 23 February 2021).
7. Instituto Nacional de Estadística y Geografía, Territorio de México. Available online: http://www.cuentame.inegi.org.mx/territorio/extension/default.aspx?tema=T (accessed on 6 February 2021).
8. Pruski, J.F. Asteraceae. In *Flora Mesoamericana*; Davidse, G., Sousa, S.M., Knapp, S., Chiang, F., Eds.; Misssouri Botanical Garden Press: St. Louis, MO, USA, 2018; Volume 5, Part 2, pp. 1–608.
9. Redonda-Martínez, R.; Villaseñor, J.L. Asteraceae, Vernonieae. In *Flora del Valle de Tehuacán-Cuicatlán*; Medina Lemos, R., Sánchez Ken, J.G., García Mendoza, A., Arias Montes, S., Eds.; Instituto de Biología, Universidad Nacional Autónoma de México: Mexico City, Mexico, 2009; Volume 72, pp. 1–23.
0. Redonda-Martínez, R.; Villaseñor, J.L. Asteraceae, Senecioneae. In *Flora del Valle de Tehuacán-Cuicatlán*; Medina Lemos, R., Sánchez Ken, J.G., García Mendoza, A., Arias Montes, S., Eds.; Instituto de Biología, Universidad Nacional Autónoma de México: Mexico City, Mexico, 2011; Volume 89, pp. 1–64.
1. Redonda-Martínez, R. Asteraceae, Gochnatieae. In *Flora del Valle de Tehuacán-Cuicatlán*; Medina Lemos, R., García Mendoza, A., Arias Montes, S., Grether González, R., Fonseca Juárez, R.M., Eds.; Instituto de Biología, Universidad Nacional Autónoma de México: Mexico City, Mexico, 2019; Volume 155, pp. 1–16.
2. Turner, B.L. Taxonomy of *Neurolaena* (Asteraceae-Heliantheae). *Plant Syst. Evol.* **1982**, *140*, 119–139. [CrossRef]
3. Redonda-Martínez, R.; Villaseñor, J.L. El género *Lepidaploa* (Familia Asteraceae, Tribu Vernonieae) en México. *Rev. Mex. Biodivers.* **2011**, *82*, 782–797. [CrossRef]
4. Botany Collection Search, Smithsonian National Museum of Natural History. Available online: http://collections.nmnh.si.edu/search/botany/?v=s1#new-search (accessed on 24 November 2020).
5. Tropicos.org. Missouri Botanical Garden. Available online: http://www.tropicos.org (accessed on 28 November 2020).
6. SEinet Arizona-New Mexico Chapter. Available online: https://swbiodiversity.org/seinet/collections/harvestparams.php (accessed on 6 December 2020).
7. Hinojosa-Espinosa, O.; Villaseñor, J.L.; Ortiz, E. On the identity of two Mexican species of *Ageratina* (Eupatorieae, Asteraceae): *A. grandifolia* and *A. rivalis*. *Bot. Sci.* **2019**, *97*, 250–259. [CrossRef]
8. Robinson, H. Generic and subtribal clasification of American Vernonieae. *Smithsonian Contr. Bot.* **1999**, *89*, 1–116. [CrossRef]
9. Schilling, E., Panero, J.L. The Eupatorieae Website. Available online: http://schillinglab.utk.edu/Danielweb/Eup/genusindex.html (accessed on 22 September 2020).
00. GBIF, Global Biodiversity Information Facility. Available online: https://www.gbif.org/ (accessed on 21 September 2020).

101. Bachmann, S.; Moat, J.; Hill, A.W.; de la Torre, J.; Scott, B. Supporting Red List threat assessments with GeoCAT: Geospatia conservation assessment tool. e-Infrastructures for data publishing in biodiversity science. *ZooKeys* **2011**, *150*, 117–126. [CrossRe [PubMed]
102. Google Earth Pro. Available online: https://www.google.com/intl/es/earth/download/gep/agree.html (accessed on 4 Decem ber 2020).
103. Instituto Nacional de Estadística y Geografía, Archivo Histórico de Localidades Geoestadísticas. Available online: http //www.inegi.org.mx/app/geo2/ahl/ (accessed on 2 December 2020).
104. Mapcarta, El Mapa Libre. Available online: https://mapcarta.com/es/ (accessed on 2 December 2020).
105. PueblosAmerica.com. Pueblos de México. Available online: https://mexico.pueblosamerica.com/ (accessed on 28 Novem ber 2020).
106. Phillips, S.J.; Anderson, R.P.; Schapire, R.E. Maximum entropy modeling of species geographic distributions. *Ecol. Model.* **200** *190*, 231–259. [CrossRef]
107. Van Proosdij, A.S.J.; Sosef, M.S.M.; Wieringa, J.J.; Raes, N. Minimum required number of specimen records to develop accurat species distribution models. *Ecography* **2016**, *39*, 542–552. [CrossRef]
108. WorldClim, Maps, Graphs, Tables, and Data of the Global Climate. Available online: https://www.worldclim.org/ (accessed o 5 January 2021).
109. Mendes, P.; Elías Velasco, S.J.; Alves de Andrade, A.F.; De Marco Júnior, P. Dealing with overprediction in species distributio models: How adding distance constraints can improve model accuracy. *Ecol. Model.* **2020**, *431*, 109180. [CrossRef]
110. Jiménez-Valverde, A.; Lobo, J.M. Threshold criteria for conversion of probability of species presence to either-or presence-absenc *Acta Oecol.* **2007**, *31*, 361–369. [CrossRef]
111. Tatebe, H.; Ogura, T.; Nitta, T.; Komuro, Y.; Ogochi, K.; Takemura, T.; Sudo, K.; Sekiguchi, M.; Manabu, A.; Saito, F.; et al. De scription and basic evaluation of simulated mean state, internal variability, and climate sensitivity in MIROC6. *Geosci. Model De* **2019**, *12*, 2727–2765. [CrossRef]
112. Fasullo, J.T. Evaluating Simulated Climate Patterns from the CMIP Archives Using Satellite and Reanalysis Dataset *Geosci. Model Dev.* **2020**, *13*, 3627–3642. [CrossRef]
113. Provincias Biogeográficas de México, Comisión Nacional para el Conocimiento y Uso de la Biodiversidad (CONABIO). 199 Available online: http://conabio.gob.mx/informacion/metadata/gis/rbiog4mgw.xml?_xsl=/db/metadata/xsl/fgdc_html.xsl _indent=no (accessed on 25 January 2021).
114. International Union for Conservation of Nature. *Red List Categories and Criteria, Version 3.1*, 2nd ed.; IUCN: Gland, Switzerlan Cambridge, UK, 2012; p. 32.

plants

MDPI

Article

Advancing Timberline on Mt. Fuji between 1978 and 2018

Hitoshi Sakio [1,*] and Takehiro Masuzawa [2]

1 Sado Island Center for Ecological Sustainability, Niigata University, Sado 952-2206, Japan
2 Department of Biology, Faculty of Science, Shizuoka University, Shizuoka 422-8529, Japan; masuzawa.takehiro@shizuoka.ac.jp
* Correspondence: sakio@agr.niigata-u.ac.jp

Received: 29 September 2020; Accepted: 7 November 2020; Published: 10 November 2020

check for updates

Abstract: Climate change is a major cause of changes in alpine and polar vegetation, particularly at the edges of distributions. In temperate regions, these changes are expected to occur at the timberline of alpine zones. On Mt. Fuji, the highest mountain in Japan, the timberline is located 2400–2500 m above sea level. Over a 40-year period (1978–2018), we researched changes in the timberline vegetation of Mt. Fuji. A permanent belt transect extending from the upper timberline to subalpine zones was set up in August 1978. Tree diameters and heights were recorded at the establishment of the transect and every 20 years afterwards. Over the 40 years of the study, the timberline advanced rapidly upwards, and the degree of vegetation cover above the timberline increased remarkably. Notably, the expansion of *Salix reinii* into the upper part of the timberline facilitated the subsequent spread of *Larix kaempferi* into this zone. Seedlings of *L. kaempferi* were particularly abundant at the upper timberline and became established on the uppermost part of the slope. The shape of *L. kaempferi* at the upper timberline changed from a prostrate form to an upright tree form. We conclude that the upward advance of the alpine timberline observed on Mt. Fuji is due to climate change.

Keywords: alpine timberline; global warming; *Larix kaempferi*; long-term ecological research; Mt. Fuji; seedling

1. Introduction

Global climate change has been analyzed using long-term meteorological and oceanographic data. Over the period from 1880 to 2012, globally averaged combined land and ocean surface temperatures followed a linear trend, increasing by 0.85 °C [1]. At the end of the 21st century (2081–2100), the change in the global surface temperature relative to 1850–1900 is projected with high confidence to exceed 1.5 °C under RCP (Representative Concentration Pathway) 4.5, RCP6.0, and RCP8.5 scenarios [1]. Breeding and species selection in agriculture have had to adapt to global warming, and the spread of infectious plant diseases associated with global warming has become a major problem [2]. In addition, glacial retreat has had considerable effects on ecosystems [3] and on the distribution of organisms in both polar regions [4] and alpine zones [5]. The movement of plant communities to the north has been confirmed in polar regions in the Northern Hemisphere, and the early arrival of spring has caused a mismatch in plant pollination, even in temperate regions [6]. Plant communities have also been spreading upwards in alpine zones [7]. Long-term monitoring is needed to confirm whether these changes are temporary or permanent, and to verify whether the simulations based on past data are correct. Long-term monitoring research is ongoing worldwide [8]. In Japan, long-term monitoring observations have been conducted in forests, grasslands, lakes, marshes, and oceans nationwide since 2003 through the "Monitoring Site 1000" project of the Ministry of the Environment [9]. Because a

20-year period is too short to capture the effects of global warming on long-term changes in natural ecosystems, these projects are expected to continue in the future.

The alpine timberline is a forefront of struggle for tree survival [10]. In this zone, the timberline migrates upwards or downwards in response to plant-limiting factors, such as low air temperature, frost damage, carbon limitation, winter desiccation, and strong wind [11]. In particular, the area called the "kampfzone" is the place with the most dynamic changes in the timberline ecosystem [10]. This kampfzone is characterized by extreme ecological conditions for survival, growth, and competition [12], and is very sensitive to changing climatic conditions. Long-term observation of ecosystem changes in such places is therefore considered useful for clearly understanding the effects of global climate change on ecosystems.

Mt. Fuji is the highest mountain with a timberline in Japan. Timberline zones of most high mountains in Japan are dominated by *Pinus pumila* communities [13,14]. In Europe and North America, coniferous trees in the upper part of the timberline exhibit a high degree of phenotypic plasticity in reaction to environmental factors in the kampfzone [10], and the forest structure around the timberline in these regions is clearly different from that of Japan. Unlike other high mountains in Japan, Mt. Fuji lacks *P. pumila* and has a forest structure similar to the timberlines of Europe and North America. Conducting a survey on Mt. Fuji is important to allow comparisons of timberline dynamics in mid-latitude Japan with those in Europe [15–20] and North America, and to analyze their relationship with global warming. In many locations in Europe, humans have had a long-term impact on the timberline, especially in the 17th to 19th centuries when high mountain meadows were extensively used for grazing and haymaking [21–23]. However, the timberline on Mt. Fuji has always been maintained in a natural state, with the exception of some low-impact activities such as mountain climbing. For these reasons, investigating the timberline of Mt. Fuji is especially important compared with other high mountains in Japan.

Many research reports have appeared on forest vegetation on Mt. Fuji, and the upward movement of the timberline has been pointed out in previous studies [24–26]. Oka [26] confirmed this phenomenon based on field surveys and an annual ring analysis, and Maruta and Masuyama [25] reported similar observations from a time series analysis using aerial photographs. These studies were short term, however, and did not clarify the mechanism and dynamics of forest change. No long-term detailed studies of the timberline of Mt. Fuji had thus been conducted. In 1978, we installed a permanent quadrat at the timberline of Mt. Fuji and have been continuously tracking the dynamics and mechanisms of the forest vegetation [27,28]. In the early years of our study, we found that the timberline had expanded upwards considerably between 1978 and 1999 [28], possibly because of climate change.

The purpose of this study was to clarify how vegetation at the timberline of Mt. Fuji changed during the 40 years from 1978 to 2018. In particular, we aimed to determine (1) whether the observed upward movement of the timberline of Mt. Fuji is continuing and (2) whether the forest structure of the timberline has changed over this period. Our results may be useful for predicting how the recent temperature rise will affect vegetation in the timberline of Mt. Fuji and how global warming will impact forest vegetation in extreme environments.

2. Results

2.1. Change in Timberline Vegetation over Time

The timberline vegetation of Mt. Fuji fluctuated drastically during 1978–2018. The number of trunks with a height of 130 cm or more varied greatly in *Alnus alnobetula* subsp. *maximowiczii*, *Salix reinii*, and *Larix kaempferi* (Figure 1). *S. reinii* decreased sharply in plots 8 and 9, and all trunks disappeared in plots 10–12 in 2018. Conversely, this species was not seen at all in plots 3 and 4 in 1978, but increased in these plots to 2 and 17 trunks, respectively, in 2018. *A. alnobetula* subsp. *maximowiczii* showed little change in plots 8–11 between 1978 and 1999 but had declined sharply by 2018. This species also increased slightly in plots 4–6 between 1978 and 2018 (Friedman's test, $p < 0.05$). *L. kaempferi* gradually increased in plots 5 and 6 from 1978 to 2018. This species first appeared in plots 3 and 4 in 1988 and in plot 2 in 2018. The number of *L. kaempferi* individuals rapidly increased in plots 3 and 4 between 1999 and 2018, and significantly increased in plots 2–6 between 1978 and 2018 (Friedman's test, $p < 0.01$). This species remained constant in plots 7–22. Few changes were observed in *Abies veitchii* and *Picea jezoensis* var. *hondoensis* in any plots during the research period.

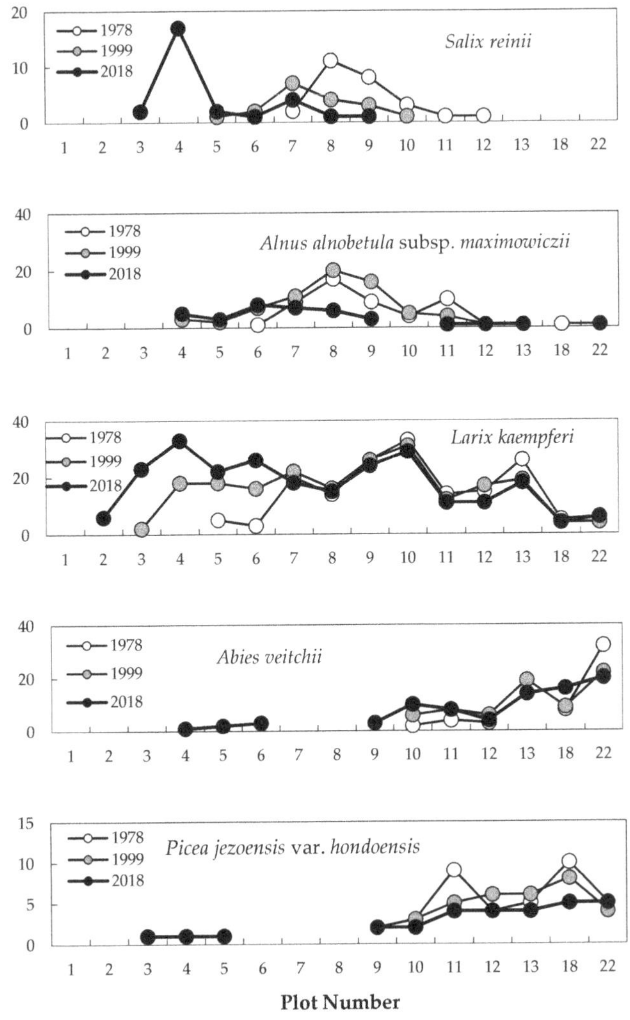

Figure 1. Distribution of the number of trees across the slope gradient between 1978 and 2018. White, gray, and black circles are the number of trees in 1978, 1999, and 2018, respectively.

The average heights of *S. reinii* and *A. alnobetula* subsp. *maximowiczii* increased in plots 7–9 from 1978 to 2018 (Figure 2). The average heights of *S. reinii* and *A. alnobetula* subsp. *maximowiczii* significantly increased between 1978 and 2018 in plots 3–6 and 4–6, respectively (Friedman's test, $p < 0.05$). In the case of *L. kaempferi*, average tree heights increased continuously in all study plots, with significant increases observed in plots 2–6 from 1978 to 2018 (Friedman's test, $p < 0.01$). In 2018, new trunks of *A. veitchii* and *P. jezoensis* var. *hondoensis* appeared in plots 4–6 and 3–5, respectively.

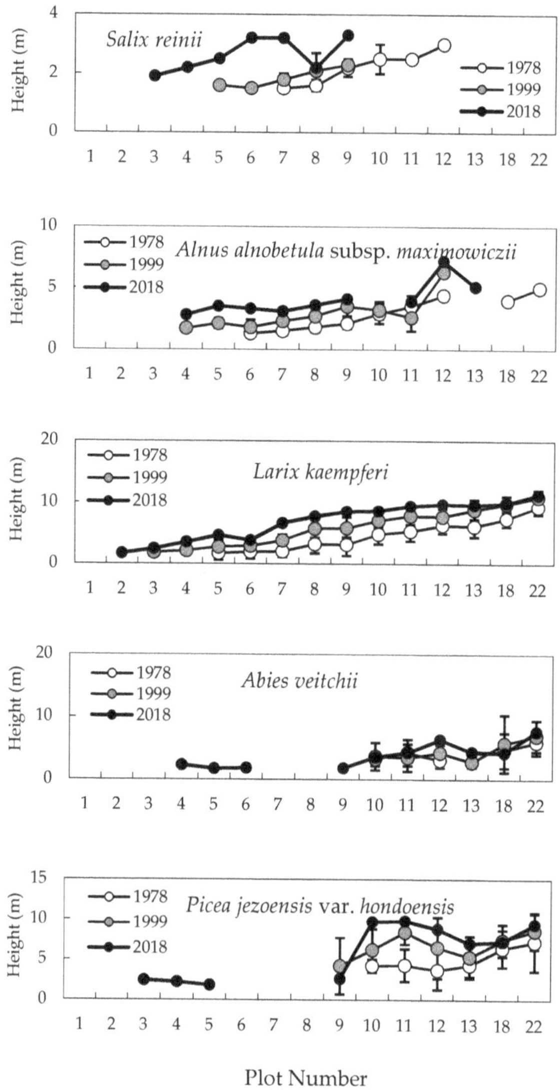

Figure 2. Distribution of tree heights across the slope gradient between 1978 and 2018. White, gray, and black circles are tree heights in 1978, 1999, and 2018, respectively.

The total basal areas (BAs) for all species significantly increased over the 40-year study period, except in plots 7 and 18 (Figure 3; Friedman's test, $p < 0.001$). The BA of *A. alnobetula* subsp. *maximowiczii* and *S. reinii* increased at elevations above plot 7 (at the timberline) and decreased below plot 8 from 1978 to 1999. *A. alnobetula* subsp. *maximowiczii* and *S. reinii* exhibited 6.7- and 2.1-fold increases, respectively, in plot 7 between 1978 and 1999, but the BA of these two species decreased sharply in all plots in 2018. The BA of *L. kaempferi* significantly increased in all plots from 1978 to 2018 (Friedman's test, $p < 0.0001$). The BA of *L. kaempferi* increased at a similar rate between 1978–1999 vs. 1999–2018 in plots 8, 9, 10, and 11 but increased more rapidly in plots 4–7 between 1999 and 2018.

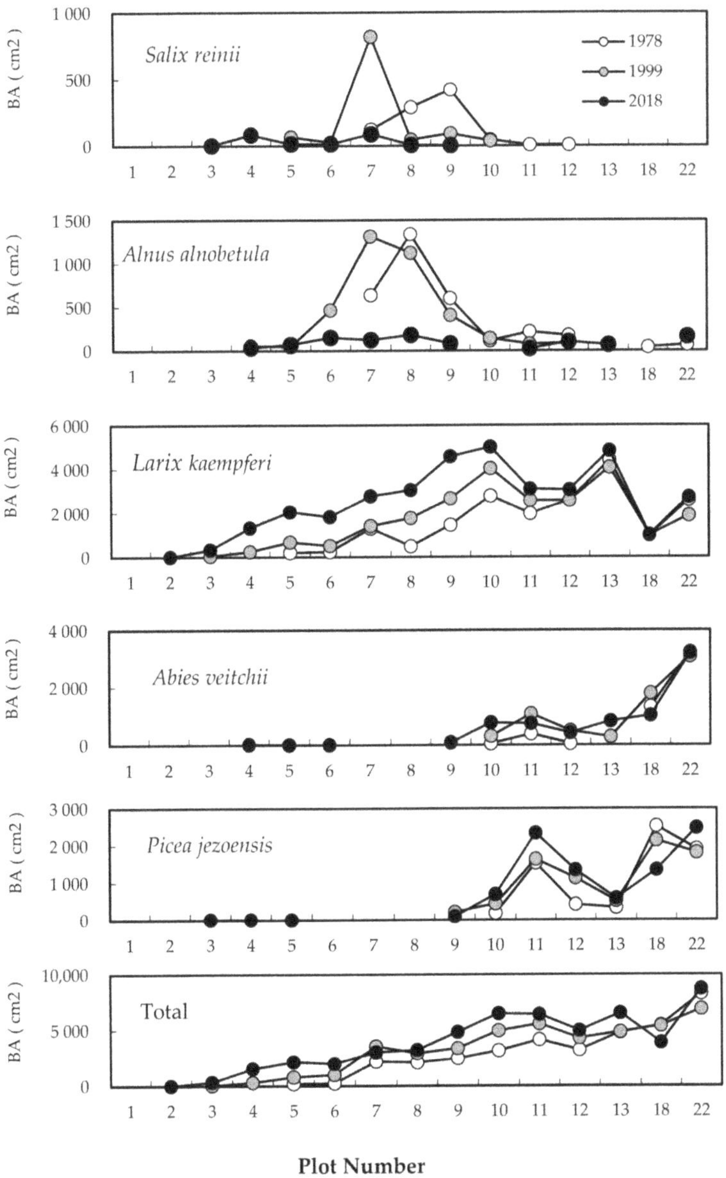

Figure 3. Distribution of tree basal areas (BAs) across the slope gradient between 1978 and 2018. White, gray, and black circles are BAs in 1978, 1999, and 2018, respectively.

2.2. Difference in Changes between the Two 20-Year Periods

Changes between the two 20-year periods are illustrated by the second-order difference between 1978, 1999, and 2018 (Figure 4). The number of trunks of *S. reinii* and *L. kaempferi* increased in plots 4 and 3, respectively, between the two periods, while *A. alnobetula* subsp. *maximowiczii* decreased in plots 8 and 9. The BA of *S. reinii* and *A. alnobetula* subsp. *maximowiczii* decreased in plot 7 between the two periods, whereas that of *L. kaempferi* increased markedly in plots 4–7.

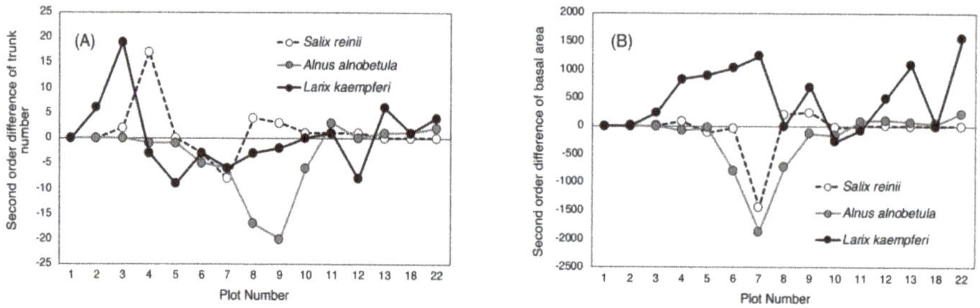

Figure 4. Second-order differences among three measurements (1978, 1999, and 2018) of trunk number (**A**) and basal area (**B**) of major tree species.

2.3. Establishment of Seedlings at the Upper Timberline

L. kaempferi seedlings were widely distributed throughout the upper area of the timberline. In particular, seedlings had colonized plots 3–6 in 1999 but decreased over the next 20 years (Figure 5A). Between 1999 and 2018, *L. kaempferi* disappeared from plots 7 and 8, and the total number of seedlings decreased by 19% in plots 3–6. In contrast, the number of *L. kaempferi* seedlings in plot 2 increased from 6 to 14 between 1999 and 2018, and seedlings were established in plot 1 for the first time during this period. Seedlings of *A. veitchii* invaded vegetation patches in plots 3–7 in 1999; this species had spread to plots 8 and 9 by 2018, and the number of seedlings in plots 3–6 had increased (Figure 5B).

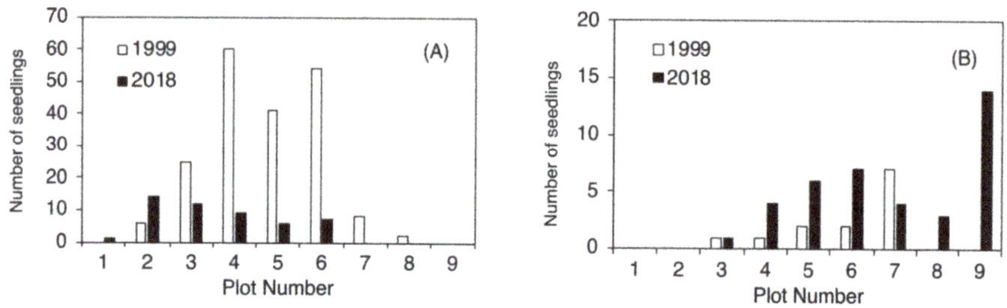

Figure 5. Number of established seedlings of *Larix kaempferi* (**A**) and *Abies veitchii* (**B**) as of 1999 and 2018.

Many seedlings were established in the upper timberline area. In the case of *L. kaempferi*, 196 and 12 new seedlings were established from 1978–1999 and 1999–2018, respectively (Table 1). The number of new seedlings during the first 20-year period was significantly higher than during the latter (Wilcoxon signed-rank test, $p < 0.05$). The range of newly invaded plots was moving upwards. No seedlings were established in plots 6–8 between 1999 and 2018. Seedlings recorded in plot 2 between 1999 and 2018 were clearly taller than those established between 1978 and 1999. In the case of *A. veitchii*, 17 and 12 new seedlings were established from 1978–1999 and 1999–2018, respectively (Table 2). The range of established new seedlings was the same in both periods.

Table 1. No. and height of *Larix* seedlings established.

Plot Number	No. of New Seedlings		Seedling Height (cm)	
	1978–1999	1999–2018	1978–1999	1999–2018
1	0	1	-	39
2	6	7	12 ± 8	36 ± 14
3	25	2	32 ± 38	18
4	60	1	52 ± 60	80
5	41	1	74 ± 79	11
6	54	0	77 ± 74	-
7	8	0	44 ± 23	-
8	2	0	11 ± 9	-
9	0	0	-	-

Table 2. No. and height of *Abies* seedlings established.

Plot Number	No. of New Seedlings		Seedling Height (cm)	
	1978–1999	1999–2018	1978–1999	1999–2018
1	0	0	-	-
2	0	0	-	-
3	1	1	16	96
4	1	2	115	75 ± 45.3
5	2	4	29 ± 16	51.3 ± 36.7
6	2	1	23 ± 9.9	50
7	7	3	57.7 ± 76.3	43.3 ± 20.7
8	4	1	40 ± 23.6	78
9	0	0	-	-

In the period 1978–1999 and 1999–2018, *L. kaempferi* individuals newly established above the timberline varied greatly in size (Figure 6). The heights of all newly established seedlings were less than 20 cm between 1978–1999 but up to 90 cm between 1999–2018. Similar to seedling heights, the diameter at ground surface of seedlings established between 1999 and 2018 was larger than that of those established between 1978 and 1999 (Welch's *t*-test, $p < 0.01$).

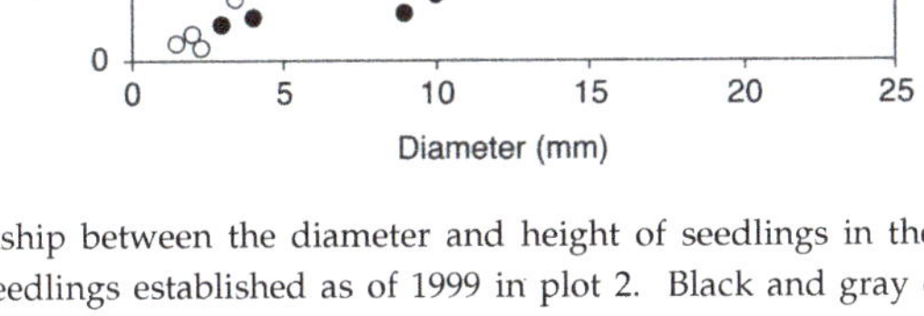

Figure 6. Relationship between the diameter and height of seedlings in the upper kampfzone. White circles are seedlings established as of 1999 in plot 2. Black and gray circles are seedlings established as of 2018 in plots 2 and 1, respectively.

2.4. Degree of Vegetation Cover

Changes in the degree of vegetation cover of the upper timberline area are shown in Figure 7. The total degree of vegetation cover increased from the top to the bottom of the upper timberline in 1978. During the 40 years, this value did not change in plot 1 but increased greatly in plots 2–5.

The degree of tree cover also increased during the 40 years. No change was observed in plot 1, but the degree of tree cover increased in plots 2–5, especially in plots 3–5. In 2018, trees accounted for approximately 90% of the total vegetation cover in plots 3–5. As a result, most of the vegetation in plots 3–5 was dominated by trees in 2018.

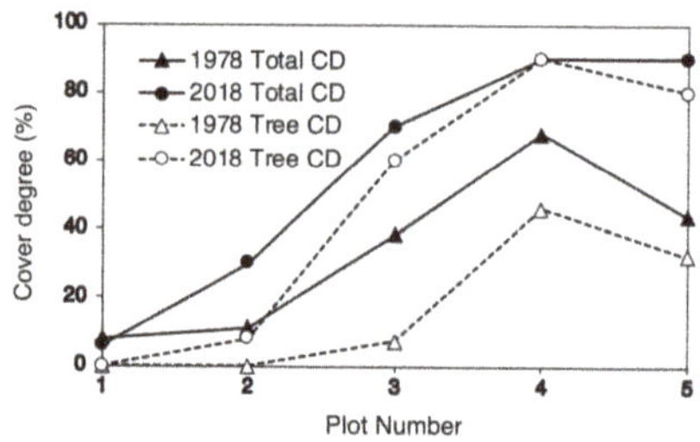

Figure 7. Degree of vegetation cover in 1978 and 2018. The total degree of cover (CD) includes the degree of tree cover.

3. Discussion

According to our earlier findings, the timberline of Mt. Fuji moved upwards between 1978, when the research site was set up, and 1999 [27,28]. In the present study, we found that the timberline of Mt. Fuji continued to considerably advance upwards until by 2018.

The vegetation around the Mt. Fuji timberline varies according to elevation, with the change in vegetation type from upper to lower elevations following this order: herbaceous plant patches, deciduous shrubs (*S. reinii*, *L. kaempferi*, and *A. alnobetula subsp. maximowiczii*), deciduous *L. kaempferi* forests, and evergreen coniferous forests of *A. veitchii* and *P. jezoensis* var. *hondoensis* [27]. The most striking change over the 40 studied years was that of the upper timberline vegetation (plots 2–6) above the deciduous shrubs (Figures 2 and 3). In plot 1, at the top, no change was observed in vegetation cover over 40 years, whereas vegetation cover in plots 2–5 increased considerably, especially that due to woody plants (Figure 7). This change was the result of an increase in the number of *S. reinii* and *L. kaempferi* individuals. Both species have pioneering properties and can invade bare land, but they have different life forms. *S. reinii* is a bush with multiple stems and a maximum height of 3 m. In contrast, *L. kaempferi* can form forests more than 10 m high below the timberline (Figure 4) and shade out *S. reinii* individuals during growth. Although *S. reinii* was able to invade the upper timberline and increase in height (Figure 4), this species was suppressed below the timberline by *L. kaempferi*, and the population therefore decreased sharply (Figures 3 and 5). Conversely, *L. kaempferi* did not decrease in population size after invading and establishing itself at higher elevations, and its BA increased with increasing tree height (Figures 4 and 5). In our earlier study, *L. kaempferi* seedlings were found to be established very close to the edges of vegetation patches [28]. Patches of *S. reinii* may play an important role in the establishment of *L. kaempferi* at the krummholz limit on Mt. Fuji [29,30] and shrubs provide safe sites through creating a more favorable microclimate [31,32]. *S. reinii* may also contribute to tree succession by providing adjacent late colonizers (*L. kaempferi*) with compatible ectomycorrhizal (ECM) symbionts [33].

In contrast to *S. reinii*, *A. alnobetula* subsp. *maximowiczii* markedly decreased, both in terms of population size and BA, over the 40 years without invading upper elevations (Figures 3 and 5). Previous studies have shown that individuals of *A. alnobetula* subsp. *maximowiczii* on Mt. Fuji have a high production rate because of their high photosynthetic rate [34] and the high nitrogen content of leaf litter [35]. In one study, in addition, the amount of annual nitrogen fixation by nodules was found to be almost the same as that of nitrogen used for annual growth [36]. This species has therefore been considered to contribute to the upward movement in nitrogen supply at the timberline of Mt. Fuji [28]. The rapid decline of *A. alnobetula* subsp. *maximowiczii* dwarf forests over the past 40 years, however, suggests that factors other than improvements in soil nitrogen have had a major effect on the advance of the timberline on Mt. Fuji.

In the case of evergreen conifers, seedlings of *A. veitchii* and *P. jezoensis* var. *hondoensis* were present in the upper part of the timberline in 1999. By 2018, some individuals with a height of approximately 2 m were observed at this higher elevation (plots 3–6). These individuals had developed from the seedlings established in 1999 (Figure 3).

Seedling establishment is an important factor in the expansion of plant and forest distributions. Many seedlings of species such as *L. kaempferi*, *A. veitchii*, and *P. jezoensis* var. *hondoensis* were found at the upper timberline (plots 3–6) between 1978 and 1999 [28]. In particular, *L. kaempferi* invasion and establishment was extensive, with a total of 196 seedlings found in 1999 in plots 2–8 (Table 1). By 2018, however, many seedlings had died as a result of an increase in the vegetation cover at the upper timberline (plots 3–6). Although 12 new seedlings were established between 1999 and 2018 (Table 1), the overall population decreased to 49 in 2018 (Figure 7). The seedling population continued to increase in the uppermost areas (plots 1 and 2), however, as *L. kaempferi* was first established in plot 1 and doubled in plot 2 in 1999 (Figure 7). The sizes of *L. kaempferi* individuals established between 1978–1999 and 1999–2018 were clearly larger in the latter period in plot 2 (Table 1). As described above, seedlings of *L. kaempferi* were steadily advancing above the timberline, which is considered to be a more severe environment.

Seedlings of *A. veitchii*, which has a higher shade tolerance than *L. kaempferi*, were found to be distributed throughout the timberline ecotone (plots 3–9), and the number of *A. veitchii* seedlings had increased in 2018 compared with 1999 (Figure 7). The number of individuals established between 1999 and 2018 was almost the same as between 1978 and 1999 (Table 2). In other words, *A. veitchii* is a recent invader of locations of previous *L. kaempferi* invasion and growth. *L. kaempferi* thus acts as a facilitator for *A. veitchii*, as the former has deeper roots than the latter and can avoid desiccation [37,38].

Tree forms are shaped by physical forces, such as strong wind and heavy snow, under severe environments [10,21,39–41]. The area around the timberline is strongly affected by strong winds in winter. Life forms with highly variable physiognomy predominate among woody plants, ranging from bushes with "flagged" leaders to cushions of krummholz pressed close to the ground (table shape) [10]. Prostrate *L. kaempferi* at the upper timberline in 1978 (plots 3 and 4) are shown in Figure 8A, while Figure 8B shows prostrate *L. kaempferi* with erect stems and an estimated age of 150 years (plot 5) at that time. In 2018, however, a different landscape of tree shapes was evident (Figure 8C). In particular, *L. kaempferi* that had newly invaded the upper portion of the krummholz limit were growing with erect trunks without dwarfing (plots 2 and 3; Figure 8C). Maruta and Masuyama [25] have reported that the first step in advancing timberline is the establishment of the dwarf type of *L. kaempferi*, which contrasts with individuals at lower elevations that gradually form erect trunks. The conflict between the results of their study and our findings may be due to differences in topography between the respective research sites as well as factors related to climate change, such as an increase in temperature.

Figure 8. Tree shape of *L. kaempferi* at the timberline of Mt. Fuji. (**A**) Prostrate trees at the upper timberline in plots 2 and 3 in 1978. (**B**) Prostrate trees with erect stems in plot 5 in 1978. (**C**) Seedlings with erect stems in plots 2 and 3 in 2018.

Air temperature and CO_2 concentration are important determinants of plant growth. Global warming has recently become a problem [1]. The significant warming occurring in recent years may have changed timberline ecosystems in Europe [7,42–44], China [45], and Japan [28]. During the last 40 years, the average maximum temperature has continued to rise during the plant growth period on Mt. Fuji (Figure 9). Higher temperatures will extend the plant growth period and elevate photosynthetic rates. As the photosynthetic period lengthens and the photosynthetic rate increases, the annual growth rate may increase, and shoots may be formed that can better withstand the winter environment. In addition to air temperature, the CO_2 concentration is rising. The mean CO_2 concentration at the summit of Mt. Fuji was approximately 335 ppm in August 1981 [46], 388 ppm in 2010 [47], and above 400 ppm in 2015 [48]. The saturation limit for CO_2 assimilation in *L. kaempferi* is at an intercellular CO_2 concentration of 600 ppm, regardless of mineral nutrient supply [49]. The photosynthetic rate may therefore continue to rise. As mentioned above, the temperature rise and the increase in CO_2 concentration are considered to be factors that increase the biomass production of trees at the timberline. As a result, the annual growth of *L. kaempferi* may have increased, and erect shoots may be able to survive, even in the severe winter environment, without unusual phenotypic response.

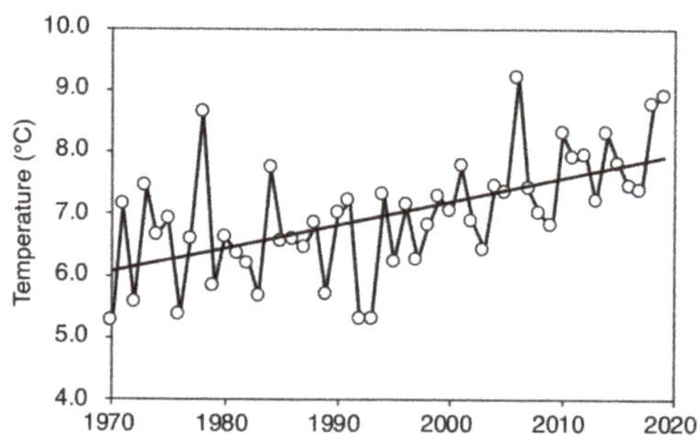

Figure 9. The average maximum temperature between June and September at the summit of Mt. Fuji over 50 years. The data in this figure are from the Japan Meteorological Agency (http://www.jma.go.jp/jma/menu/report.html).

The mechanism responsible for timberline rise on Mt. Fuji has been thought to entail the invasion of deciduous shrub trees, such as *S. reinii* and *A. alnobetula* subsp. *maximowiczii*, into herbaceous patches to form shrub forests, with *L. kaempferi* also invading to form a table-shaped shrub forest that eventually stands upright. However, *L. kaempferi* has invaded the upper part of the timberline and continued to grow upright without forming dwarf shrubs with a prostrate form. This phenomenon is thought to be due to changes in external factors in addition to natural succession occurring after the eruption in 1707. One such external factor is an increase in annual growth due to temperature rise. This temperature increase promotes an increase in photosynthetic rate and an expansion in the photosynthetic period. In addition to the rise in air temperature, the increase in CO_2 concentration accelerates the growth rate.

Previous studies have pointed out the consequences of the imbalance between rapid climate change and slow biological responses. Even among the most mobile species such as butterflies, these pollinators have been unable to extend their ranges as fast as required to keep pace with climate change [50,51]. On the other hand, results of our research over forty years and global warming forecasts [1], has suggested that the timberline of Mt. Fuji will continue to advance upwards. These results may indicate that monitoring of the alpine ecosystems may be effective in capturing the sensitive impact of climate change on forests. The "Monitoring site 1000" project of the Ministry of the Environment in Japan, which began in 2003, has yielded results for many forests, but its impact on rapid climate change in recent years has been less apparent. Therefore, long-term monitoring in various climate zones, including alpine ecosystems, will be necessary to assess the effects of global warming on organisms.

4. Study Site and Methods

4.1. Study Site

Mt. Fuji (3776 m) is the highest volcano in Japan. A stratovolcano mainly composed of basalt, the mountain spans both Yamanashi and Shizuoka prefectures. Mt. Fuji is still a young adolescent volcano and is believed to have begun on a submarine volcano one million to 700,000 years ago, but its exact origins remain unclear. The hillside over 2500 m above sea level is covered with volcanic products, with vegetation distributed on the slope below. The research site (35°21′ N, 138°45′ E) was located at the timberline (ca. 2400 m) on the southeastern slope, where the vegetation is recovering from damage caused by the 1707 volcanic eruption of Hoei-Zan, a parasitic crater. This area is a special protection zone of Fuji-Hakone-Izu National Park. Special protection areas, which feature the most outstanding natural scenery and pristine conditions, are the most strictly regulated areas of the park.

Most of the timberline vegetation in Japan comprises of *Pinus pumila*, whereas the timberline on Mt. Fuji is dominated by *L. kaempferi*. The vegetation at the timberline of Mt. Fuji changes dramatically as one proceeds down the slope, with variations in herbs, deciduous shrubs, deciduous conifers, and evergreen conifers (Figure 10). Perennial herb vegetation occurs above the timberline and includes *Astragalus laxmannii* var. *adsurgens*, *Arabis serrata* var. *serrata*, *Aconogonon weyrichii* var. *alpinum*, *Carex doenitzii*, and *Artemisia pedunculosa* [27]. The timberline vegetation comprises of deciduous dwarf trees: *A. alnobetula* subsp. *maximowiczii*, *S. reinii*, *L. kaempferi*, and *Betula ermanii*. Downslope of the dwarf vegetation, the forest composition changes from *L. kaempferi* forest to coniferous evergreen forest, of which *A. veitchii* and *P. jezoensis* var. *hondoensis* are dominant [27].

Figure 10. Timberline vegetation on Mt. Fuji in 1980.

The timberline weather on Mt. Fuji is very cold and windy [52], but little snow cover is present (ca. 30 cm in depth from November to February). The annual mean air temperature is 1.1 °C, with the highest and lowest monthly means of 11.8 °C in August and −9.5 °C in February [27]. The annual precipitation is approximately 4500 mm [53]. The precipitation level is high throughout the year, especially during the summer growing season because of the rainy and typhoon season. Relative humidity is very high because of frequent fog from June to September (mean >80%) [27].

The surface substratum at the research site consists of basalt scoria from the volcanic eruptions of Hoei in 1707. This scoria is easily moved downward by repeated freezing and thawing, and by strong wind or heavy rain. The ground surface is therefore very unstable. The nitrogen and carbon content of the soil at the upper timberline is very low, 0.02% and 0.3%, respectively [27].

4.2. Methods

We established a 220-m-long permanent belt transect (10 m wide) extending from the upper timberline zone to the subalpine forest dominated by coniferous evergreen trees in August 1978. The transect comprised of 22 numbered contiguous plots (10 × 10 m; Figure 11). All living trees

(≥130 cm tall) were identified to species level. The diameter at breast height (DBH; diameter 130 cm above ground level) and height of trees were recorded in 1978 in the uppermost plots at 0–130 m along the transect and in lower plots at 180 and 220 m [27]. For deciduous shrub tree such as *A. alnobetula* subsp. *maximowiczii* and *S. reinii*, the longest trunks of individual plants were selected for tree height measurements. Basal area (BA) was calculated from the DBH data for all trunks and summarized as the total area per plot for each tree species. A second census taken in 1999 confirmed that the timberline of Mt. Fuji was advancing upwards [28]. At that time, climate change was proposed to be one of the causes of this expansion. Since the beginning of the 21st century, extreme weather events have occurred around the world, and rising temperatures have been observed in Japan. To be able to compare changes in the timberline over the next 20 years, we repeated the measurements in 2018. Friedman's test (followed by a least significant difference test) was used to detect significant differences in the number of trees, tree height, and BA among the three measurements obtained from the very uppermost plots (plots 1–6). In addition, we compared changes in the number of trees and BA between the first half of the study period (1987–1999) and the second half (1999–2018). We first determined the first-order difference between 1987 and 1999 and that between 1999 and 2018. The second-order difference, obtained by subtracting the former from the latter, was then calculated, and the change between the two periods was compared.

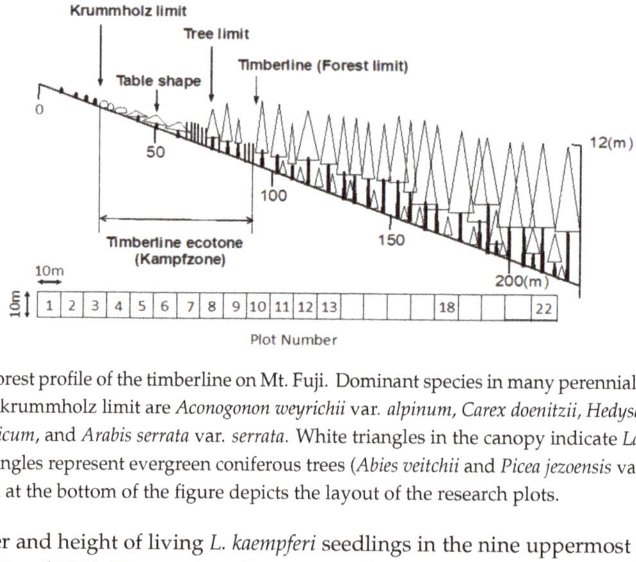

Figure 11. Forest profile of the timberline on Mt. Fuji. Dominant species in many perennial herb patches at the upper krummholz limit are *Aconogonon weyrichii* var. *alpinum*, *Carex doenitzii*, *Hedysarum vicioides* subsp. *japonicum*, and *Arabis serrata* var. *serrata*. White triangles in the canopy indicate *Larix kaempferi*, and gray triangles represent evergreen coniferous trees (*Abies veitchii* and *Picea jezoensis* var. *hondoensis*). The diagram at the bottom of the figure depicts the layout of the research plots.

The number and height of living *L. kaempferi* seedlings in the nine uppermost plots (1–9) were measured in 1999 and 2018. The number, diameter, and height of new seedlings established between 1978–1999 and 2000–2018 were measured in the nine uppermost plots (1–9) in 1999 and 2018. These data were used to investigate the upwards advance of vegetation.

The total degree of vegetation cover and degree of tree cover in each plot were measured in 1978 and 2018. The degree of vegetation cover, an index of the degree of foliage overgrowth, is defined as the vegetation area obtained by orthographically projecting the foliage of the vegetation onto a horizontal plane, that is, the occupancy degree of vegetation per unit area of each plot. The degree of tree cover is defined in the same way.

Author Contributions: Conceptualization, H.S. and T.M.; methodology, H.S. and T.M.; formal analysis, H.S.; investigation, H.S. and T.M.; resources, T.M.; data curation, H.S.; writing—original draft preparation, H.S.; writing—review and editing, H.S. and T.M.; visualization, H.S.; supervision, T.M.; project administration, T.M.; funding acquisition, T.M. All authors have read and agreed to the published version of the manuscript.

Funding: A part of this research was funded by a Grant-in-Aid for Scientific Research (B) (No. 19310008) from the Ministry of Education, Culture, Sports, Science and Technology, Japan.

Acknowledgments: The authors are indebted to F. Konta for his advice, and to members of the Laboratory of Plant Ecology, Shizuoka University and to member of the Niigata University Forest, for their kind assistance during field work.

Conflicts of Interest: The authors declare no conflict of interest.

References

1. IPCC. *Climate Change 2014: Synthesis Report. Contribution of Working Groups I, II and III to the Fifth Assessment Report of the Intergovernmental Panel on Climate Change*; IPCC: Geneva, Switzerland, 2014.
2. Evans, N.; Baierl, A.; Semenov, M.A.; Gladders, P.; Fitt, B.D.L. Range and severity of a plant disease increased by global warming. *J. R. Soc. Interface* **2008**, *5*, 525–531. [CrossRef] [PubMed]
3. Shakun, J.D.; Clark, P.U.; He, F.; Lifton, N.A.; Liu, Z.; Otto-Bliesner, B.L. Regional and global forcing of glacier retreat during the last deglaciation. *Nat. Commun.* **2015**, *6*, 8059. [CrossRef] [PubMed]
4. Bjorkman, A.; Criado, M.G.; Myers-Smith, I.H.; Ravolainen, V.; Jónsdóttir, I.S.; Westergaard, K.B.; Lawler, J.P.; Aronsson, M.; Bennett, B.; Gardfjell, H.; et al. Status and trends in Arctic vegetation: Evidence from experimental warming and long-term monitoring. *Ambio* **2019**, *49*, 678–692. [CrossRef] [PubMed]
5. Rixen, C.; Wipf, S. Non-equilibrium in Alpine Plant Assemblages: Shifts in Europe's Summit Floras. In *High Mountain Conservation in a Changing World*; Catalan, J., Ninot, J.M., Aniz, M.M., Eds.; Springer: Cham, Switzerland, 2017; pp. 285–303.
6. Gérard, M.; Vanderplanck, M.; Wood, T.; Michez, D. Global warming and plant–pollinator mismatches. *Emerg. Top. Life Sci.* **2020**, *4*, 77–86. [CrossRef] [PubMed]
7. Vitasse, Y.; Hoch, G.; Randin, C.F.; Lenz, A.; Kollas, C.; Körner, C. Tree recruitment of European tree species at their current upper elevational limits in the Swiss Alps. *J. Biogeogr.* **2012**, *39*, 1439–1449. [CrossRef]
8. Mirtl, M.; Borer, E.T.; Djukic, I.; Forsius, M.; Haubold, H.; Hugo, W.; Jourdan, J.; Lindenmayer, D.; McDowell, W.H.; Muraoka, H.; et al. Genesis, goals and achievements of Long-Term Ecological Research at the global scale: A critical review of ILTER and future directions. *Sci. Total Environ.* **2018**, *626*, 1439–1462. [CrossRef]
9. Biodiversity Center of Japan. Monitoring Sites 1000. Available online: http://www.biodic.go.jp (accessed on 20 September 2020).
10. Tranquillini, W. *Physiological Ecology of the Alpine Timber-Line—Tree Existence at High Altitude with Special Reference to the European Alps*; Springer: Berlin/Heidelberg, Germany, 1979. [CrossRef]
11. Holtmeier, F.K. *Mountain timberlines. Ecology, patchiness, and dynamics. Advances in Global Change Research 14*; Kluwer Academic Publishers: Dordrecht, The Netherlands, 2003.
12. Skre, O. Northern treelines as indicators of climate and land use changes—A literature review. *Agrotechnology* **2019**, *9*, 190. [CrossRef]
13. Okitsu, S. Comparative studies on the Japanese alpine zone with special reference to the ecology of *Pinus pumila* thickets. *Geogr. Rev. Jpn.* **1984**, *57*, 791–802. [CrossRef]
14. Okitsu, S. Consideration of vegetational zonation based on the establishment process of a *Pinus pumila* zone in the Hokkaido, northern Japan. *Jap. J. Ecol.* **1985**, *35*, 113–121, (In Japanese with English summary). [CrossRef]
15. Payette, S.; Eronen, M.; Jasinski, P. The circumboreal tundra-taiga interface: Late Pleistocene and Holocene changes. *Ambio* **2002**, *12*, 15–22.
16. Skre, O.; Baxter, R.; Crawford, R.M.M.; Challaghan, T.V.; Fedorkov, A. How will the tundra-taiga interface respond to climate change? *Ambio* **2002**, *12*, 37–46.
17. Dalen, L.; Hofgaard, A. Differential Regional Treeline Dynamics in the Scandes Mountains. *Arct. Antarct. Alp. Res.* **2005**, *37*, 284–296. [CrossRef]
18. Juntunen, V.; Neuvonen, S. Natural regeneration of Scots pine and Norway spruce close to the timberline in northern Finland. *Silva Fenn.* **2006**, *40*, 443–458. [CrossRef]
19. Hofgaard, A.; Tømmervik, H.; Rees, G.; Hanssen, F. Latitudinal forest advance in northernmost Norway since the early 20th century. *J. Biogeogr.* **2012**, *40*, 938–949. [CrossRef]
20. Bryn, A.; Potthoff, K. Elevational treeline and forest line dynamics in Norwegian mountain areas—A review. *Landsc. Ecol.* **2018**, *33*, 1225–1245. [CrossRef]

21. Holtmeier, F.K. Geoecological aspects of timberlines in Northern and Central Europe. *Arct. Antarct. Alp. Res.* **1973**, *5*, 45–54.
22. Rochefort, R.; Little, R.L.; Woodward, A.; Peterson, D.L. Changes in sub-alpine tree distribution in western North America: A review of climatic and other causal factors. *Holocene* **1994**, *4*, 89–100. [CrossRef]
23. Treml, V.; Migoń, P. Controlling factors limiting timberline position and shifts in the Sudetes: A review. *Geogr. Pol.* **2015**, *88*, 55–70. [CrossRef]
24. Kobayashi, T.; Nashimoto, M.; Takeuchi, T.; Nakano, T. Distribution and ages of larch (*Larix kaempferi*) in different altitude on the west slope of Mt. Fuji. *Mt. Fuji Res.* **2012**, *6*, 55–60. (In Japanese)
25. Maruta, E.; Masuyama, K. Elevation mechanism of timberline ecotone on the southern slope of Mt. Fuji. *Mt. Fuji Res.* **2009**, *3*, 1–12.
26. Oka, S. The tree limit and its dynamics on the western and northwestern slopes of Mount Fuji, central Japan. *Geogr. Rep. Tokyo Metrop. Univ.* **1992**, *27*, 1–28.
27. Masuzawa, T. Ecological studies on the timberline of Mt. Fuji I. Structure of plant community and soil development on the timberline. *Bot. Mag. Tokyo.* **1985**, *98*, 15–28. [CrossRef]
28. Sakio, H.; Masuzawa, T. The advancing timberline on Mt. Fuji: Natural recovery or climate change? *J. Plant Res.* **2012**, *125*, 539–546. [CrossRef] [PubMed]
29. Endo, M.; Yamamura, Y.; Tanaka, A.; Nakano, T.; Yasuda, T. Nurse-plant effects of a dwarf shrub on the establishment of tree seedlings in a volcanic desert on Mt. Fuji, central Japan. *Arct. Antarct. Alp. Res.* **2008**, *40*, 335–342. [CrossRef]
30. Nabeta, K.; Yamamura, Y.; Nakano, T.; Yasuda, T. Effects of dwarf shrub *Salix reinii* on the establishment, survival and growth of tree seedlings *Larix kaempferi* in a volcanic desert on Mt. Fuji: Verification by monitoring for 14 years. *Mt. Fuji Res.* **2015**, *9*, 17–24. (In Japanese)
31. Aradottir, Á.L.; Eysteinsson, T. Restoration of birch woodlands in Iceland. In *Restoration of Boreal and Temperate Forests*; Stanturf, J.A., Madsen, P., Eds.; CRC Press: Boca Raton, FL, USA, 2005; pp. 195–209.
32. Chen, J.; Yang, Y.; Wang, S.; Sun, H.; Schöb, C. Shrub facilitation promotes selective tree establishment beyond the climatic treeline. *Sci. Total Environ.* **2020**, *708*, 134618. [CrossRef]
33. Nara, K. Pioneer dwarf willow may facilitate tree succession by providing late colonizers with compatible ectomycorrhizal fungi in a primary successional volcanic desert. *New Phytol.* **2006**, *171*, 187–198. [CrossRef]
34. Sakio, H.; Masuzawa, T. Ecological studies on the timberline of Mt. Fuji II. Primary productivity of *Alnus maximowiczii* dwarf forest. *Bot. Mag. Tokyo* **1987**, *100*, 349–363. [CrossRef]
35. Sakio, H.; Masuzawa, T. Ecological studies on the timber-line of Mt. Fuji III. Seasonal changes in nitrogen content in leaves of woody plants. *Bot. Mag. Tokyo* **1992**, *105*, 47–52. [CrossRef]
36. Tsutsumi, H.; Nakatsubo, T.; Ino, Y. Field measurements of nitrogen-fixing activity of intact saplings of *Alnus maximowiczii* in the subalpine zone of Mt Fuji. *Ecol. Res.* **1993**, *8*, 685–692. [CrossRef]
37. Yura, H. Comparative ecophysiology of *Larix kaempferi* (Lamb.) Carr. and *Abies veitchii* Lindle. I. Seedling establishment on bare ground on Mt. Fuji. *Ecol. Res.* **1988**, *3*, 67–73. [CrossRef]
38. Yura, H. Comparative ecophysiology of *Larix kaempferi* (Lamb.) Carr. and *Abies veitchii* Lindl. II. Mechanisms of higher drought resistance of seedlings of *L. kaempferi* as compared with *A. veitchii*. *Ecol. Res.* **1989**, *4*, 351–360. [CrossRef]
39. Heikkinen, O.; Tuovinen, M.; Autio, J. What determines the timberline? *Fennia* **2002**, *180*, 67–74.
40. Smith, W.K.; Germino, M.J.; Hancock, T.E.; Johnson, D.M. Another perspective on altitudinal limits of alpine timberlines. *Tree Physiol.* **2003**, *23*, 1101–1112. [CrossRef]
41. Nakano, Y.; Sakio, H. The regeneration mechanisms of a *Pterocarya rhoifolia* population in a heavy snowfall region of Japan. *Plant Ecol.* **2018**, *219*, 1387–1398. [CrossRef]
42. Grabherr, G.; Gottfried, M.; Pauli, H. Climate effects on mountain plants. *Nature* **1994**, *369*, 448. [CrossRef]
43. Kullman, L. 20th century climate warming and tree-limit rise in the southern Scandes of Sweden. *Ambio* **2001**, *30*, 72–80. [CrossRef]
44. Sanz-Elorza, M.; Dana, E.D.; Gonzáalez, A.; Sobrino, E. Changes in the high-mountain vegetation of the Central Iberian Peninsula as a probable sign of global warming. *Ann. Bot.* **2003**, *92*, 273–280. [CrossRef]
45. Parolo, G.; Rossi, G. Upward migration of vascular plants following a climate warming trend in the Alps. *Basic Appl. Ecol.* **2008**, *9*, 100–107. [CrossRef]
46. Nakazawa, T.; Aoki, S.; Fukabori, M.; Tanaka, M. The concentration of atmospheric carbon dioxide on the summit of Mt. Fuji (3776 m), Japan. *J. Meteorol. Soc. Jpn.* **1984**, *62*, 688–695. [CrossRef]

47. Sunaga, A.; Mukai, H.; Machida, T.; Nojiri, Y.; Conway, T.; Masarie, K.; Crotwell, A.; Dlugokenchy, E.J.; White, J.; Vaughn, B. Comparison of co-located air samples at Mauna Loa Observatory and CO2 observations at Mt. Fuji. In Proceedings of the Abstract of 39th NOAA Earth System Research Laboratory 2011 Global Monitoring Annual Conference, Boulder, CO, USA, 17 May 2011; NOAA: Boulder, CO, USA, 2011; p. 3.
48. Nomura, S.; Mukai, H.; Terao, Y.; Machida, T.; Nojiri, Y. Six years of atmospheric CO_2 observations at Mt. Fuji recorded with a battery-powered measurement system. *Atmos. Meas. Tech.* **2017**, *10*, 667–680. [CrossRef]
49. Yazaki, K.; Ishida, S.; Kawagishi, T.; Fukatsu, E.; Maruyama, Y.; Kitao, M.; Tobita, H.; Koike, T.; Funada, R. Effects of elevated CO_2 concentration on growth, annual ring structure and photosynthesis in *Larix kaempferi* seedlings. *Tree Physiol.* **2004**, *24*, 941–949. [CrossRef]
50. Menéndez, R.; Megías, A.G.; Hill, J.K.; Braschler, B.; Willis, S.G.; Collingham, Y.; Fox, R.; Roy, D.B.; Thomas, C.D. Species richness changes lag behind climate change. *Proc. R. Soc. B.* **2006**, *273*, 1465–1470. [CrossRef] [PubMed]
51. Bedford, F.E.; Whittaker, R.J.; Kerr, J.T. Systemic range shift lags among a pollinator species assemblage following rapid climate change. *Botany* **2012**, *90*, 587–597. [CrossRef]
52. Oka, S. On the deformation of larches on Mt. Fuji and their causal factors. *J. Geogr.* **1980**, *89*, 97–112. (In Japanese with English summary) [CrossRef]
53. Ito, E. Climate of Mt. Fuji. *Bull. Fac. Agric. Shizuoka Univ.* **1964**, *14*, 117–187. (In Japanese)

Publisher's Note: MDPI stays neutral with regard to jurisdictional claims in published maps and institutional affiliations.

plants

MDPI

rticle

)iversity and Relationships among Neglected Apricot Prunus armeniaca L.) Landraces Using Morphological Traits ınd SSR Markers: Implications for \gro-Biodiversity Conservation

;iandomenico Corrado [1,*], Marcello Forlani [1], Rosa Rao [1,2] and Boris Basile [1]

1 Department of Agricultural Sciences, University of Naples Federico II, 80055 Portici, NA, Italy; marcello.forlani@unina.it (M.F.); rosa.rao@unina.it (R.R.); boris.basile@unina.it (B.B.)
2 Consorzio Interuniversitario Biotecnologie (CIB), University of Naples Federico II Unit, 80055 Portici, NA, Italy
* Correspondence: giandomenico.corrado@unina.it

Abstract: Apricot (*Prunus armeniaca* L.) is an economically important tree species globally cultivated in temperate areas. Italy has an ample number of traditional varieties, but numerous landraces are abandoned and at risk of extinction because of increasing urbanization, agricultural intensification, and varietal renewal. In this work, we investigated the morphological and genetic diversity present in an ex-situ collection of 28 neglected varieties belonging to the so-called "Vesuvian apricot". Our aim was to understand the level of diversity and the possible link between the promotion of specific fruit types (e.g., by public policies) and the intraspecific variation in apricot. The combination of five continuous and seven categorical traits allowed us to phenotypically distinguish the varieties; while fruit quality-related attributes displayed high variation, both apricot size and skin colour were more uniform. The twelve fluorescent-based Simple Sequence Repeats (SSRs) markers identified cultivar-specific molecular profiles and revealed a high molecular diversity, which poorly correlated with that described by the morphological analysis. Our results highlighted the complementary information provided by the two sets of descriptors and that DNA markers are necessary to separate morphologically related apricot landraces. The observed morphological and genetic differences suggest a loss of diversity influenced by maintenance breeding of specific pomological traits (e.g., skin colour and size). Finally, our study provided evidence to recommend complementary strategies to avoid the loss of diversity in apricot. Actions should pivot on both the promotion of easily identified premium products and more inclusive biodiversity-centred on-farm strategies.

Keywords: stone fruit; local varieties; germplasm; pomological traits; DNA fingerprinting; microsatellites

:**itation:** Corrado, G.; Forlani, M.; .ao, R.; Basile, B. Diversity and .elationships among Neglected .pricot (*Prunus armeniaca* L.) .andraces Using Morphological raits and SSR Markers: Implications or Agro-Biodiversity Conservation. *lants* **2021**, *10*, 1341. https:// loi.org/10.3390/plants10071341

.cademic Editor: Gregor Kozlowski

.eceived: 22 April 2021
.ccepted: 29 June 2021
'ublished: 30 June 2021

1. Introduction

Apricot (*Prunus armeniaca* L.) is a stone-fruit tree globally appreciated for the rich flavour, fragrant aroma, and versatility of use of the drupes. The juicy, firm, and nutritious fruits can be consumed fresh, dried, in syrup, or made into jams [1]. Apricot was domesticated in Central Asia and was spread across Western Europe by the Romans [2,3]. It is no coincidence that the word 'apricot' allegedly derives from the Latin *arbor praecox* (the tree with an early production) [2]. The temperate climate and the varied orographic conditions favoured the diffusion of apricot in Italy and this species has experienced a considerable diversification, whose heritage is still visible. According to the International Plant Genetic Resources Institute, Italy has the largest collection of apricot varieties [4].

Italy is one of the top world producers of apricots [5] and the Campania region (Southern Italy) has the largest cultivated area, providing around one-third of the entire national production [6]. In this region, the empirical selection of the growers has created a wealth of landraces that are distributed mainly in the Naples Province [7], mostly because apricot has been one of the typical cultivations of the South-facing, rain-fed Vesuvian slopes since Roman times [8]. This rich germplasm is often referred to as Vesuvian apricots [9,10]. Nonetheless, the Neapolitan word from apricot (*cresommela*) originates from Greek [χρυσό-μήλο; golden fleshy (tree) fruit], suggesting that the introduction of this tree in this area could date back to the settlers of the Magna Graecia era [3].

In the last decade, apricot has provided satisfactory economic results for the entire supply chain (including sweet manufacturing) when compared to other stone fruits. The apricot sector was sustained by a strong varietal innovation driven by consumer demand for fruits with strongly coloured skin [11]. Therefore, the preference of apricot growers has experienced a transition from varieties that are also suitable for processing to the ones employed only for the fresh market. Specifically, vast attention has been given to new cultivars, often introduced from foreign countries, that have fruits with an intense red skin over colour (red blush), and to those that allow the extension of the harvesting period [12]. On the other hand, apricot is considered a species with reduced environmental adaptability, and the introduction of exotic germplasm may also result in fluctuating or limited yield. This is associated with differences in fertilization, chilling requirements, late-frost resistance, cold-hardiness, and in certain instances, the need for specific cultural techniques [13]. Especially in Southern Italy, there are problems in introducing contemporary self-incompatible, freestone varieties in areas where chilling requirements cannot be always fully satisfied, an issue of rising importance in the face of climate change [14,15]. For all these reasons, traditional cultivars in the Campania region still provide interesting economic results in local markets, remaining a popular option in small farms (<5 ha) and, more recently, in agritourism and farm stay enterprises [16]. Although yield and resistance to mechanical injury may not always be ideal, the locally adapted landraces are appreciated for their superior flavour and aroma [17], leading to the request of the EU Protected Geographical Indication (PGI) "Albicocca Vesuviana" (Vesuvian apricot) label. Moreover, apricots cannot be stored for a long period, leaving room for Short Food Supply Chains as alternative promoters of agricultural, social, and economic sustainability [18].

The potential of traditional or neglected varieties to diversify the apricot sector, to support local producers, and to promote traditional gastronomic products that use dried and candied fruits has not gained momentum because of the lack of information on available plant material [17,19,20]. Knowledge of the characteristics and variability of the apricot landraces is central for the selection and promotion of premium products [21–23]. It is therefore necessary to fill the gap between the available diversity and the folk names recognised by local consumers [24]. In addition, the evaluation of the apricot landraces is also a measure to indirectly support and acknowledge farmers in return for their precious role in promoting agro-biodiversity and, specifically for the Vesuvian germplasm, in sustaining rural areas in the most densely populated volcanic region in the world.

The objective of this study was to characterize and evaluate the diversity of 28 traditional apricots cultivated in the Campania region. Specifically, our work aimed at addressing the potential impact of a pattern of specialization in the local apricot market over landraces diversity. We used 12 morphological traits and 12 fluorescent based-SSR molecular markers to offer a more comprehensive view of the variation present in the germplasm. Morphological descriptors allow a technically undemanding evaluation of the diversity and represent an easily adaptable classification approach, while DNA fingerprinting is an indispensable tool to assess genetic diversity, discriminate varieties, identify possible synonyms and homonyms, and genetically trace plant varieties in food chains [25–27].

2. Results

2.1. Morphological Analysis

To assess the morphological diversity in the germplasm collection, we scored five multistate categorical traits and seven quantitative traits of the fruit (Supplementary Table S1). All morphological traits were polymorphic, presenting two or more different forms (Tables 1 and 2). The most variable qualitative pomological trait was the colour of the flesh, considering the scored phenotypes, their distribution, and the Simpson Index of Diversity (SDI). Little variation was present for the ground colour of the skin. We scored only two and similar ground colour of the seven possible categories listed in the UPOV guidelines. A single predominant phenotype was not evident, and the SDI was relatively high (0.48), indicating a good distribution of the abundance of this trait. Very little variability was present for the adherence of the stone to the fruit and the kernel bitterness (SDI: 0.07). Nine varieties presented a unique combination of qualitative traits; nonetheless, the most common phenotype (a strongly vigorous tree producing fruits with a yellowish ground skin colour and a medium-orange pulp, and with a bitter, free stone) was present only in four varieties.

Table 1. Frequency (*f*) and relative frequency (rf) of the categorically scored traits in the germplasm collection. For each trait, phenotypes (Phen.) are ranked in decreasing order. For each qualitatively scored trait, the Simpson Index of Diversity (SDI) is also reported.

Tree Vigour (SDI: 0.52)			Fruit: Ground Colour of Skin (SDI: 0.48)			Fruit: Colour of Flesh (SDI: 0.70)			Kernel Bitterness (SDI: 0.07)			Fruit: Adherence of Stone to Flesh (SDI: 0.07)		
Phen. [1]	*f*	rf	Phen. [2]	*f*	rf	Phen. [3]	*f*	rf	Phen. [4]	*f*	rf	Phen. [4]	*f*	rf
S	17	0.61	Y	18	0.64	LO	13	0.46	P	27	0.96	A-VW	27	0.96
M	3	0.11	LO	10	0.36	MO	7	0.25	A-VW	1	0.04	P	1	0.04
NA	3	0.11				C	6	0.21						
VS	3	0.11				DO	2	0.07						
W	2	0.07												

[1] S: strong; M: medium; NA: not available/not consistent; VS: very strong: W: weak. [2] Y: yellowish; LO: light orange; R: red; O: orange; YG: yellowish-green; OR: orange-reddish; G: green. [3] LO: light orange; MO: medium orange; C: cream; DO: dark orange. [4] A-VW: absent-very weak; P: present.

Table 2. Selected descriptive statistics for the quantitative traits of the 28 apricot landraces. For each trait, the table reports the coefficient of variation (CV) and the maximum (max), average (mean), and minimum (min) value.

Trait (Abbreviation)	Unit	CV	Max	Average	Min
Fruit fresh weight (FFW)	g/fruit	18.4%	62.3	46.7	30.0
Fruit length (FL)	mm	9.9%	53.7	45.1	38.2
Fruit width (FW)	mm	8.9%	47.3	41.1	34.9
Fruit volume (FV)	cm^3	26.3%	62.8	40.7	25.2
Solid Soluble Content (SSC)	°Brix	19.4%	22.7	15.9	10.4
Titratable acidity (TA)	g/L	38.0%	2.5	1.3	0.5
Flesh firmness (FF)	N	34.3%	54.0	29.0	17.7

The range of variation of the quantitative traits is presented in Table 2. The average coefficient of variation (CV) was 22.2, indicating the presence of considerable differences in the pomological traits.

Traits displayed substantial differences in their range of variation. For instance, fruits varied slightly in length and width. The fruit volume was the most variable morphological feature, but its variation (~27%) was proportionate to the extent of the linear measurements (~9%) in three dimensions. The greatest differences were evident for fruit quality, with the titratable acidity having the highest CV, followed by the flesh firmness and the soluble solid content. The presence of a rather uniform fruit shape is indirectly indicated by the strong positive correlation between width, length, and volume of the fruit, while quality-related pomological traits displayed non-significant correlations, with the notable exception of the negative value for SSC and TA (Figure 1).

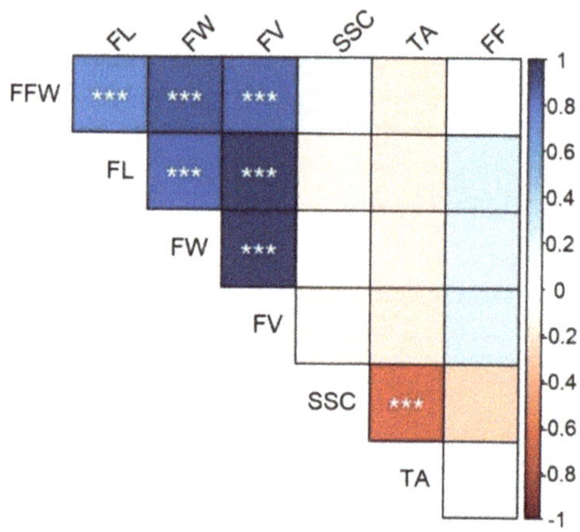

Figure 1. Correlogram (Pearson) of the quantitative variables of the fruits. Pairwise correlation between variables (see Table 2. for the code) are colour-mapped according to the colour scale of the bar on the right-hand side. Asterisks indicate statistically significant correlations (***: $p < 0.001$).

To visualize the relationships among the apricot landraces under investigation, we performed a multivariate cluster analysis using both qualitative and quantitative traits. At a high hierarchical node (k = 2), the top cluster (PAZ-MON) was associated with a yellowish skin colour (13 varieties out of 14), while the bottom (BOC-SCI) mainly with the light orange one (9 out of 14) (Figure 2).

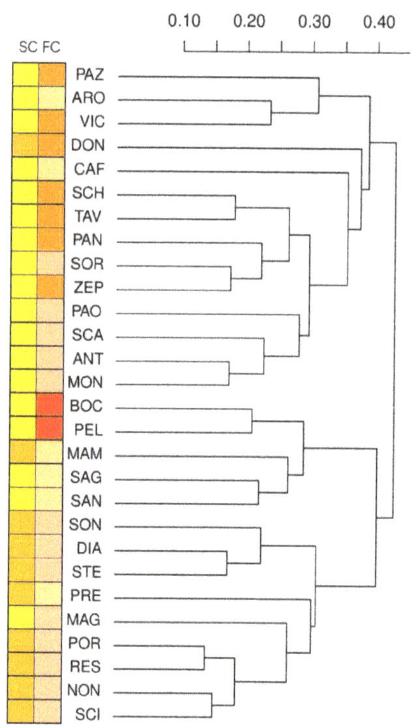

Figure 2. Hierarchical cluster analysis of the 28 apricot landraces based on the 12 scored traits. Distances were computed with the Gower's coefficient. The coloured squares on the left indicate the ground skin colour (SC) and flesh colour (FC) of the fruits (data are reported in Supplementary Table S1).

This explorative analysis also indicated that cutting the tree at k = 11 highlights groups of landraces with similar skin and fruit flesh colour. For instance, the cluster SCH-ZEP (five varieties) was characterised mostly by yellowish skin colour and a medium orange flesh, while the cluster PAO-MON (four varieties) differed because of light orange flesh. Other phenotypically distinct clusters were the BOC-PEL (two varieties; yellowish skin and dark orange flesh), the SON-STE (three varieties; light orange skin and flesh colour) and the MAG-SCI cluster (five varieties; mostly with light orange skin and with light orange pulp). Overall, the morphological analysis indicated that the germplasm under investigation presented some distinctive features, considering the trait variation described in apricot. Albeit the varieties could be distinguished based on both quantitative and qualitative traits, the germplasm under investigation was largely characterized by medium size, obovate fruits with not strongly coloured fruits and flesh. Moreover, the association of a fruit phenotype with agglomerative clusters suggests a possible similar origin for at least some varieties.

2.2. Analysis of Genetic Diversity by SSR Markers

The genetic diversity was assessed using 12 SSRs selected from the literature as specific to apricot. The loci were all polymorphic, and the main genetic parameters are presented in Table 3. We detected in total 76 alleles and their length (from 79 to 300 bp), was consistent with the literature [19,28,29]. Differences among loci were in the number of alleles, which ranged from 2 (AMPA111) to 10 (UDAp-446). Considering the effective number of alleles (i.e., the number of alleles weighted for their frequencies), the most diverse locus was AMPA112, followed by UDAp-410 and UDAp-446. Not surprisingly for an agamically propagated species, the observed heterozygosity (Ho) was high (mean value: 63%) but large variations were present among loci. The number of alleles was positively associated with the Ho, but the correlation was not significant ($r_s = 0.32$, p (two-tailed) = 0.16, Spearman's rho). Specifically, heterozygotes were not present for one locus (AMPA111), which was fixed in our population. The transcript in the *P. armeniaca* genome (Seq_id: tig00008589_30087 in the assembly 1.0) closest to AMPA111 (distant approximately 400 bp) putatively codes for a protein that has the highest similarity (blastx e-value: 7×10^{-37}; similarity: 47%) with a zeaxanthin epoxidase from *P. mume* (XP_008224462.2), which is involved in carotenoid accumulation [30]. On the other hand, UDAp-419 also displayed a substantial positive Fixation Index (0.63) yet this locus had eight different alleles. As expected, the Polymorphic Information Content of the loci significantly correlated with the number of alleles ($r_s = 0.96$, p (two-tailed) < 0.001, Spearman's rho). Three loci (UDAp-446, AMPA112, and UDAp-410) were almost equally highly informative, considering the Information index, the number of alleles, and PIC. Finally, the rate of proportional abundance homogeneity of individual alleles in the population was high for all loci, as indicated by the Evenness values, which were negatively correlated with the number of alleles ($r_s = -0.63$, p (two-tailed) < 0.05, Spearman's rho).

To evaluate the genetic relationship between varieties, we built a UPGMA dendrogram (Figure 3). Genetic distances are reported in Supplementary Table S2. All the varieties could be discriminated, and the average (± standard deviation) genetic distance was 0.44 ± 0.11. Moreover, a clear tendency in grouping phenotypically similar varieties was not evident. For instance, varieties with similar fruit or flesh colour did not clearly agglomerate according to the genetic analysis. Overall, the dendrogram based on molecular data did not identify groups shown by the morphological analysis.

Table 3. Main genetic indices of the apricot landraces obtained by SSR analysis. Na: number of different alleles; I: Shannon's information index; Ho: observed heterozygosity; PIC: polymorphic information content; ENA: effective number of alleles; Ev: evenness F: Wright fixation index.

Locus	Na	I	Ho	PIC	ENA	Ev	F
AMPA095	4	1.06	0.75	0.61	2.55	0.82	−0.23
AMPA111	2	0.68	0.00	0.49	1.96	0.98	1.00
AMPA112	9	1.71	0.61	0.77	4.43	0.76	0.22
AMPA113	4	0.99	0.58	0.57	2.34	0.79	−0.01
AMPA124	7	1.32	0.48	0.64	2.79	0.65	0.25
UDAp-401	7	1.57	0.85	0.76	4.17	0.83	−0.11
UDAp-410	8	1.72	0.86	0.77	4.31	0.72	−0.12
UDAp-414	5	1.15	0.81	0.62	2.67	0.77	−0.30
UDAp-415	5	1.15	0.81	0.61	2.58	0.73	−0.33
UDAp-419	8	1.67	0.28	0.76	4.21	0.74	0.63
UDAp-420	7	1.49	0.68	0.72	3.52	0.73	0.05
UDAp-446	10	1.78	0.89	0.77	4.28	0.66	−0.16
Mean	6.33	1.36	0.63	0.67	3.32	0.77	0.07
Standard error	0.68	0.10	0.08	0.03	0.27	0.03	0.12

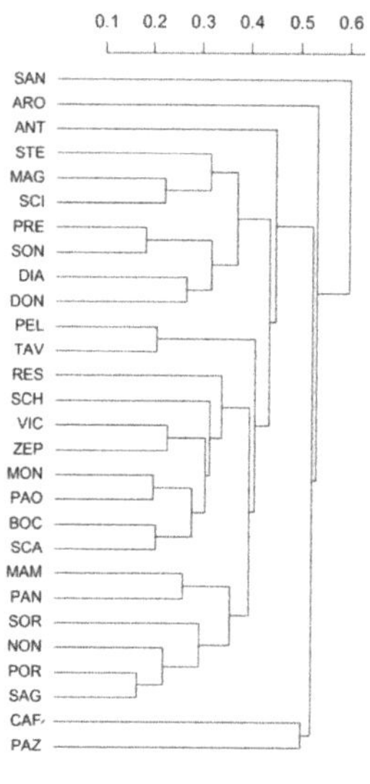

Figure 3. Cluster analysis (UPGMA algorithm based on Bruvo's distances from SSR data) of the 28 apricot landraces. The scale bar for the genetic distance is presented on the top.

To better visualize the differences between morphological and molecular data in describing diversity, we compared dendrograms. The correlation between the distances obtained with the two datasets (r = 0.18; $p > 0.05$; Mantel test with 9999 permutations) and the cophenetic correlation between the dendrograms (r = 0.19) were not significant. The tanglegram indicated that the resemblance depicted with the morphological traits did not largely correlate with that obtained with SSR markers (Figure 4).

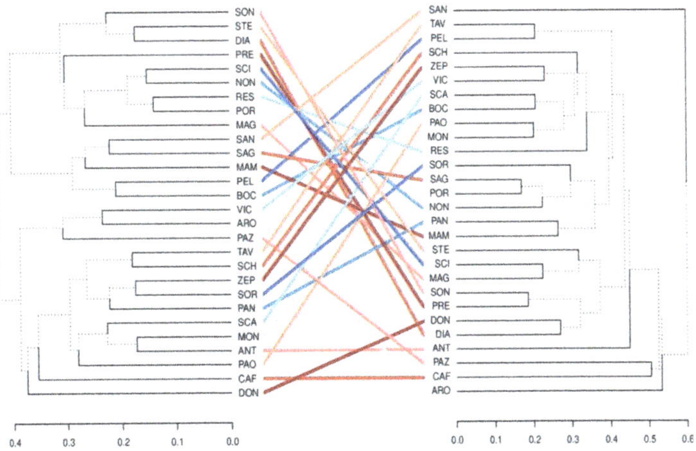

Figure 4. A comparison of the hierarchical clustering of the apricot varieties using morphological (left tree) or SSR marker (right tree) data. Clustering was performed with the UPGMA algorithm for both dendrograms. To ease the comparison, coloured lines connect identical names. The different line types in the dendrograms highlight distinct edges in a tree (compared to the other one).

3. Discussion

The market promotion of the traditional plant varieties necessitates a two-way transfer of knowledge between farmer–custodians and extension specialists [31], sharing the common goal of enhancing landraces' symbolic and socio-cultural value and their profitability. To raise stakeholder engagement, it is needed a thorough description and discrimination of the germplasm's features to go beyond, and hopefully reduce, the still essential strategy of providing economic incentives to maintain crop landraces on-farm [32,33]. Moreover, the recognition of the existing diversity is a prerequisite to define policies for agricultural biodiversity [34,35] and to increase consumer loyalty, at least in local markets and for selected traditional products [36].

Our study revealed the phenotypic and genetic diversity in the virtually neglected apricot germplasm under investigation. Several phenotypes were scored, but there was limited variability in the ground skin colour of the fruits considering the UPOV categories and the literature [37–39]. Different colours in local germplasm (e.g., green-yellowish colour for 'Persicara'; dark orange for 'Cerasona' and 'Parrocchiana') have been reported in non-scientific and grey literature, suggesting a possible loss of diversity. In our germplasm, higher variation was evident for the flesh colour and tree habit, while the bitterness of the kernel and the adherence of the stone displayed a similar frequency (e.g., a largely predominant phenotype) than in other apricot collections [38,40]. All this can be explained considering that the observable pomological characteristics have clear commercial importance, and therefore, these traits are more likely to be objects of maintenance breeding [41]. Traditional local varieties of the Campania region, and especially those considered typical of Mount Vesuvius, are associated with a light orange-yellowish colour, medium weight, and an elliptic/obovate shape [42]. The recognition of the typical phenotype of the Vesuvian apricot, and related market demand, could have been important drivers for maintenance breeding in the face of biodiversity erosion. The apricot cultivation in the Campania region largely changed in the second half of the XIX century, where in a rentier agricultural economy, there was a shift towards the production of large quantities of standardized apricots to satisfy the demand of a rising, low-income, urban population of Naples, at that time, the third largest European city [43,44]. More recently, the cultivation of apricot has been limited by the anthropization of a vast portion of the peri-urban hilly areas of the Naples Province [45].

Hierarchical analysis indicated that different clusters could be associated with som qualitative pomological traits, although the quantitative traits under investigation wer higher in number and with good variability. For instance, the variation of TSC and T was higher compared to traditional apricot landraces and similar to that reported i contemporary cultivars [39,46], implying that the traditional germplasm can be a potentia source of variation of quality-related fruit traits [12]. This is relevant because, different fror other works [37,46,47], we analysed plants growing in the same environment and wit the same cultural practices [39], allowing us to untangle the genetic and environmenta contribution to quantitative trait variation.

Another difference from previously published works is that our cluster analysis jointl considered qualitative and quantitative traits [37,47]. Moreover, in our work, quantitativ traits were first standardized (but not weighted), not only to share a common scale bu also for their different biological meaning and unit of measurement [48]. The morpho logically similar subclusters prompted us to test a possible common ancestry by usin DNA molecular markers. It is known that traditional apricot varieties in small farms ar also selected from seedlings, leaving open the possibility of a common origin of at leas some of the clusters that show a similar pomological appearance. The molecular analysi indicated a high level of molecular diversity, comparable to that of other reports on large numbers of landraces [49], or higher than in similarly sized collections [50]. For instanc the SSR loci showed an average level of heterozygosity that was in line with previou work, but locus-specific differences in some genetic parameters were also evident [51]. On locus (AMPA111) was fixed and another (UDAp-419) had a fixation index higher than 0. While the former had two evenly distributed alleles, the latter was characterized by a hig number of alleles. It is difficult to speculate on the reasons for these locus-specific diffe ences and, specifically, on the AMPA111 fixation. The agamic propagation preserves bot adaptive and neutral genetic diversity, thus affecting both the distribution of alleles at loc under selection and that of neutral loci. The limited variability in fruit skin colour make it tempting to speculate a possible selection-driven fixation of a locus tightly associate with a carotenogenic gene [30] however, further genetic and biochemical studies will hav to clarify the adaptive significance of these variations, along with the possible presenc of a founder effect, or of genetic drift. For all SSR loci, the number of alleles was withi if not higher than the range reported in the literature [28,29,52,53]. Distinct DNA profile could be identified for each variety, without possible cases of synonymy and/or duplicate accessions that can be present in traditional germplasm [52,54]. This may be explaine considering that the analysed germplasm belongs to an ex-situ collection, whose materia was previously classified based on farmers' description. For all these reasons, the molecula analysis does not favour the possibility that the similar-looking varieties have a recen common origin, deriving for instance, from the selection of seedlings of open-pollinate plants from neighbouring farms [51]. The data suggested that the frequency of plant with a similar pomological appearance is due to the selection of aesthetically importan indicators of marketability.

Finally, the comparison between morphological and molecular diversity indicated tha morphological descriptors in apricot provide different classifications from the molecula ones, consistent with studies in other tree crops [55,56]. Our results imply that in aprico landraces, an estimation of the variability exclusively based on morphological traits ca misrepresent the level of diversity, and therefore, DNA markers can be very useful fo building core collections in apricot [57]. Moreover, the assignment of trees to a variety, whe based exclusively on morphological similarity, would require scoring a high number o both qualitative and quantitative traits, especially to avoid possible spurious homonymies

4. Materials and Methods

4.1. Plant Material

The work was carried out on 28 *Prunus armenicaca* L. landraces present in the germplas collection of the Azienda Agricola Sperimentale Regionale Improsta, Centro per la Ricerc

Applicata in Agricoltura (C.R.A.A.). The varieties (abbreviation) were: 'Antonaniello' (ANT), 'Aronzo' (ARO), 'Boccuccia liscia II' (BOC), 'Cafona' (CAF), 'Diavola' (DIA), 'Don Aniello' (DON), 'Magnalona' (MAG), 'Mammana' (MAM), 'Montedoro' (MON), 'Nonno' (NON), 'Panzona' (PAN), 'Paolona' (PAO), 'Pazza' (PAZ), 'Pelese Correale' (PEL), 'Portuallara' (POR), 'Presidente' (PRE), 'Resina' (RES), 'San Giorgio' (SAG), 'Sant'Antonio' (SAN), 'Scassulillo' (SCA), 'Schiavona' (SCH), 'Scialo' (SCI), 'Sonacampana' (SON), 'Sorrentino' (SOR), 'Stella' (STE), 'Taviello' (TAV), 'Vicienzo (syn: Vicienzo 'e maria)' (VIC), 'Zeppa (syn: Zeppa 'e sisco)' (ZEP). These locally cultivated varieties were selected because they are considered at risk of extinction since they were not included in the proposal for the EU Protected Geographical Indication (PGI) "Albicocca Vesuviana" (Vesuvian apricot) (Gazzetta Ufficiale della Repubblica Italiana, Serie Generale n.66, 19-03-2002). These crop varieties can be considered "neglected" according to the literature [58], because in the past they were of greater importance in traditional agriculture, in the diet of local communities, and in local food processing activities [59]. Subsequently, they have been marginalized because of the introduction of contemporary varieties and gradual change in consumer demand, as well as for economic, societal, and cultural factors that contribute to the disappearance of social groups that cultivated this material [58].

4.2. Analysis of Morphological Data

Tree vigour was visually evaluated in the field and assigned to one of the following categories: very weak, weak, medium, strong, very strong, according to the document TG/70/4 Rev. (proj.2) of the International Union for the Protection of new Varieties of Plants (UPOV, Geneva, Switzerland). Not available (NA) was assigned in the cases of off-types and lack of uniformity. A sample of 30 fruits per variety (six fruits from five plants) was collected and used to assess the following morphological traits: fresh weight, height, ventral width, lateral width, volume, skin ground colour, flesh colour, flesh firmness, soluble solids content, titratable acidity, kernel bitterness, and stone adherence to the flesh. Fruit fresh weight was measured with a digital scale, whereas fruit height, ventral width, and lateral width were measured with a digital caliper. For each fruit, mean fruit width was calculated averaging ventral and lateral widths. Single fruit volume was estimated assuming the apricot fruit had an ellipsoidal shape and considering the height and the ventral and lateral widths as axes. Ground skin and flesh colour were evaluated visually, and the fruits were assigned to one of the following possible categories: (a) ground colour of the skin: not visible, white, yellowish, yellow green, light orange, medium orange, and dark orange; (b) flesh colour: whitish green, white cream, light orange, medium orange, and dark orange. Flesh firmness (N) and the soluble solids content in fruit (°Brix) were measured with a digital fruit firmness tester (model #53205, TR, Forlì, Italy) fitted with an 8-mm diameter plunger and a digital refractometer (HI96811, Hanna Instruments, TX, USA), respectively. Titratable acidity was measured adding a 0.1 N NaOH solution to filtered fruit juice until reaching a pH of 8.2. During titration, pH was measured continuously with a laboratory pH-meter (GLP 21; Crison, Alella, Barcelona, Spain). Titratable acidity was expressed in grams per liter of malic acid (g/L). Before juice extraction, fruits were split along the suture to visually evaluate the adherence of the stone to the flesh, classified as present or absent–very weak. Bitterness of ground kernels was also dichotomously classified as present or absent–very weak. The Simpson Index of Diversity (SDI) of each qualitative trait was calculated as 1-D, where D is the sum of $n_i(n_i - 1)/N(N - 1)$; n_i is the number of varieties having the ith-phenotype, and N is the total number of varieties. Hierarchical cluster analysis, using the Unweighted Pair Group Method with Arithmetic Mean (UPGMA) algorithm based on Gower's distance of the unweighted and scaled variables, was carried out as already reported [60]. We did not assign a weight to each trait (or each category of traits), and quantitative data were normalized (Z-score) because variables are on different scales.

4.3. DNA Isolation and SSR Analysis

Genomic DNA was isolated from young leaves stored at −80 °C. Powdered leave (approx. 500 mg) were mixed with 15 mL of warm (65 °C) extraction buffer [61] and afte the first precipitation of nucleic acids [61], the pellet was solubilised in 400 μL Buffer AP (Qiagen). The DNA was then purified according to the instructions of the DNeasy Plan Mini Kit (Qiagen). Polymerase Chain Reaction (PCR) amplifications were carried out usin 12 apricot SSRs primer pairs selected from the literature and reported in the Supplementar Table S1 [19,28,29]. We genotyped two plants per variety, which provided an identica profile for each landrace. PCR reactions (25 μL final volume containing 100 ng of genomi DNA) were performed as previously described [62] using the annealing temperatures listec in Supplementary Table S3. Amplicons were resolved in an agarose gel-electrophoresis tc verify the presence of and to quantify the amplified fragments. Allelic discrimination wa performed by fluorescence-based capillary electrophoresis using an ABI PRISM 3130 Avan (Applied Biosystems, Milan, Italy) and the POP4 polymer (Applied Biosystems). Allel sizes were calculated with GeneScan 4.0 (Applied Biosystems) software using the loca Southern algorithm, as previously described [62]. The raw size was rounded to an intege and scaled considering the number of bases of the repeated core motif (Supplementar Table S3). Alleles were binned, minimizing the mean offset of the allelic size for each SSI within the resolution of the instrument (±1 bp).

4.4. Molecular Data Analysis

We calculated, for each SSR locus, the number of different alleles (N), the Shannon's information index (I), the observed heterozygosity (Ho), and the Polymorphic Informatior Content (PIC), equivalent to the expected Heterozygosity) as previously reported [56] The effective number of alleles (ENA), Evenness (Ev), and pairwise genetic distances between varieties using Bruvo's coefficient were computed with the poppr R-library [63] Hierarchical clustering (UPGMA algorithm) was carried out as previously described [60] To test the correlation between the morphological and molecular data, the two paralle matrices were compared by a Mantel test (9999 permutations) [60].

5. Conclusions

Our work indicated that the combination of molecular and morphological data for the classification of traditional apricot varieties is probably necessary to separate morphologi cally related accessions, an important issue particularly for local germplasms subject to a possible genetic erosion because of the selection for specific fruit types. Our results imply that the biodiversity preservation and promotion of apricot landraces should not only rely on *ex-situ* strategies. It is worth developing a good balance between the commercia exploitation of specific fruit types and brands (important to strengthen socio-economic structures) and more inclusive biodiversity-based sustainable agriculture that can provide ecosystem services through an increased attractiveness of varietal mixtures [64].

Supplementary Materials: The following are available online at https://www.mdpi.com/article/10.3390/plants10071341/s1, Table S1: Phenotypic trait analysis. The full name of the variety is reported in the main article. Table S2: Matrix of the pairwise genetic distance between the landraces under investigation. See the main text for the landraces' code. Table S3: SSR loci employed, primer sequences (in 5′ to 3′ order), and their main features.

Author Contributions: Conceptualization, G.C.; methodology, G.C. and B.B.; formal analysis, G.C and B.B.; investigation, G.C. and B.B.; resources, M.F. and R.R., writing—original draft preparation G.C.; writing—review and editing, G.C. and B.B.; visualization, G.C. and B.B.; funding acquisition M.F. and R.R. All authors have read and agreed to the published version of the manuscript.

Funding: This work was supported by the SALVE project (Salvaguardia della Biodiversità Vegetale della Campania-Regione Campania, PSR 2007–2013, misura 214 az. f2).

Data Availability Statement: Data are contained within the article or supplementary material. The raw data generated during and/or analysed during the current study are available from the corresponding author on reasonable request.

Acknowledgments: We thank Matteo Giaccone for the assistance during field data collection. We also wish to thank the two anonymous reviewers for the useful comments.

Conflicts of Interest: The authors declare no conflict of interest.

eferences

Sinha, N.; Sidhu, J.; Barta, J.; Wu, J.; Cano, M.P. *Handbook of Fruits and Fruit Processing*; John Wiley & Sons: Hoboken, NJ, USA, 2012.

Faust, M.; Suranyi, D.; Nyujto, F. Origin and dissemination of apricot. *Hortic. Rev. Westport N. Y.* **1998**, *22*, 225–260.

Yu, Y.; Fu, J.; Xu, Y.; Zhang, J.; Ren, F.; Zhao, H.; Tian, S.; Guo, W.; Tu, X.; Zhao, J. Genome re-sequencing reveals the evolutionary history of peach fruit edibility. *Nat. Commun.* **2018**, *9*, 1–13. [CrossRef] [PubMed]

Ledbetter, C.A. Apricots. In *Temperate Fruit Crop Breeding: Germplasm to Genomics*; Hancock, J.F., Ed.; Springer: Dordrecht, The Netherlands, 2008; pp. 39–82. [CrossRef]

FAO. Available online: http://www.fao.org/ (accessed on 1 March 2021).

Istat. Available online: http://dati.istat.it/ (accessed on 1 March 2021).

Pugliano, G.; Forlani, M. Two-year observations on the biology and fructification of apricot. *Acta Hortic.* **2019**, *192*, 383–400.

Jashemski, W.F. Ancient Roman gardens in Campania and Tunisia: A comparison of the evidence. *J. Gard. His.* **1996**, *16*, 231–243. [CrossRef]

Basile, B.; Cirillo, C.; Santin, A.; Forlani, M. Fruit quality of Vesuvian apricots harvested at different ripening stages after a cold-storage period. In Proceedings of the V International Postharvest Symposium, Verona, Italy, 6–11 June 2005; Volume 682, pp. 1443–1450.

0. Cirillo, C.; Basile, B.; Hernandez, G.; Pannico, A.; Giaccone, M.; Forlani, M. Influence of fruiting shoot on flowering pattern and fruit quality of Vesuvian apricot cultivars. *Acta Hortic.* **2010**, *862*, 557–564. [CrossRef]

1. Mennone, C. Effect of climate on apricot productivity. *Informatore Agrario* **2016**, *72*, 47–50.

2. Campoy, J.A.; Audergon, J.M.; Ruiz, D.; Martínez-Gómez, P. Genomic designing for new climate-resilient apricot varieties. In *Genomic Designing of Climate-Smart Fruit Crops*; Kole, C., Ed.; Springer: Cham, Switzerland, 2019; pp. 73–89.

3. Pennone, F.; Abbate, V. Prospettive per la valorizzazione della coltura dell'albicocco nel Mezzogiorno attraverso il miglioramento genetico. *Rivista di Frutticoltura e di Ortofloricoltura* **2004**, *66*, 36–39.

4. Campoy, J.A.; Ruiz, D.; Egea, J. Dormancy in temperate fruit trees in a global warming context: A review. *Sci. Hortic.* **2011**, *130*, 357–372. [CrossRef]

5. Bartolini, S.; Massai, R.; Viti, R. The influence of autumn-winter temperatures on endodormancy release and blooming performance of apricot (*Prunus armeniaca* L.) in central Italy based on long-term observations. *J. Hortic. Sci. Biotechnol.* **2020**, *95*, 794–803. [CrossRef]

6. Cirillo, O. Fashion and tourism in Campania in the middle of the twentieth century: A story with many protagonists. *Almatour. J. Tour. Cult. Territ. Dev.* **2018**, *9*, 23–46.

7. Mainolfi, A.; Abbate, V.; Buccheri, M.; Damiano, C. Molecular characterization of local ecotypes and commercial varieties of apricot (*Prunus armeniaca* L.) [Campania]. *Italus Hortus* **2006**, *13*, 262–265.

8. Enjolras, G.; Aubert, M. Short food supply chains and the issue of sustainability: A case study of French fruit producers. *Int. J. Retail Distrib. Manag.* **2018**, *46*, 194–209. [CrossRef]

9. Rao, R.; Bencivenni, M.; Mura, L.; Araujo-Burgos, T.; Corrado, G. Molecular characterisation of Vesuvian apricot cultivars: Implications for the certification and authentication of protected plant material. *J. Hortic. Sci. Biotechnol.* **2010**, *85*, 42–347. [CrossRef]

0. Rao, R.; Bencivenni, M.; Corrado, G.; Basile, B.; Forlani, M. Molecular characterization of apricot varieties included in the "Albicocca Vesuviana" PGI Regulation. *Acta Hortic.* **2010**, *862*, 61–66. [CrossRef]

1. Krishna, V.V.; Pascual, U. Can greening markets help conserve landraces in situ? In *Agrobiodiversity Conservation and Economic Development*; Routledge: London, UK, 2008; Volume 10, p. 267.

2. Dwivedi, S.; Goldman, I.; Ortiz, R. Pursuing the potential of heirloom cultivars to improve adaptation, nutritional, and culinary features of food crops. *Agronomy* **2019**, *9*, 441. [CrossRef]

3. Cirillo, C.; Pannico, A.; Basile, B.; Rivera, C.; Giaccone, M.; Colla, G.; De Pascale, S.; Rouphael, Y. A simple and accurate allometric model to predict single leaf area of twenty-one European apricot cultivars. *Eur. J. Hortic. Sci* **2017**, *82*, 65–71. [CrossRef]

4. Vermeulen, S.; Koziell, I. *Integrating Global and Local Values: A Review of Biodiversity Assessment*; IIED: London, UK, 2002.

5. Hagen, L.; Khadari, B.; Lambert, P.; Audergon, J.-M. Genetic diversity in apricot revealed by AFLP markers: Species and cultivar comparisons. *Theor. Appl. Genet.* **2002**, *105*, 298–305. [CrossRef] [PubMed]

6. Geuna, F.; Toschi, M.; Bassi, D. The use of AFLP markers for cultivar identification in apricot. *Plant Breed.* **2003**, *122*, 526–531. [CrossRef]

7. Corrado, G.; Imperato, A.; La Mura, M.; Perri, E.; Rao, R. Genetic diversity among olive varieties of Southern Italy and the traceability of olive oil using SSR markers. *Hortic. Sci. Biotechnol.* **2011**, *86*, 461–466. [CrossRef]

28. Hagen, L.; Chaïb, J.; Fady, B.; Decroocq, V.; Bouchet, J.; Lambert, P.; Audergon, J. Genomic and cDNA microsatellites from apricc (*Prunus armeniaca* L.). *Mol. Ecol. Notes* **2004**, *4*, 742–745. [CrossRef]
29. Messina, R.; Lain, O.; Marrazzo, M.; Cipriani, G.; Testolin, R. New set of microsatellite loci isolated in apricot. *Mol. Ecol. Note* **2004**, *4*, 432–434. [CrossRef]
30. Kita, M.; Kato, M.; Ban, Y.; Honda, C.; Yaegaki, H.; Ikoma, Y.; Moriguchi, T. Carotenoid accumulation in Japanese apricot (*Prunu mume* Siebold & Zucc.): Molecular analysis of carotenogenic gene expression and ethylene regulation. *J. Agric. Food Chem.* **200** *55*, 3414–43420. [PubMed]
31. Tuxill, J.D.; Nabhan, G.P. *People, Plants, and Protected Areas: A Guide to in Situ Management*; Routledge: London, UK, 2001; Volume
32. Smith, E.A.; Wishnie, M. Conservation and subsistence in small-scale societies. *Annu. Rev. Anthropol.* **2000**, *29*, 493–452 [CrossRef]
33. Emerton, L. *Using Economic Incentives For Biodiversity Conservation*; IUCN: Gland, Switzerland, 2000.
34. Bellon, M.R.; Gotor, E.; Caracciolo, F. Conserving landraces and improving livelihoods: How to assess the success of on-farn conservation projects? *Int. J. Agric. Sustain.* **2015**, *13*, 167–4182. [CrossRef]
35. Brooks, J.S.; Franzen, M.A.; Holmes, C.M.; Grote, M.N.; Mulder, M.B. Testing hypotheses for the success of different conservatio strategies. *Conserv. Biol.* **2006**, *20*, 1528–41538. [CrossRef]
36. Chrysochou, P.; Krystallis, A.; Giraud, G. Quality assurance labels as drivers of customer loyalty in the case of traditional foo products. *Food Qual. Prefer.* **2012**, *25*, 156–4162. [CrossRef]
37. Krichen, L.; Audergon, J.-M.; Trifi-Farah, N. Variability of morphological characters among Tunisian apricot germplasm. *Sc Hortic.* **2014**, *179*, 328–4339. [CrossRef]
38. Salari, H.; Samim, A.K.; Ahadi, S.; Etemadi, S.A. Preliminary evaluation of morphological and pomological characters to illustrat genetic diversity of apricots (*Prunus domestica* L.) in Afghanistan. *Eur. J. Agric. Food Sci.* **2020**, *2*, 104. [CrossRef]
39. Drogoudi, P.D.; Vemmos, S.; Pantelidis, G.; Petri, E.; Tzoutzoukou, C.; Karayiannis, I. Physical characters and antioxidant, suga and mineral nutrient contents in fruit from 29 apricot (*Prunus domestica* L.) cultivars and hybrids. *J. Agric. Food Chem.* **2008**, *5* 10754–10760. [CrossRef] [PubMed]
40. Milošević, T.; Milošević, N.; Glišić, I.; Glišić, I.S. Determination of size and shape properties of apricots using multivariate analysi *Acta Sci. Pol. Hortorum Cultus* **2014**, *13*, 77–90.
41. Zeven, A.C. Traditional maintenance breeding of landraces: 2. Practical and theoretical considerations on maintenance o variation of landraces by farmers and gardeners. *Euphytica* **2002**, *123*, 147–158. [CrossRef]
42. Pennone, F.; Abbate, V. Apricot breeding in Caserta: New perspectives of apricot growing in Southern Italy. In Proceedings of th XIII International Symposium on Apricot Breeding and Culture, Murcia, Spain, 13–17 June 2005; Volume 717, pp. 157–162.
43. Izzo, P. *L'alimentazione del Popolo Minuto di Napoli (Al Tempo dei Borbone)*; Stamperia del Valentino: Naples, Italy, 2017.
44. Tino, P. Napoli ei suoi dintorni. Consumi alimentari e sistemi colturali nell'Ottocento. *Meridiana* **1993**, 47–99.
45. Ricciardi, M.; Mazzoleni, S.; La Valva, V. The flora and vegetation of the Somma-Vesuvius volcanic complex. In *Elementi d Biodiversità del Parco Nazionale del Vesuvio*; Picariello, O., Di Fusco, N., Frassinet, M., Eds.; Ente Nazionale Parco del Vesuvic Naples, Italy, 2000; pp. 51–65.
46. Kafkaletou, M.; Kalantzis, I.; Karantzi, A.; Christopoulos, M.V.; Tsantili, E. Phytochemical characterization in traditional anc modern apricot (*Prunus armeniaca* L.) cultivars—Nutritional value and its relation to origin. *Sci. Hortic.* **2019**, *253*, 195–202 [CrossRef]
47. Wani, A.A.; Zargar, S.A.; Malik, A.H.; Kashtwari, M.; Nazir, M.; Khuroo, A.A.; Ahmad, F.; Dar, T.A. Assessment of variability ir morphological characters of apricot germplasm of Kashmir, India. *Sci. Hortic.* **2017**, *225*, 630–637. [CrossRef]
48. Mohamad, I.B.; Usman, D. Standardization and its effects on K-means clustering algorithm. *Res. J. Appl. Sci. Eng. Technol.* **2013**, *6* 3299–3303. [CrossRef]
49. Lamia, K.; Hedia, B.; Jean-Marc, A.; Neila, T.-F. Comparative analysis of genetic diversity in Tunisian apricot germplasm using AFLP and SSR markers. *Sci. Hortic.* **2010**, *127*, 54–63. [CrossRef]
50. Cheng, Z.; Huang, H. SSR fingerprinting Chinese peach cultivars and landraces (*Prunus persica*) and analysis of their genetic relationships. *Sci. Hortic.* **2009**, *120*, 188–193. [CrossRef]
51. Maghuly, F.; Fernandez, E.B.; Ruthner, S.; Pedryc, A.; Laimer, M. Microsatellite variability in apricots (*Prunus armeniaca* L.) reflects their geographic origin and breeding history. *Tree Genet. Genomes* **2005**, *1*, 151–165. [CrossRef]
52. Zhebentyayeva, T.; Reighard, G.; Gorina, V.; Abbott, A. Simple sequence repeat (SSR) analysis for assessment of genetic variability in apricot germplasm. *Theor. Appl. Genet.* **2003**, *106*, 435–444. [CrossRef]
53. Hormaza, J. Molecular characterization and similarity relationships among apricot (*Prunus domestica* L.) genotypes using simple sequence repeats. *Theor. Appl. Genet.* **2002**, *104*, 321–328. [CrossRef]
54. Trujillo, I.; Ojeda, M.A.; Urdiroz, N.M.; Potter, D.; Barranco, D.; Rallo, L.; Diez, C.M. Identification of the worldwide olive germplasm Bank of Córdoba (Spain) using SSR and morphological markers. *Tree Genet. Genomes* **2014**, *10*, 141–155. [CrossRef]
55. Corrado, G.; La Mura, M.; Ambrosino, O.; Pugliano, G.; Varricchio, P.; Rao, R. Relationships of Campanian olive cultivars Comparative analysis of molecular and phenotypic data. *Genome* **2009**, *52*, 692–700. [CrossRef]
56. Manco, R.; Basile, B.; Capuozzo, C.; Scognamiglio, P.; Forlani, M.; Rao, R.; Corrado, G. Molecular and phenotypic diversity of traditional european plum (*Prunus domestica* L.) germplasm of southern Italy. *Sustainability* **2019**, *11*, 4112. [CrossRef]

7. Wang, Y.; Zhang, J.; Sun, H.; Ning, N.; Yang, L. Construction and evaluation of a primary core collection of apricot germplasm in China. *Sci. Hortic.* **2011**, *128*, 311–319. [CrossRef]
8. FAO. *Neglected Crops: 1492 from A Different Perspective*; FAO: Rome, Italy, 1994; Volume 26.
9. Pugliano, G.; Forlani, M.; Giofrrè, D.; Pasquarella, C.; Rotundo, A.; Sonnino, G. Individuazione di cv idonee alla trasformazione industriale. *Agrcultura e Ricerca* **1980**, *14*, 3–57.
0. Manco, R.; Chiaiese, P.; Basile, B.; Corrado, G. Comparative analysis of genomic-and EST-SSRs in European plum (*Prunus domestica* L.): Implications for the diversity analysis of polyploids. *3 Biotech* **2020**, *10*, 1–9. [CrossRef]
1. Melchiade, D.; Foroni, I.; Corrado, G.; Santangelo, I.; Rao, R. Authentication of the 'Annurca'apple in agro-food chain by amplification of microsatellite loci. *Food Biotechnol.* **2007**, *21*, 33–43. [CrossRef]
2. Verdone, M.; Rao, R.; Coppola, M.; Corrado, G. Identification of zucchini varieties in commercial food products by DNA typing. *Food Control* **2018**, *84*, 197–204. [CrossRef]
3. Kamvar, Z.N.; Tabima, J.F.; Grünwald, N.J. Poppr: An R package for genetic analysis of populations with clonal, partially clonal, and/or sexual reproduction. *PeerJ* **2014**, *2*, e281. [CrossRef]
4. Barot, S.; Allard, V.; Cantarel, A.; Enjalbert, J.; Gauffreteau, A.; Goldringer, I.; Lata, J.-C.; Le Roux, X.; Niboyet, A.; Porcher, E. Designing mixtures of varieties for multifunctional agriculture with the help of ecology. *Rev. Agron. Sustain. Dev.* **2017**, *37*, 13. [CrossRef]

MDPI
St. Alban-Anlage 66
4052 Basel
Switzerland
Tel. +41 61 683 77 34
Fax +41 61 302 89 18
www.mdpi.com

Plants Editorial Office
E-mail: plants@mdpi.com
www.mdpi.com/journal/plants

www.ingramcontent.com/pod-product-compliance
Lightning Source LLC
Chambersburg PA
CBHW061610210926
51850CB00009B/2274
9783036520568